Hyphenated and Alternative Methods of Detection in Chromatography

CHROMATOGRAPHIC SCIENCE SERIES

A Series of Textbooks and Reference Books

Editor:
NELU GRINBERG

Founding Editor:
JACK CAZES

1. Dynamics of Chromatography: Principles and Theory,
   J. Calvin Giddings
2. Gas Chromatographic Analysis of Drugs and Pesticides,
   Benjamin J. Gudzinowicz
3. Principles of Adsorption Chromatography: The Separation of Nonionic
   Organic Compounds, Lloyd R. Snyder
4. Multicomponent Chromatography: Theory of Interference,
   Friedrich Helfferich and Gerhard Klein
5. Quantitative Analysis by Gas Chromatography, Josef Novák
6. High-Speed Liquid Chromatography, Peter M. Rajcsanyi
   and Elisabeth Rajcsanyi
7. Fundamentals of Integrated GC-MS (in three parts),
   Benjamin J. Gudzinowicz, Michael J. Gudzinowicz,
   and Horace F. Martin
8. Liquid Chromatography of Polymers and Related Materials, Jack Cazes
9. GLC and HPLC Determination of Therapeutic Agents (in three parts),
   Part 1 edited by Kiyoshi Tsuji and Walter Morozowich, Parts 2 and 3
   edited by Kiyoshi Tsuji
10. Biological/Biomedical Applications of Liquid Chromatography,
    edited by Gerald L. Hawk
11. Chromatography in Petroleum Analysis, edited by Klaus H. Altgelt
    and T. H. Gouw
12. Biological/Biomedical Applications of Liquid Chromatography II,
    edited by Gerald L. Hawk
13. Liquid Chromatography of Polymers and Related Materials II,
    edited by Jack Cazes and Xavier Delamare
14. Introduction to Analytical Gas Chromatography: History, Principles,
    and Practice, John A. Perry
15. Applications of Glass Capillary Gas Chromatography, edited by
    Walter G. Jennings
16. Steroid Analysis by HPLC: Recent Applications, edited by
    Marie P. Kautsky
17. Thin-Layer Chromatography: Techniques and Applications,
    Bernard Fried and Joseph Sherma
18. Biological/Biomedical Applications of Liquid Chromatography III,
    edited by Gerald L. Hawk
19. Liquid Chromatography of Polymers and Related Materials III,
    edited by Jack Cazes

20. Biological/Biomedical Applications of Liquid Chromatography, edited by Gerald L. Hawk

21. Chromatographic Separation and Extraction with Foamed Plastics and Rubbers, G. J. Moody and J. D. R. Thomas

22. Analytical Pyrolysis: A Comprehensive Guide, William J. Irwin

23. Liquid Chromatography Detectors, edited by Thomas M. Vickrey

24. High-Performance Liquid Chromatography in Forensic Chemistry, edited by Ira S. Lurie and John D. Wittwer, Jr.

25. Steric Exclusion Liquid Chromatography of Polymers, edited by Josef Janca

26. HPLC Analysis of Biological Compounds: A Laboratory Guide, William S. Hancock and James T. Sparrow

27. Affinity Chromatography: Template Chromatography of Nucleic Acids and Proteins, Herbert Schott

28. HPLC in Nucleic Acid Research: Methods and Applications, edited by Phyllis R. Brown

29. Pyrolysis and GC in Polymer Analysis, edited by S. A. Liebman and E. J. Levy

30. Modern Chromatographic Analysis of the Vitamins, edited by André P. De Leenheer, Willy E. Lambert, and Marcel G. M. De Ruyter

31. Ion-Pair Chromatography, edited by Milton T. W. Hearn

32. Therapeutic Drug Monitoring and Toxicology by Liquid Chromatography, edited by Steven H. Y. Wong

33. Affinity Chromatography: Practical and Theoretical Aspects, Peter Mohr and Klaus Pommerening

34. Reaction Detection in Liquid Chromatography, edited by Ira S. Krull

35. Thin-Layer Chromatography: Techniques and Applications, Second Edition, Revised and Expanded, Bernard Fried and Joseph Sherma

36. Quantitative Thin-Layer Chromatography and Its Industrial Applications, edited by Laszlo R. Treiber

37. Ion Chromatography, edited by James G. Tarter

38. Chromatographic Theory and Basic Principles, edited by Jan Åke Jönsson

39. Field-Flow Fractionation: Analysis of Macromolecules and Particles, Josef Janca

40. Chromatographic Chiral Separations, edited by Morris Zief and Laura J. Crane

41. Quantitative Analysis by Gas Chromatography, Second Edition, Revised and Expanded, Josef Novák

42. Flow Perturbation Gas Chromatography, N. A. Katsanos

43. Ion-Exchange Chromatography of Proteins, Shuichi Yamamoto, Kazuhiro Naka-nishi, and Ryuichi Matsuno

44. Countercurrent Chromatography: Theory and Practice, edited by N. Bhushan Man-dava and Yoichiro Ito

45. Microbore Column Chromatography: A Unified Approach to Chromatography, edited by Frank J. Yang

46. Preparative-Scale Chromatography, edited by Eli Grushka

47. Packings and Stationary Phases in Chromatographic Techniques, edited by Klaus K. Unger

48. Detection-Oriented Derivatization Techniques in Liquid Chromatography, edited by Henk Lingeman and Willy J. M. Underberg

49. Chromatographic Analysis of Pharmaceuticals, edited by John A. Adamovics

50. Multidimensional Chromatography: Techniques and Applications, edited by Hernan Cortes
51. HPLC of Biological Macromolecules: Methods and Applications, edited by Karen M. Gooding and Fred E. Regnier
52. Modern Thin-Layer Chromatography, edited by Nelu Grinberg
53. Chromatographic Analysis of Alkaloids, Milan Popl, Jan Fähnrich, and Vlastimil Tatar
54. HPLC in Clinical Chemistry, I. N. Papadoyannis
55. Handbook of Thin-Layer Chromatography, edited by Joseph Sherma and Bernard Fried
56. Gas–Liquid–Solid Chromatography, V. G. Berezkin
57. Complexation Chromatography, edited by D. Cagniant
58. Liquid Chromatography–Mass Spectrometry, W. M. A. Niessen and Jan van der Greef
59. Trace Analysis with Microcolumn Liquid Chromatography, Milos Krejcl
60. Modern Chromatographic Analysis of Vitamins: Second Edition, edited by André P. De Leenheer, Willy E. Lambert, and Hans J. Nelis
61. Preparative and Production Scale Chromatography, edited by G. Ganetsos and P. E. Barker
62. Diode Array Detection in HPLC, edited by Ludwig Huber and Stephan A. George
63. Handbook of Affinity Chromatography, edited by Toni Kline
64. Capillary Electrophoresis Technology, edited by Norberto A. Guzman
65. Lipid Chromatographic Analysis, edited by Takayuki Shibamoto
66. Thin-Layer Chromatography: Techniques and Applications: Third Edition, Revised and Expanded, Bernard Fried and Joseph Sherma
67. Liquid Chromatography for the Analyst, Raymond P. W. Scott
68. Centrifugal Partition Chromatography, edited by Alain P. Foucault
69. Handbook of Size Exclusion Chromatography, edited by Chi-San Wu
70. Techniques and Practice of Chromatography, Raymond P. W. Scott
71. Handbook of Thin-Layer Chromatography: Second Edition, Revised and Expanded, edited by Joseph Sherma and Bernard Fried
72. Liquid Chromatography of Oligomers, Constantin V. Uglea
73. Chromatographic Detectors: Design, Function, and Operation, Raymond P. W. Scott
74. Chromatographic Analysis of Pharmaceuticals: Second Edition, Revised and Expanded, edited by John A. Adamovics
75. Supercritical Fluid Chromatography with Packed Columns: Techniques and Applications, edited by Klaus Anton and Claire Berger
76. Introduction to Analytical Gas Chromatography: Second Edition, Revised and Expanded, Raymond P. W. Scott
77. Chromatographic Analysis of Environmental and Food Toxicants, edited by Takayuki Shibamoto
78. Handbook of HPLC, edited by Elena Katz, Roy Eksteen, Peter Schoenmakers, and Neil Miller
79. Liquid Chromatography–Mass Spectrometry: Second Edition, Revised and Expanded, Wilfried Niessen
80. Capillary Electrophoresis of Proteins, Tim Wehr, Roberto Rodríguez-Díaz, and Mingde Zhu
81. Thin-Layer Chromatography: Fourth Edition, Revised and Expanded, Bernard Fried and Joseph Sherma

82. Countercurrent Chromatography, edited by Jean-Michel Menet and Didier Thiébaut
83. Micellar Liquid Chromatography, Alain Berthod and Celia García-Alvarez-Coque
84. Modern Chromatographic Analysis of Vitamins: Third Edition, Revised and Expanded, edited by André P. De Leenheer, Willy E. Lambert, and Jan F. Van Bocxlaer
85. Quantitative Chromatographic Analysis, Thomas E. Beesley, Benjamin Buglio, and Raymond P. W. Scott
86. Current Practice of Gas Chromatography–Mass Spectrometry, edited by W. M. A. Niessen
87. HPLC of Biological Macromolecules: Second Edition, Revised and Expanded, edited by Karen M. Gooding and Fred E. Regnier
88. Scale-Up and Optimization in Preparative Chromatography: Principles and Bio-pharmaceutical Applications, edited by Anurag S. Rathore and Ajoy Velayudhan
89. Handbook of Thin-Layer Chromatography: Third Edition, Revised and Expanded, edited by Joseph Sherma and Bernard Fried
90. Chiral Separations by Liquid Chromatography and Related Technologies, Hassan Y. Aboul-Enein and Imran Ali
91. Handbook of Size Exclusion Chromatography and Related Techniques: Second Edition, edited by Chi-San Wu
92. Handbook of Affinity Chromatography: Second Edition, edited by David S. Hage
93. Chromatographic Analysis of the Environment: Third Edition, edited by Leo M. L. Nollet
94. Microfluidic Lab-on-a-Chip for Chemical and Biological Analysis and Discovery, Paul C.H. Li
95. Preparative Layer Chromatography, edited by Teresa Kowalska and Joseph Sherma
96. Instrumental Methods in Metal Ion Speciation, Imran Ali and Hassan Y. Aboul-Enein
97. Liquid Chromatography–Mass Spectrometry: Third Edition, Wilfried M. A. Niessen
98. Thin Layer Chromatography in Chiral Separations and Analysis, edited by Teresa Kowalska and Joseph Sherma
99. Thin Layer Chromatography in Phytochemistry, edited by Monika Waksmundzka-Hajnos, Joseph Sherma, and Teresa Kowalska
100. Chiral Separations by Capillary Electrophoresis, edited by Ann Van Eeckhaut and Yvette Michotte
101. Handbook of HPLC: Second Edition, edited by Danilo Corradini and consulting editor Terry M. Phillips
102. High Performance Liquid Chromatography in Phytochemical Analysis, edited by Monika Waksmundzka-Hajnos and Joseph Sherma
103. Hydrophilic Interaction Liquid Chromatography (HILIC) and Advanced Applications, edited by Perry G. Wang and Weixuan He
104. Hyphenated and Alternative Methods of Detection in Chromatography, edited by R. Andrew Shalliker

Hyphenated and Alternative Methods of Detection in Chromatography

Edited by
R. ANDREW SHALLIKER

CRC Press
Taylor & Francis Group
Boca Raton London New York

CRC Press is an imprint of the
Taylor & Francis Group, an **informa** business

Published 2012 by CRC Press
Taylor & Francis Group
6000 Broken Sound Parkway NW, Suite 300
Boca Raton, FL 33487-2742

First issued in paperback 2019

No claim to original U.S. Government works

ISBN-13: 978-0-367-45220-9 (pbk)
ISBN-13: 978-0-8493-9077-7 (hbk)

This book contains information obtained from authentic and highly regarded sources. Reasonable efforts have been made to publish reliable data and information, but the author and publisher cannot assume responsibility for the validity of all materials or the consequences of their use. The authors and publishers have attempted to trace the copyright holders of all material reproduced in this publication and apologize to copyright holders if permission to publish in this form has not been obtained. If any copyright material has not been acknowledged please write and let us know so we may rectify in any future reprint.

Visit the Taylor & Francis Web site at
http://www.taylorandfrancis.com

and the CRC Press Web site at
http://www.crcpress.com

Contents

Preface..xi
Editor xiii
Contributors xv

Chapter 1   Mass Spectrometry and Separation Science 1

              Xavier Conlan

Chapter 2   Analytical Aspects of Modern Gas Chromatography:
              Mass Spectrometry...31

              Shin Miin Song and Philip J. Marriott

Chapter 3   Coupling Liquid Chromatography and Other Separation
              Techniques to Nuclear Magnetic Resonance Spectroscopy..............61

              Cristina Daolio and Bernd Schneider

Chapter 4   Application of Infrared and Raman Spectroscopy for Detection
              in Liquid Chromatographic Separations 99

              Peter J. Mahon

Chapter 5   Evaporative Light Scattering and Charged Aerosol Detector..........145

              Pierre Chaminade

Chapter 6   Detection and Determination of Heteroatom-Containing
              Molecules by HPLC: Inductively Coupled Plasma Mass
              Spectrometry 161

              Sandra Mounicou, Kasia Bierla, and Joanna Szpunar

Chapter 7   HPLC with Electrochemical Detection..187

              Fumiyo Kusu and Akira Kotani

Chapter 8   Liquid-Phase Chemiluminescence Detection for HPLC 221

              Paul S. Francis and Jacqui L. Adcock

Chapter 9   Multidimensional High-Performance Liquid Chromatography....... 251

Coleen S. Milroy, Paul G. Stevenson,
Mariam Mnatasakyan, and R. Andrew Shalliker

Index..287

Preface

This book is intended for analytical chemists, particularly those involved in the analysis of natural, biological, and environmental samples. Perhaps, more specifically, the collection of techniques presented here is intended for the scientist who is interested in the separation, analysis, and identification of sample constituents in complex mixtures. This book will serve as a detailed reference text for those who are attempting to start out in the field, while at the same time the specific details of the application in separation and detection will suit those who are more experienced yet require assistance in new directions. The book is also an ideal reference for senior college and university students, particularly at the postgraduate level, wishing to supplement their learning experience. Containing over 800 references, this book is an excellent source of literature information within the field of hyphenated methods of chromatographic analysis.

Separation techniques are widely employed in almost every laboratory. This field of science serves to separate species into, ideally, individual constituents, following which, components within the sample can be quantified and with the assistance of other methods of analysis, the identity of these components can be discovered. In a modern laboratory, the manual transport of sample constituents from the chromatographic system to other analytical instrumentation is now almost unheard of. Instead, hyphenated methods of analysis dominate the analytical laboratory. Perhaps the simplest application of hyphenation was LC coupled with diode array UV/Vis detection. Nowadays such methodology is taken for granted, and few would consider this a hyphenated method of analysis. Since then, more sophisticated methods have been developed, such as GC-MS, which is now a routine application in almost all laboratories. LC-MS and LC-MS/MS are techniques that require substantial expertise in operation. Nevertheless, there is a growing acceptance of these techniques, and they have found widespread application. In this book, we have detailed the development and application of mass spectral detection coupled with both liquid phase (Chapter 1) and gas phase (Chapter 2) chromatographies, and we also include a chapter dedicated to the coupling of LC-ICP-MS. The difference in the development of these coupled systems is an interesting factor that results primarily from the vacuum requirements of the mass spectrophotometer and the limitations associated with buffer systems.

Together with mass spectral analysis, NMR is the other technique that is the most important in relation to the identification of unknown constituents. This technique is discussed in Chapter 3. For many years, the limitations in the sensitivity of the NMR have inhibited its application as a hyphenation mode in LC. However, as methodologies develop so, too, has the application of LC-NMR. This chapter details the development of the LC-NMR and illustrates its application in chromatographic science, in particular in relation to the analysis of natural and environmental samples, and in the study of reaction monitoring, biosynthetic analysis, and structural elucidation.

Although FTIR is one of the most important spectroscopic analytical techniques, it finds relatively limited application as a detector in LC. This is largely related to the problems associated with solvent compatibility between the most common types of liquid chromatography (i.e., water) and the detection cell. Even so, solvent elimination procedures have been developed that allow these systems to be coupled. The most significant application of LC-FTIR has come from the area of polymer analysis, where the ability of FTIR to offer fingerprint spectral information has proven to be a valuable aid. Chapter 4 provides a detailed discussion on LC-Raman and LC-FTIR.

Evaporative light scattering detection is another important method that is seeing widespread growth. This technique is covered in Chapter 5. Originally finding employment in the area of polymer analysis, this form of detector is finding application as a universal detector, not limited to species that have, for example, chromophores, provided they are more volatile than the mobile phase. It is therefore fitting that this technique is discussed after that of FTIR detection. The ramifications of the nonlinear concentration response are discussed in detail throughout this chapter. Charged aerosol detectors are also covered in this chapter.

Chapter 6 details aspects of LC-ICP, also coupled to MS. The application of LC-ICP is a growing method of chemical speciation, although it has yet to receive widespread recognition. Here, the technique is highlighted through application in biological samples for the speciation of organoarsenic and organoselenium compounds; metal complexes in microorganisms, plants, and foods of plant origin; human body fluids and tissues; and in the detection of chemical warfare agents.

Chapter 7 discusses electrochemical detection in liquid chromatography. It elaborates on system optimization and highlights its applications through the analysis of flavonoids.

While techniques such as MS and NMR provide a means for identification, chemiluminescence is a valuable tool in that it is simple and cheap to operate, requires little expertise, and can be relatively selective in the type of analytes it detects. For instance, chemiluminescence has been shown to be a powerful means of detecting antioxidants in foods. This is especially useful in the wine industry. In addition, chemiluminescence is a technique that is ideally suited to microcolumn and capillary methods of separation. These issues are covered in Chapter 8.

The last chapter, Chapter 9, deals with the process of employing multidimensional separation methods as a means of identifying components in complex samples. Retention times in one-dimensional chromatographic systems are not unique. However, in a two-dimensional separation, there is a higher probability of "unique" component displacement in the two-dimensional plane. This means that conventional detectors, such as the UV/Vis, that cannot identify components can be employed in these types of systems and the sample identity can be confirmed through two-dimensional retention times. The technique of multidimensional HPLC can also be used as a means of selectivity screening for complex samples, yielding information not available in one-dimensional systems for complex samples. This technique is illustrated in this chapter by the analysis of café expresso.

R. Andrew Shalliker

Editor

Dr. R. Andrew Shalliker received his PhD in 1992 and his BSc (Hons) in 1987 from Deakin University, Geelong, Australia. He joined the University of Western Sydney (UWS) in 1999. He is currently an associate professor of analytical chemistry at the same university and teaches second- and third-year undergraduates analytical chemistry units. Prior to joining UWS, he was a research fellow in Professor Guiochon's group at the University of Tennessee, Knoxville, Tennessee, and Oak Ridge National Laboratories, Oak Ridge, Tennessee (1997–1999). From 1995 to 1997, he held a Queensland University of Technology (QUT) Postdoctoral Research Award at QUT Brisbane. In 1994, he was an associate lecturer of analytical chemistry at the University of Tasmania, Hobart. Between 1992 and 1994, he was a research chemist at Faulding Pharmaceuticals/David Bull Laboratories, Melbourne, Victoria, Australia.

Dr. Shalliker's research interests are in the field of liquid chromatography. He has published over 100 scientific papers in this field, focusing on the areas of column technology, stationary phase design, multidimensional HPLC, polymer and oligomer separations, and the analysis of natural products employing chromatographic separations.

In 2007, his research team joined the research groups at UTAS and RMIT as a member of the Australian Centre for Research on Separation Science (ACROSS) consortium.

Contributors

Jacqui L. Adcock
School of Life and Environmental
    Sciences
Deakin University
Geelong, Victoria, Australia

Kasia Bierla
Unité Mixte de Recherche
Centre National de la Recherche
    Scientifique
Pau, France

Pierre Chaminade
EA4041 Groupe de Chimie
Analytique de Paris-Sud
Faculté de Pharmacie
University of Paris-Sud
Châtenay-Malabry, France

Xavier Conlan
School of Life and Environmental
    Sciences
Deakin University
Geelong, Victoria, Australia

Cristina Daolio
Bruker Biospin GmbH
Rheinstetten, Germany

Paul S. Francis
School of Life and Environmental
    Sciences
Deakin University
Geelong, Victoria, Australia

Akira Kotani
School of Pharmacy
Tokyo University of Pharmacy and Life
    Sciences
Tokyo, Japan

Fumiyo Kusu
School of Pharmacy
Tokyo University of Pharmacy and Life
    Sciences
Tokyo, Japan

Peter J. Mahon
Faculty of Life and Social Sciences
Swinburne University of Technology
Hawthorn, Victoria, Australia

Philip J. Marriott
Department of Applied Chemistry
School of Applied Sciences
RMIT University
Melbourne, Victoria, Australia

Coleen S. Milroy
School of Natural Sciences
and
Australian Centre for Research on
    Separation Science
and
Nanoscale Organisation and Dynamics
    Group
University of Western Sydney
Parramatta, New South Wales,
    Australia

Mariam Mnatasakyan
School of Natural Sciences
and
Australian Centre for Research on
    Separation Science
University of Western Sydney
Parramatta, New South Wales,
    Australia

Sandra Mounicou
Unité Mixte de Recherche
Centre National de la Recherche
    Scientifique
Pau, France

Bernd Schneider
Max Planck Institute for Chemical
    Ecology
Jena, Germany

R. Andrew Shalliker
School of Natural Sciences
and
Australian Centre for Research on
    Separation Science
and
Nanoscale Organisation and Dynamics
    Group
University of Western Sydney
Parramatta, New South Wales,
    Australia

Shin Miin Song
Department of Applied Chemistry
School of Applied Sciences
RMIT University
Melbourne, Victoria, Australia

Paul G. Stevenson
School of Natural Sciences
and
Australian Centre for Research on
    Separation Science
and
Centre for Complimentary Medicine
University of Western Sydney
Parramatta, New South Wales,
    Australia

Joanna Szpunar
Unité Mixte de Recherche
Centre National de la Recherche
    Scientifique
Pau, France

1 Mass Spectrometry and Separation Science

Xavier Conlan

CONTENTS

1.1 Introduction 2
1.2 Brief History 2
1.3 Principle 3
    1.3.1 Electron Ionization 4
    1.3.2 Chemical Ionization...5
    1.3.3 Atmospheric Pressure Ionization..5
    1.3.4 Electrospray Ionization..5
    1.3.5 Atmospheric Pressure Chemical Ionization 6
    1.3.6 Matrix-Assisted Laser Desorption Ionization 6
    1.3.7 Fast Atom Bombardment...7
    1.3.8 Inductively Coupled Plasma 8
1.4 Ionization Energy and Ionization Efficiency 8
1.5 Mass Spectral Data..9
    1.5.1 Nitrogen Rule...9
    1.5.2 Isotopic Information 10
    1.5.3 Rings and Double Bonds 11
    1.5.4 Fragmentation.. 11
1.6 Mass Analyzers.. 14
    1.6.1 Magnetic Sector... 14
    1.6.2 Quadrupole 15
    1.6.3 Time of Flight.. 15
    1.6.4 Ion Trap... 17
    1.6.5 Tandem Mass Spectrometry 18
        1.6.5.1 Triple Quadrupole... 19
        1.6.5.2 Quadrupole Time of Flight.. 19
1.7 Resolution 20
1.8 Ion Enhancement/Suppression Solvent Considerations.............................. 21
1.9 Contamination and Practical Considerations 23
1.10 Quantitation 25
1.11 Data Handling: Chemometrics 25
References..27

1.1   INTRODUCTION

Mass spectrometry has become an integral tool for the analytical scientist enabling advances in diverse fields, such as metabolomics, proteomics, drug discovery, surface science, materials science, food science, and forensic science. The importance of mass spectrometry to science in the early twenty-first century is clearly noticeable when viewing the increasing number of publications where mass spectrometry is prominent (see in Figure 1.1).

Due to its sensitivity, limits of detection, resolution, mass range, accuracy, and flexibility, mass spectrometry has become increasingly important as a detection system coupled to both gas- and liquid-phase separation technologies. It is the intention of this chapter to discuss the theoretical and practical aspects of mass spectrometry that are relevant to separation science. There are several good texts that detail the fundamentals of mass spectrometry, for example, deHoffmann and Stroobant [1] and McLafferty and Turecek [2], and the reader is urged to review these.

1.2   BRIEF HISTORY

J.J. Thomson is regarded as the father of mass spectrometry upon his discovery in 1897 of the electron and his subsequent determination of its mass to charge ratio [3]. Prior to this, in 1886, E. Goldstien had discovered positively charged gas-phase ions, anode rays, from within a gas discharge tube [4]. Wilhelm Wein went on to use a magnet to maneuver these charged rays determining that they carried a positive charge [5]. Several years later, in 1912, J.J. Thomson developed the first mass spectrometer where he observed several charged ions [6]. Nearly simultaneously, in 1918 and 1919, A.J. Dempster [7] and F.W. Aston [8], respectively, developed mass spectrometers that started to resemble modern day instruments based on magnetic sectors and velocity focusing systems. The next major development occurred in 1948 when A.E. Cameron and D.F. Eggers [9] designed a linear time-of-flight mass spectrometer based on a concept proposed by W. Stephens in 1946 [10].

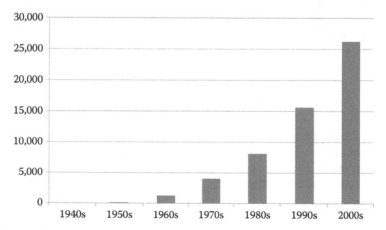

FIGURE 1.1   The increasing number of publications regarding mass spectrometry from the 1940s to 2008.

To this point in time, mass spectrometry was not in an applicable form to aid chromatographic problems, and several key developments over the latter half of the twentieth century enabled the utility of chromatography and mass spectrometry. F.W. McLaffery and R.S. Gohlke brought separation science and mass spectrometry together for the first time in the late 1950s when they coupled a mass spectrometer and a gas chromatography instrument [11,12]. Chemical ionization (CI) was established in 1966 by M.S.B. Munson and F.H. Field [13] and has subsequently led to soft ionization techniques that are conducive to producing large stable molecular ions. The reflectron was developed by Karataev and coworkers in 1972 [14], leading to advancement in the resolution capabilities of mass spectrometry. By 1974, Arpino, Baldwin, and McLaffer had coupled HPLC to mass spectrometry with some success [15]. The first coupling of capillary electrophoresis and mass spectrometry was presented by Smith in 1987 [16]; the significant challenge in this case surrounded the liquid to gas phase interface of the system. The final frontier and potentially one of the most important developments was the introduction of electrospray in 1989 by Fenn et al. [17]. Electrospray has proved pivotal in the analysis of proteins, polymers, and small molecules and offers an ideal mode of sample introduction from liquid chromatography to the mass spectrometer.

While these are only some of the historical highlights in the development of separation science and mass spectrometry, many researchers have been involved in the optimization of the various techniques and instruments for specific scientific purposes. The mass spectrometry systems that have been developed include atmospheric pressure chemical ionization (APCI), thermospray (TSP) and electrospray (ESI) liquid chromatography (LC-MS), gas chromatography (GC-MS), continuous flow fast atom bombardment (FAB), capillary electrophoresis (CE), continuous flow matrix-assisted laser desorption ionization (MALDI), and nano liquid chromatography. Gelpi [18] aptly describes the possibilities for processing liquid chromatography eluent in mass spectrometry as shown in Figure 1.2. Because of the intricacies of mass spectrometry, we need to further discuss the fundamental principles and practicalities involved in coupling such systems.

1.3 PRINCIPLE

Inherent to mass spectrometry is the ionization of the molecule(s) of interest and by this process either directly or indirectly producing a gas-phase ion, subsequently leading to the possibility to manipulate the ion via a magnetic or electric field prior to detection. Ionization is a process leading to the charged state either positively or negatively of a molecule, which can be initiated in a variety of ways. The charged state exists due to the difference between the number of protons and electrons. A positive electric charge is produced when an electron bond to an atom or molecule absorbs enough energy from an external source to escape from the electric potential barrier that originally confined it, where the amount of energy required is called the ionization potential. A negative electric charge is produced when a free electron collides with an atom and is subsequently caught inside the electric potential barrier, releasing any excess energy.

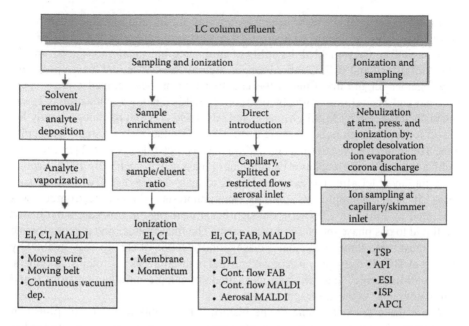

FIGURE 1.2 In LCMS, the LC column effluent can be sampled by various devices based on preferential solvent removal or effluent splitting and then the fraction sampled into the ion source of the mass spectrometer is ionized by different ionization techniques under vacuum, or ionized first by various atmospheric pressure ionization methods prior to the sampling of the ions into the vacuum of the mass spectrometer. Abbreviations: API, atmospheric pressure ionization; APCI, atmospheric pressure chemical ionization; EI, electron ionization; CI, chemical ionization; DLI, direct liquid introduction; ESI, electrospray ionization; FAB, fast atom bombardment; ISP, ion spray; MALDI, matrix-assisted laser desorption/ionization; TSP, thermospray. (Adapted from Gelpi, E., *J. Mass Spectrom.*, 37, 241, 2002.)

1.3.1   ELECTRON IONIZATION

Electron ionization (EI) is performed by impacting an electron (usually from a hot filament) with some energy into the gas-phase molecule of interest. The energy contained within the impacting electron is sufficient enough to remove an electron (with the lowest bonding energy) from the molecule. Generally, the electron has more than enough energy to dislodge the electron and the excess energy goes into the fragmentation of the molecule. It is because of this process that EI is considered a hard ionization technique (Figure 1.3).

$$M + e^- \rightarrow M^{+\bullet} + e + e^-$$

Molecule   ~70 eV   Molecular ion   ~55 eV   0.1 eV

FIGURE 1.3 Electron ionization.

$$CH_4^{+\cdot} + M \rightarrow + CH_4 + M^{+\cdot} \text{ (molecular ion)}$$

$$CH_5^+ + M \rightarrow + CH_4 + MH^+ \text{ (protonated molecular)}$$

FIGURE 1.4 Chemical ionization formation of the molecular ion and the protonated or pseudo-molecular ion.

When coupling an EI system to a gas-phase/volatile molecule separation, the sampling line can be directly connected. In the case of liquid chromatographic interfaces, the vapor pressure of the liquid is generally increased.

1.3.2 CHEMICAL IONIZATION

CI is a softer ionization technique than EI imparting less energy into the molecule. In principle, CI occurs by the transfer of energy from an interaction between an initial ion and the molecule of interest. Often reagent gases such as methane, isobutene, and ammonia are used for this purpose. Alongside the production of the molecular ion, one of the most common interactions is the transfer of a proton to the molecule of interest (Figure 1.4).

1.3.3 ATMOSPHERIC PRESSURE IONIZATION

Atmospheric pressure ionization fundamentally refers to any ionization process that occurs prior to the ultrahigh vacuum region of the mass spectrometer. These are primarily the interface used when coupling liquid-phase separation systems to mass spectrometry. Several techniques have been established with electrospray ionization (ESI) and atmospheric pressure chemical ionization (APCI) the most commonly employed.

1.3.4 ELECTROSPRAY IONIZATION

ESI is the predominant process applied to liquid chromatography offering several advantages over other techniques. This soft ionization technique offers good control over the fragmentation of the molecule of interest and is conducive to producing multiply charged molecules. Multiple charged molecules can be useful as they allow the detection of molecules of a molecular weight that falls outside the normal working range of the mass spectrometer. In electrospray, ions are formed in the liquid phase, which flow from the tip of a fine needle due to an induced electrostatic condition that produces a Taylor cone, which was first described by Geoffrey Taylor in 1964 (see Figure 1.5). The charged liquid forms the Taylor cone where upon the tip a fine filament of flow extends, this filament then forms a spray. This occurs due to solvent evaporation until the excess charge equals the cohesive force of surface tension; at this point the droplets are in the order of 1 μm. These droplets subsequently undergo further solvent evaporation until the ions exist in the gas phase. It is this reproducible conversion of ions from the liquid phase to the gas phase that have allowed for ESI to become one of the most useful forms for coupling liquid chromatographic systems to mass spectrometry.

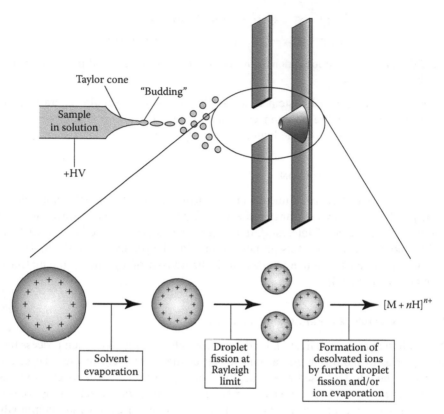

FIGURE 1.5 Droplet production in the electrospray interface. (Adapted from Gaskell, S., *J. Mass Spectrom.*, 32, 1378, 1997.)

1.3.5 ATMOSPHERIC PRESSURE CHEMICAL IONIZATION

APCI is generally applicable to liquid chromatography as it is capable of perform-ing with high solvent flow rates. Ions are produced via a corona discharge electrode that is placed in the solvent spray, one that is generated in a similar fashion to the aforementioned ESI system. The droplets of solvent are heated directly within a quartz heating tube prior to entering a corona discharge region where ioniza-tion occurs in the gas phase. APCI has become a popular ionization technique for LC-MS (Figure 1.6).

1.3.6 MATRIX-ASSISTED LASER DESORPTION IONIZATION

MALDI was famously developed by Karas and Hillenkamp [21]. The technique exploits the use of a laser for the ionization process applied to the sample in an indirect manner. In short, a sample is either sandwiched in layers or directly mixed with a matrix and placed on a sample stub, this matrix has significant UV absorption bands enabling absorption of a laser pulse that is subsequently applied. The energy absorbed by the matrix is transferred to the sample molecules in a process that forms

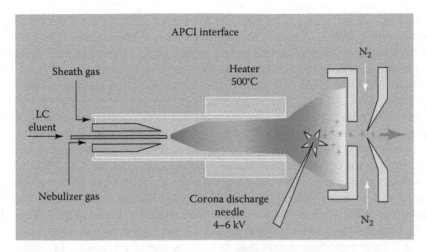

FIGURE 1.6 Schematic of APCI source. (From Abian, J., *J. Mass Spectrom.*, 35, 157, 1999.)

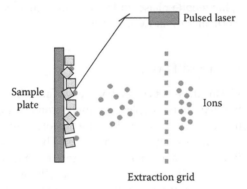

FIGURE 1.7 MALDI ion generation. (From Aebersold, R. and Mann, M., *Nature*, 442, 198, 2003.)

a gas-phase plume. While not fully understood, the ionization process occurs after this laser ablation. This ionization process is particularly gentle, enabling large ionized molecules to remain intact, which can be in the order of 10^6 Da. MALDI generates less multiply charged ions than ESI and is commonly employed for the analysis of large labile molecules such as proteins, polymers, and large inorganic species (Figure 1.7).

1.3.7 Fast Atom Bombardment

Fast atom bombardment (FAB) is an efficient technique used for the analysis of polar, ionic, thermally labile, and high-molecular-weight compounds that cannot be analyzed by EI or CI techniques. FAB is commonly employed in the analysis of natural products, surfactants, and biomolecules. The technique was first introduced in 1981 by two research groups based in Manchester [23,24]. By 1983, investigations into the coupling of liquid chromatography and FAB occurred [25]. LC-FAB has been successfully applied to a variety of biological sample matrices including plasma and serum, antibiotics [26], and foodstuffs including honey [27]. Several attributes of LC-FAB-MS lend itself to the determination of steroidal compounds offering benefits over the more common GC techniques, an example of which is presented by Watson et al. [28]. Watson states that in the electron-impact mode, the methyoxime-trimethylsilyl (MO-TMS) derivatives of corticosteroids produce significant ions of very low intensity, if at all; however, negative-ion chemical ionization mass spectrometry (NICI-MS) produces mass spectra from the MO-TMS derivatives that contain a few high-intensity significant ions. In the case of corticosteroids bearing an ester group at the 17-position, GC/MS is less satisfactory, since thermal elimination of the 17-ester group occurs in the injection port and at the head of the GC column and this produces broad peaks with mass spectra containing molecular ions at a lower molecular weight than that of the parent compound. Frit-fast atom bombardment provides a relatively simple interface between a liquid chromatograph and a mass spectrometer and the commonly used glycerol matrix can be introduced into the chromatographic mobile phase, in low concentration, without affecting the chromatographic performance of an LC column.

1.3.8 INDUCTIVELY COUPLED PLASMA

Inductively coupled plasma mass spectrometry is a useful technique for the determination of trace elements and can be coupled to liquid chromatography. The technique allows for the detection of both metals and nonmetals offering excellent sensitivity and selectivity across a wide variety of samples [29]. The solution stream from the LC is sprayed into flowing argon and passed into an area that is inductively heated to approximately 10,000°C. The gas and almost everything in it is atomized and ionized at this temperature, a plasma providing both excited and ionized atoms with an ionization efficiency of nearly 100% is produced. The positive ions produced are detected by quadrupole mass spectrometer enabling the detection of nearly all the elements in the periodic table at detection limits below 1 ug/L for most elements. ICP-MS is an expensive detection system; however, the ease with which it can be coupled to liquid chromatography systems, which offer a wide range of separation processes, has enabled LC-ICP-MS to be the leading system for the detection and determination of metallospecies from complex organic sample matrices [30]. Further detailed discussion of this technique follows in Chapter 6.

1.4 IONIZATION ENERGY AND IONIZATION EFFICIENCY

Regardless of the ionization process, there are two simple concepts that are handy for any mass spectroscopist to have a handle on, ionization potential and ionization efficiency. The ionization energy E_i (which historically was referred to as the ionization potential) is defined by IUPAC conventions as; the minimum energy required to eject an electron out of a neutral atom or molecule in its ground state. The adiabatic

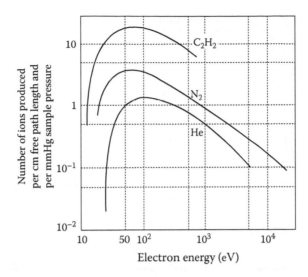

FIGURE 1.8 Number of ions produced as a function of electron energy. (From deHoffmann, E. and Stroobant, V., *Mass Spectrometry Principles and Applications*, John Wiley & Sons Ltd., Chichester, U.K., 2007.)

ionization energy refers to the formation of the molecular ion in its ground vibrational state and the vertical ionization energy applies to the transition to the molecular ion without change in geometry. It was this quantity that was formerly called ionization potential. The second ionization energy of an atom is the energy required to eject the second electron from the atom [31]. The ionization efficiency is simply the ratio of the number of ions formed to the number of electrons or photons used in an ionization process [31].

deHoffmann and Stroobant [1] highlight with regards to electron ionization the relationship between electron impact energy and ionization efficiency particularly well. It can be noted from Figure 1.8 that a maximum around 70 eV appears, which has previously been mentioned as the optimum electron energy for electron ionization processes.

1.5 MASS SPECTRAL DATA

The mass spectrum presents a great deal of chemical information to the analyst and care must be taken for the optimum interpretation of the data. In order to do this, a logical interpretation procedure must be applied. The general characteristics of a mass spectrum includes several aspects, such as a molecular ion (M^+) or pseudo-molecular ion ($M \pm X^\pm$) and daughter or fragment peaks, of these, the peak with the greatest intensity is known as the base peak (Figure 1.9).

1.5.1 NITROGEN RULE

The *nitrogen rule* is particularly useful in the interpretation of mass spectral data. If a compound has an odd number of nitrogen atoms, irrespective of the number

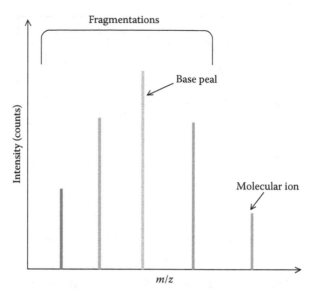

FIGURE 1.9   A typical mass spectrum.

of C, H, halogens, O, S, Si, or P, then $M^{+\bullet}$ will have an odd nominal mass. For a compound with an even number of nitrogen atoms, the $M^{+\bullet}$ will have an even nominal mass.

1.5.2   ISOTOPIC INFORMATION

Isotopic information can be obtained from a mass spectrum as the resolving power of modern mass spectrometers allows for differentiation of the peaks associated with the molecule that consists of the heavier isotopes. When investigating hydrocarbons, the relative abundance on ^{12}C 98.93% and ^{13}C 1.07% and hydrogen ^{1}H 99.98% and ^{2}H 0.012% can be taken into account for a molecule with the empirical formula of C_nH_m. In this instance, the intensity of the A + 1 peak or the next highest observable peak one mass unit higher than the molecular ion can be defined by the following equation:

$$\text{Intensity} = n \times 1.08\% + m \times 0.012\% \tag{1.1}$$

McLafferty, a leader in the interpretation of mass spectrometry [2], describes C and N in terms of their isotopic nature as A + 1 type elements and O, Si, S, Cl, Br as A + 2 type elements. The A + 1 peak can be used to generate an estimation of the number of carbons in a compound; this can be achieved by dividing the percentage abundance of the A + 1 peak by 1.1.

Distinctive mass spectral isotopic patterns can be observed for molecular ions containing chlorine and bromine, some of which are presented in Figure 1.10.

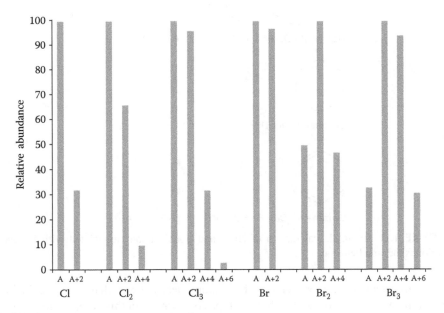

FIGURE 1.10 Chlorine and bromine isotopes.

The ratio of A and A + 2 of approximately 3:1 and 1:1 are indicative of the presence of Cl and Br, respectively.

1.5.3 RINGS AND DOUBLE BONDS

The number of rings and double bonds in the molecule of interest can be calculated if the elemental makeup of the molecule has been determined by the exact mass. The equation for this calculation is as follows:

$$R + DB = x - \frac{y}{2} + \frac{z}{2} + 1 \qquad (1.2)$$

where
  R is the number of rings
  DB is the number of double bonds
  x is the number of atoms that make four bonds, i.e., C, Si, etc.
  y is the number of hydrogen and halogen atoms
  z is the number of atoms that make three bonds, i.e., N, P, etc.

The comparison of hexane, cyclohexane, and benzene presents a good example for this calculation (Figure 1.11).

1.5.4 FRAGMENTATION

The mass spectrum enables the observation of fragment ions resulting from the distribution of internal energy induced by intermolecular collisions that occur in the gas phase, which is inherently presented by the chemical ionization processes. In the case

Hexane: $R + DB = 6 - 14/2 + 0/2 + 1 = 6 - 7 + 1 = 0$

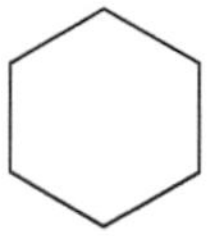

Cyclohexane: $R + DB = 6 - 12/2 + 0/2 + 1 = 6 - 6 + 1 = 1$

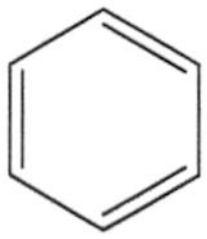

Benzene: $R + DB = 6 - 6/2 + 0/2 + 1 = 6 - 3 + 1 = 4$

FIGURE 1.11 Example calculation of rings and double bonds associated with hexane, cyclohexane, and benzene.

of electron ionization, there tends to be a dramatically lower chance of a collision due to the vacuum associated with these types of sources. In this case, the ionization occurs by the interaction with an electron, a process that occurs in a very quick time frame. The energy in this case is distributed via vibrations within the ion and any fragmentation occurs within 10^{-8} s. The daughter ions within a mass spectrum enable the fragmentation patterns to be determined due to the loss of logical neutral molecules.

Fragmentation patterns can be quite complex, take, for example, the positive ion mass spectrum generated for the polyethylene terephthalate (PET) monomer with a molecular ion at m/z 193.

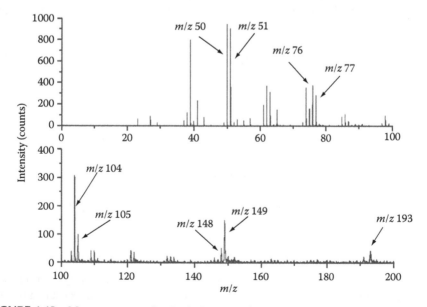

FIGURE 1.12 Mass spectrum of polyethylene terephthalate (PET) monomer.

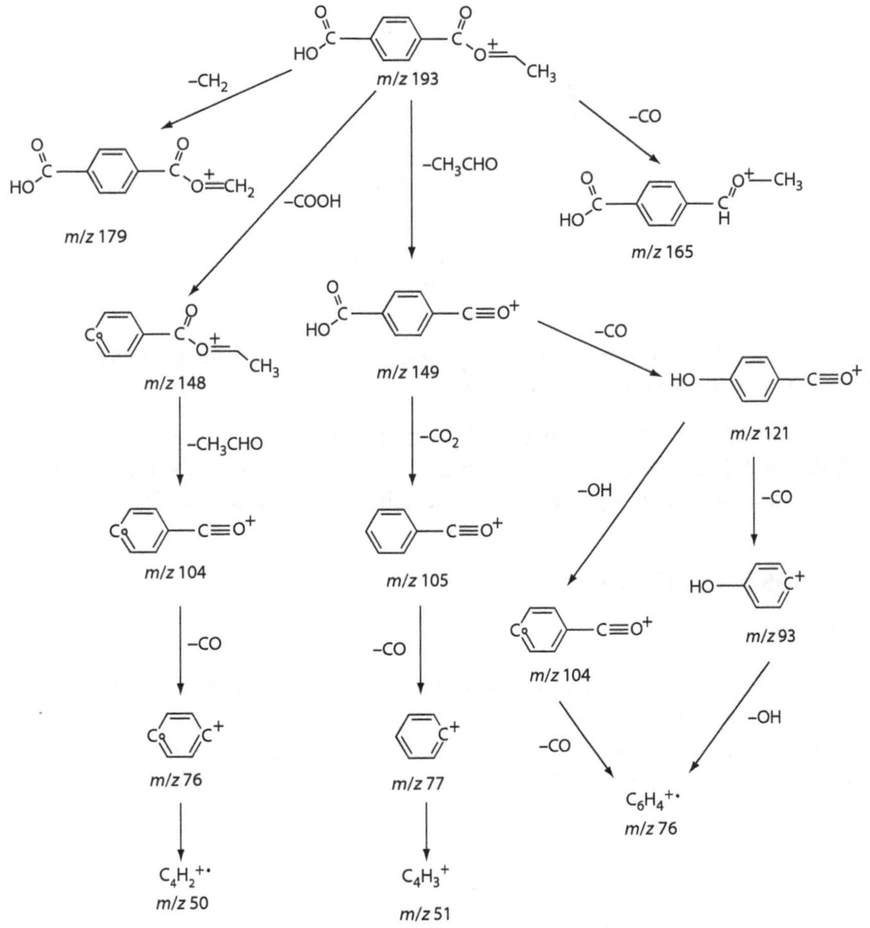

FIGURE 1.13 Fragmentation pathways of PET from the molecular ion [M + H]+ *m/z* 193 showing both the positively charged ion and radical ion pathways.

The pathways in which the molecular ion can fragment is quite complex. One fragmentation process occurs via a positively charged ion pathway and the other via the formation of positively charged radical species. The fragmentation pathways are outlined in Figures 1.12 and 1.13.

Analysis of the fragmentation patterns enables the molecular structure to be confirmed and is particularly useful if the molecular ion is not present in the mass spectrum. These principles can be applied to large molecules and have proven to be particularly fruitful in the protein sciences enabling the amino acid sequencing of proteins with the aid of ESI and MALDI mass spectrometry. The amino acids fragment or cleave at distinct points along the protein backbone; the mass spectral peaks associated with these cleavages can be determined and processed against a known database for identification purposes. Henzel et al. [32,33] were the first to present the use of this technique, however, around this time several research groups were perusing similar research [34–36].

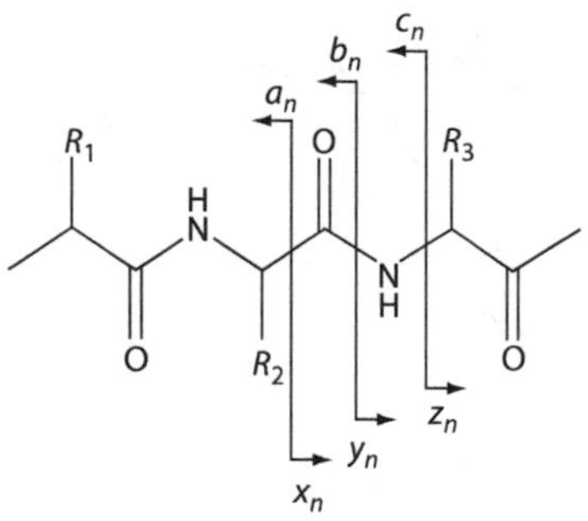

FIGURE 1.14 The distinct peptide cleavage points and nomenclature. (From Haroudin, J., *Mass Spectrom. Rev.*, 26, 672, 2007.)

In Figure 1.14, the distinct peptide cleavage points are identified.

A detailed discussion of the fragmentation characteristics is beyond the scope of this text, however, several good texts on the subject are available [1,2].

1.6 MASS ANALYZERS

1.6.1 MAGNETIC SECTOR

The use of magnets in mass analyzers has been common since the early days of mass spectrometry, all the way back to Wein's era [5]. In general, a magnet is positioned over one region of the ions' flight path. An ion of charge z moving with velocity v that transverses a magnetic field B at right angles to the direction of the field will experience a centrifugal force:

$$F = zevB \tag{1.3}$$

When this force is equal to the centripetal force, the ions adopt a circular path or radius r:

$$zevB = \frac{(mv^2)}{r} \tag{1.4}$$

$$r = \frac{(mv)}{zeB} \tag{1.5}$$

Therefore, ions of a particular charge z moving through a fixed magnetic field B, the radius of their path is dependant only upon their momentum mv. Since the initial kinetic energy of the ions $(1/2mv^2)$ equals z eV, the initial velocity of the ions is dependent on the potential V through which they are accelerated. If we rearrange to solve for v^2, we get

$$v^2 = \frac{(2z\,eV)}{m} \tag{1.6}$$

From Equation 1.4, the following can be derived:

$$v = (zeBr)m \qquad (1.7)$$

Squaring both sides of Equation 1.7 gives

$$v^2 = \frac{(zeBr)^2}{m^2} \qquad (1.8)$$

Following this, we can combine Equations 1.6 and 1.8 to represent the relationship for m/z:

$$\frac{m}{z} = \frac{eB^2r^2}{2V} \qquad (1.9)$$

Therefore, specific values of V or B allow unique ions in mass to charge to pass through the magnetic field to the detector. This method is used to isolate molecules of a particular mass prior to the detection system.

1.6.2 QUADRUPOLE

The magnetic sector mass analyzer has limited capability for chromatographic coupling with the most common mass analyzer coupled to liquid chromatography systems the quadrupole, a system that requires ultrahigh vacuum. The analyzer is known as a quadrupole (or quad) because it is made up of four parallel metal rods upon which is applied a constant voltage and an oscillating radio frequency. This allows all ions with a resonant frequency to pass through to the detector with non-resonant ions filtered out. To monitor ions over a particular mass range, the voltages applied to the rods are varied within a short time frame.

1.6.3 TIME OF FLIGHT

The concept of linear time-of-flight mass spectrometer (TOFMS) was first proposed by W. Stephens to the American Physical Society meeting in Cambridge, Massachusetts, in 1946 [10]. The first commercial TOF instrument was marketed by the Bendix corporation in the late 1950s, the design of which was based on the Wiley and MacLaren instrument [38].

The basic ion optics of a TOF instrument is relatively straightforward, ions need to be extracted from an ion source in short pulses and then directed down a flight tube to a detector. The time taken to travel the length of the flight (or drift) tube depends on the mass of the ion and its charge. For singly charged ions ($z = 1$; $m/z = m$), the time taken to traverse the distance from the source to the detector is proportional to a function of mass. The greater the mass of the ion, the slower it is in arriving at the detector.

The first step of the secondary ion is acceleration through an electric field (E volts). With the usual nomenclature (m = mass, z = number of charges on an ion, e = the charge on an electron, and v = the final velocity reached on acceleration), the kinetic energy ($mv^2/2$) of the ion is given in the following equation:

$$\frac{mv^2}{2} = zeE \tag{1.10}$$

This equation can be rearranged to form the following equation:

$$v = \left[\frac{(2zeE)}{m}\right]^{1/2} \tag{1.11}$$

If the distance from the ion source to the detector is d, then the time (t) taken for an ion to transverse the flight tube is given by the following equation:

$$t = \frac{d}{v} = \frac{d}{[(2zeE)/m]^{1/2}} = d\left[\frac{(m/z)^{1/2}}{(2eE)^{1/2}}\right] \tag{1.12}$$

In the aforementioned equation, d is fixed, E is held constant in the instrument during analysis, and e is a universal constant. Thus, the time of flight of an ion t is directly proportional to the square root of the m/z multiplied by a constant.

The resolving power of a TOF instrument is limited, particularly at high mass, for two major reasons: one inherent in the technique and the other a practical problem. First, the flight times are proportional to the square root of the m/z. The difference in the flight times (t_m and t_{m+1}) for two ions separated by unit mass is given by Equation 1.13:

$$t_m - t_{m+1} = \Delta t = \left[\left(\frac{m}{z}\right)^{1/2} - \left(\frac{(m+1)}{z}\right)^{1/2}\right] \times c \tag{1.13}$$

where c is a constant. As m increases, Δt becomes progressively smaller; thus, the difference in arrival times at the detector becomes small making it harder to differentiate the two masses. The second practical problem arises due to not all ions of any given m/z value reach the same velocity after acceleration nor are they formed at exactly the same point from the sample. This results in a longer window of time at the detector for ions of the same m/z to arrive, resulting in an overlap of ions with very similar m/z ratios.

This problem can be resolved by using a reflectron, an example of which is shown in Figure 1.15.

A reflectron consists of a series of ring electrodes each of which is placed an electric potential. The first ring has the lowest potential and the last ring the highest to produce an electrostatic field that increases from the front end of the reflectron to the back. The field is at the end of the flight path of the ions and has a polarity the same as that of the ions (a retarding potential). In the reflectron, the ions come to a stop and are then accelerated in the opposite direction. The reflectron is often at a

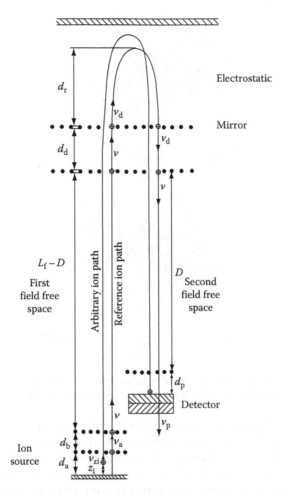

FIGURE 1.15 Reflectron diagram. (Reproduced from Ioanoviciu, D., *Int. J. Mass Spectrom.*, 206, 211, 2001.)

slight angle to the line of the flight of the ions so that the reflected ions travel on a slightly deflected line toward the detector. If we consider ions of any given *m/z* value, the faster ions, having greater kinetic energy, travel further into the reflectron before being reflected than the slower ions. As a result, the faster ions spend slightly more time in the reflectron; therefore, the faster ions have further to travel to the detector than the slower ones. The faster ions catch up to the slower ions over the second drift region (from reflectron to detector), so the ions of the same mass arrive together.

1.6.4 ION TRAP

The ion trap was originally invented in 1953, by the 1989 winner of the Nobel Prize for physics, Wolfgang Paul and Steinwedel [40]. Modern day three-dimensional (3D) quadrupole ion traps contain three electrodes, a hyperbolically shaped ring electrode and two hyperbolically shaped end caps electrodes. The end cap electrodes have

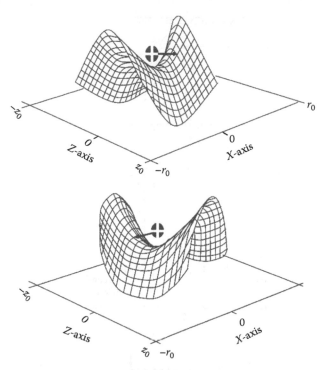

FIGURE 1.16   The quadrupole potential is depicted for a plane that intersects the trap's symmetry axis. (From Fernández, L.E.M., *Carbohydr. Polym.*, 68, 797, 2007.)

orifices for the injection and extraction of ions. The ions are trapped or stored by applying a high-voltage oscillating potential known as the fundamental radio frequency (RF) to the ring electrode with the end caps in an electrically grounded state.

Fernández [41] describes the confinement of the ions within the trap nicely as illustrated by the quadrupole potential that is represented in Figure 1.16. Ions in the central part of the trap are confined in the axial z-direction; however, in the radial r-direction, ions are accelerated toward the end caps and are not confined. Simultaneous confinement (trapping) of the ion in both directions can be obtained by changing the direction of the field every time the ion is approaching the electrodes. For use as a mass analyzer, the trapped ions need to be extracted from the ion trap in a mode dependent of their m/z value. This is achieved by two possible modes of operation, the mass-selective instability mode and the mass-resonant ejection mode.

1.6.5   TANDEM MASS SPECTROMETRY

Tandem mass spectrometry requires the use of two or more stages of mass analysis, a method that aids considerably in the structural elucidation of a molecule. The most common way in which tandem mass spectrometry works is via the isolation of a precursor ion, which can be further fragmented prior to analysis by a second mass spectrometry system (Figure 1.17).

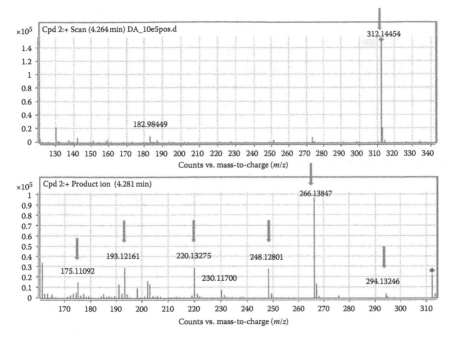

FIGURE 1.17 Highlighting the tandem mass spectrometry of domoic acid produced by QToF technology. (In the top spectrum, the molecular ion is observed with the fragment ions observed in the bottom spectrum.)

The subsequent dimensions of mass spectral stages allow the pieces of the molecular jigsaw to be established. In conjunction with high-resolution mass spectrometry instruments, multidimensional mass spectrometry has the potential to present full structural elucidation of a target molecule. Two methods of tandem mass spectrometry commonly coupled to liquid chromatography include triple quadrupole (QqQ) and a quadrupole time-of-flight systems.

1.6.5.1 Triple Quadrupole

In Melbourne in 1978, a QqQ, which until that time had only been used to study photodissociation processes, was adapted by Yost and Enke [42] for mass spectrometry investigations. In this system, the first quadrupole acts as an ion isolation source by adjusting the RF and high voltages to allow a particular ion through. The second quadrupole often denoted by a lower case q in the region in which an induced collision is brought about by reaction with gaseous ions. The final quadrupole is used in a traditional manner in order to detect the fragment ions (Figure 1.18).

1.6.5.2 Quadrupole Time of Flight

The quadrupole time of flight (QToF) isolates precursor ions and generates the fragment ions in a similar manner to the triple quad. In this instance, however, in place of the third quadrupole, a time-of-flight analyzer is present. The addition of the ToF enables high-resolution mass spectral data to be generated (Figure 1.19).

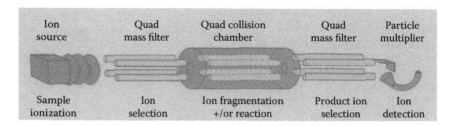

Ion source	Quad mass filter	Quad collision chamber	Quad mass filter	Particle multiplier
Sample ionization	Ion selection	Ion fragmentation +/or reaction	Product ion selection	Ion detection

FIGURE 1.18 Schematic of triple quadrupole mass spectrometer. (From Yost, R.A. and Enke, C.G., *Anal. Chem.*, 51, 1251, 1979.)

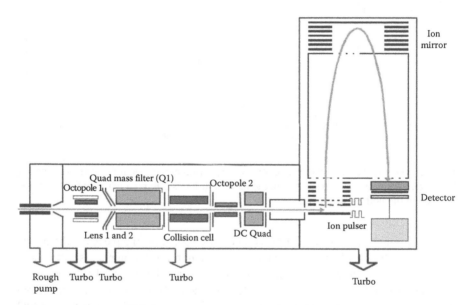

FIGURE 1.19 Schematic of a quadrupole time-of-flight mass spectrometer. (Courtesy of Agilent Technologies, 6510 QTOF Mass Spectrometers. 2004, Agilent Technologies, Australia.)

1.7 RESOLUTION

Resolution is one of the defining characteristics of a mass spectrometer, and can be described as the ability to distinguish between ions differing in the quotient mass/charge by a small increment. It may be characterized by giving the peak width, measured in mass units, expressed as a function of mass, for at least two points on the peak, specifically at 50% and at 5% of the maximum peak height [31].

It is important to note that slightly different methods for the determination of resolving power tend to be used depending on the mass spectrometer in question. The greater the level of resolving power means that a mass spectrometer can better separate any two mass spectral peaks. Historically, with older style magnetic sectors, the resolving power ($M/\Delta M$) is defined as the separating potential required to produce peaks with no lower than a 10% valley between the two peaks. Unit resolution

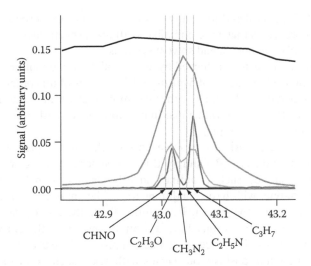

FIGURE 1.20 Peak comparisons from different mass spectrometers. (From DeCarlo, P.F. et al., *Anal. Chem.*, 78, 8281, 2006.)

is often used with respect to quadrupole mass spectrometry such that two successive integer masses will be resolved up to a given mass [44]. The modern description outlined in the previous paragraph known as full width half maximum (FWHM) is employed in ion trap, ion cyclotron, and time-of-flight mass spectrometry, those that generate high-resolution data (Figure 1.20).

The rapid development of mass spectrometry in the last 10 years has produced systems with resolution so high that very accurate characteristic information can be gained on a molecule of interest. The resolving power achievable by Thomson [6] in 1913 was a mere 13 while by 1998 Marshall and coworkers had achieved resolution in the order of 8,000,000 [46]. Due to this mass accuracy, the accuracy of the m/z generated by the mass spectrometer has greatly improved. The mass accuracy can be defined as the difference between the mass observed and the theoretical mass of the molecule. This is generally given in parts per million, however, it should technically be described in millimass units (mmu). The greater the accuracy, there is a better chance for a positive match of a chemical formula from the mass spectral data. In general, accurate mass measurement is nonambiguous up to 300 daltons (Da) due to the small number of potential molecular formulae. With increasing mass to charge ratio (m/z), the number of possible formulae dramatically increases, making identification much more difficult [47].

1.8 ION ENHANCEMENT/SUPPRESSION SOLVENT CONSIDERATIONS

ESI and APCI are the main ion sources coupled to liquid chromatography in modern laboratories. Therefore, consideration is given here to the implications of chromatography on ionization.

The ionization process is paramount to achieving good results in mass spectrometry. Upon coupling chromatographic separations to mass spectrometry, many factors need to be considered to minimize degradation of this process. This limits the separation processes that are compatible with mass spectrometry due to achievable flow rates, pH conditions, and solvent systems.

The flow rates that can be achieved by analytical scale monolithic column chromatography can reach in excess of 5 mL/min. Flows rates of this order are much too high for ESI and APCI. Therefore, either the mobile phase flow rate is tailored to suit the mass spectral system, with a maximum flow rate of 1 mL/min, or the mobile phase flow is split post separation.

In general, the best sensitivity is achieved with the analyte presented in an ionized state in the liquid phase. Put simply, this can be done by using an acidic mobile phase for basic analytes, such as amines, ideally placing the pH two units below the pK_a of the analyte, with basic conditions for acidic analytes, such as carboxylic acids and phenols with the ideal pH two units above the pK_a of the analyte [48]. The pH of a solvent will not only have a dramatic effect on the chromatographic performance, it will also influence the ionization process. The use of acid conditions favors the production of positive ion species with acids, such as formic acid, acetic acid, and trifluoroacetic acid particularly suitable, ammonium hydroxide is the primary agent used to promote negative ion generation. The pH conditions required for mass spectrometry may often not be suitable for the chromatographic task at hand. Overcoming this problem can be as simple as the adjustment of the pH of the solvent system post-chromatographic separation.

The generation of the charged droplets is the fundamental process required for successful electrospray results, a process that is driven by electrophoretic migration, which promotes the separation of the anions from the cations [49,50]. Due to this, the choice of solvent is critical with the conductivity of the solvent required to be sufficient for the process to occur. Both polar and nonpolar solvents can be used in APCI, whereas only polar or medium polar solvents are suitable for ESI [51]. There is a range of particularly applicable solvents that can be used including acetonitrile, water, methanol, ethanol, propanol, isopropanol, tetrahydrofuran, and chloroform. Regarding APCI solvents such as hydrocarbons, carbon tetrachloride and carbon disulfide may also be utilized. It is important to note that optimization of the preferential solvent system needs to be assessed on a sample to sample basis. In general, water is a less suitable solvent than the organic solvents for ESI, because of the difficulty in maintaining a stable spray as the charge separation process is reduced by the higher viscosity of water. The regular users of ESI would be aware that the production of negative ions is less efficient than positive ions, this is due to the electric discharge occurring at a lower electric field in negative ion mode [51]. There is a decrease in the ability of gas-phase ions to be produced with water, which can be attributed to high surface tension, low volatility, and the efficient solvation of ions in water. The higher sensitivity achievable in organic solvents systems is due to low surface tension, higher volatility, and less efficient solvation of ions in organic solvents [51,52]. Evaporation of the solvent plays an important role here, with the evaporation of water considerably slower than for most organic solvents employed for HPLC. The reduced potential for evaporation can also be attributed to water's higher surface tension, which enables the formation of large droplets.

Small amounts of buffers will be tolerated within mass spectrometry, however, phosphate buffers are to be avoided. Up to 10 mM levels of buffers such as ammonium acetate, ammonium formate can be used; it is important to note, however, that often buffer concentrations an order of magnitude higher than this are commonly used for chromatographic separations.

1.9 CONTAMINATION AND PRACTICAL CONSIDERATIONS

One of the greatest analytical advantages of mass spectrometry is its ability to achieve excellent sensitivity. While this is great for targeting our analyte of choice, the mass spectrometer is also inherently sensitive to contamination. Put simply, great care must be taken to ensure possible sites of contamination are controlled. Every container that comes into contact with a sample, solvent, or additive for mass spectrometry must be pristine including any traces of detergent or soaps both of which should be avoided when cleaning apparatus. Salts may be a particular problem for the mass spectrometry task at hand with vials, pipettes, paper, and glassware being common points of contamination. Glassware that is sodium free, while quite expensive, may be required. Further to the salt issue, process paper can carry a range of contaminants including industrial detergents. Plastics, including those used to make bottles, tubing, and trays, possess a series of contamination risks including phthalates, silicones, antioxidants, and UV stabilizers. Phthalates are commonly associated with solvent contamination and are often present in material that has been cleaned using a dishwasher. The phthalates have been known to concentrate on the separation column, only to be released during a gradient solvent run. A common contamination observed includes the m/z +44 series or a contamination that consists of a series of mass spectral peaks 44 mass units apart. The +44 is a common indicator of polymer contamination including; from polyethylene glycol (PEG) and other ethoxylated polymers. PEG has been shown to be a problem in the analysis of rat neuropeptides as outlined in Figures 1.21 and 1.22.

Sample carry over is a problem in all forms of chromatography and may be of even higher consideration when mass spectrometry is coupled due to the increased sensitivity of the detector. In any case, sample carryover must be limited by the use of appropriate analyte solvent combinations and routine injection needle wash cycles between sample injections.

All scientists have seen the stereotypical "bad" laboratory with rubbish, samples, and glassware strewn about the place, basically a den for sample contamination. Even in a clean, well-managed laboratory, we recommend dedicated glassware, reserved for mass spectrometry use that is regularly cleaned using detergent and chromate-free techniques. Further to this, it is good practice to burn off any organic residue that may be present in an oven or furnace. Contact of the sample matrix or important internal components of the mass spectrometer with paper should be avoided at all costs, with any tool used for maintenance cleaned thoroughly to avoid oil contamination. Solvent containers need to be kept covered as contamination can occur, including phthalates in air dissolving in organic solvents.

Metals are a source of concern for protein and peptide analyses and contact with any metal should be avoided as chelation of the metal by a protein may occur. Filtration of

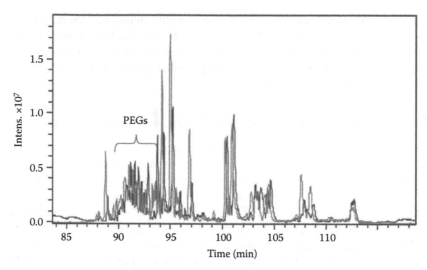

FIGURE 1.21 Overlaid base peak chromatograms (BPCs) for the 100 mM fractions from rat brain hypothalamus extracts. Interfering PEG peaks covered the range from 90 to 95 min and overlapped with the neuropeptide peaks. (From Mihailova, A. et al., *J. Separ. Sci.*, 29, 576, 2006.)

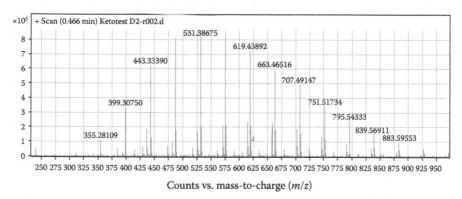

FIGURE 1.22 +44 Da series, characteristic polymer contamination.

all solvents and water prior to analysis is paramount for mass spectrometry purposes care should be taken, the use of a 0.45 μm filtration routine should be applied to all solvents. Generally, mass spectrometers contain an inline filter between the chromatographic column and the mass spectrometry source. This filter should be monitored and replaced as part of a routine maintenance program. Algae growth in solvent bottles, particularly in buffered water, needs to be monitored and such solvents should not be used any longer than 1 week to avoid a "minor mass spectrometry algal bloom." Upon the purchase of a new column or when changing the conditions of use of an existing column, it is good practice to condition the column prior to connecting the solvent stream to the mass spectrometer so as to avoid any unforeseen contamination of the mass spectrometer. DMSO and DMF are not friends with our mass spectrometry systems and where possible it is prudent to clean samples prior to analysis.

1.10 QUANTITATION

Quantitative information can be gained via mass spectrometry in a similar fashion to more common HPLC and GC systems with a correlation between mass spectral peak area and the quantity of a particular analyte. Mass spectrometry presents a particularly dynamic detection process and therefore great care needs to be taken in developing a methodology with conditions that are reproducible from sample run to sample run. Importance must be placed on the mass spectral peaks to be used with a peak of the highest mass possible, most likely to be structurally significant. The monitoring of more than one peak is helpful including monitoring the molecular ion and any important fragment ions. The sensitivity of an instrument should be taken into account and accurate notes on the mass spectrometry settings should be taken as changes to optics, etc., can have a large effect on sensitivity. The detection limit of the system needs to be determined and can be considered the signal achievable at least three times the background noise generated from the smallest sample concentration. There are several methodologies that can be employed for quantitative analysis including the external standard method, internal standard method, and isotopic dilution.

The external standard method utilizes a known quantity of the molecule of interest (often a synthetic sample) and measuring the mass spectral signal intensity. A range of concentrations of the standard are analyzed and a calibration curve can be generated. The real-world sample is then processed and the intensity of the signal used against the calibration data in order to calculate the concentration of the analyte.

The internal standard method is more robust than the external standard method and enables limitation of potential errors. An internal standard is selected based on its chemical characteristics and how closely they resemble the analyte. Due to their similarity, the internal standard will undergo very similar losses during any extraction process or during ionization. In a similar manner to the external standard method, a calibration curve is generated for comparison with the real-world sample.

The isotopic dilution method exploits the chemical characteristic requirements of the internal standard to the fullest degree. In this instance, a molecule with a variation in isotopic nature is used, most commonly deuterated molecules. This quantitative method can be very accurate, however, a standard is required for any analyte of interest.

1.11 DATA HANDLING: CHEMOMETRICS

Data handling is an issue often associated with scientific pursuits, mass spectrometry is no exception. Mass spectrometry generates a vast amount of data points in any given mass spectrum; if the analyst is looking for an unknown entity, this data set can be very daunting. Chemometrics, data processing techniques, gathered pace in the mid-1970s, subsequently it has proven to be a particularly useful tool, more so for mass spectroscopists in recent years. There are a range of mathematical procedures that have been successfully employed for mass spectrometry data analysis including principle component analysis (PCA) [54], partial least squares (PLS) [55,56], discriminant function analysis (DFA) [54], and hierarchical cluster analysis (HCA) [57].

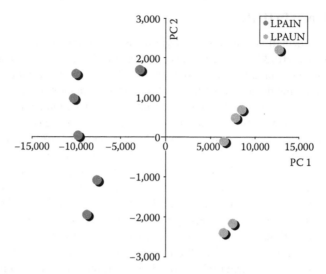

FIGURE 1.23   A typical PCA plot for mass spectrometry data.

These procedures are commonly coined "multivariate analysis" as they all deal with data sets that contain many variables. Generally, the main aim of all of these approaches is to simplify data and measure mathematically the relationship (similarity, difference, significance, variance) between one data point and another.

Principle component analysis involves a rotation and transformation of the axes of the original data into new axes. The formation of the new axes is made along the directions of maximum variance, with the axes orthogonal, which makes the new data uncorrelated. Inherently in this process, the dimensionality of the parameter space is reduced. Furthermore, chemical or physiochemical assessment of the data can be inferred as PCA can highlight structure in the data series [58]. A typical PCA plot for mass spectrometry data is presented in Figure 1.23.

Discriminant function analysis is a technique that comes under the banner of supervised learning and is used for pattern recognition. In its most simplistic terms, discriminant analysis is a statistical technique to find a set of descriptors that can be used to detect and rationalize separation between activity classes [31]. The method is used to determine which continuous variables discriminate between two or more of the groups by determining their descriptive statistics, such as mean and standard deviation [58].

The IUPAC described partial least squares as the "projection to latent structures (PLS) is a robust multivariate generalized regression method using projections to summarize multitudes of potentially collinear variables" [31]. Hierarchical cluster analysis groups gather the data that are in close proximity to each other. The procedure involves repeated calculations of distance between data points, and between the clusters of data. A dendrogram is used to visualize the data, an example of which is shown in Figure 1.24.

One could go into great detail on the methods for chemometrics that can enhance mass spectrometry experiments and several good texts are available [58]. The area of chemometrics is very large and can be a daunting proposition for a classically trained

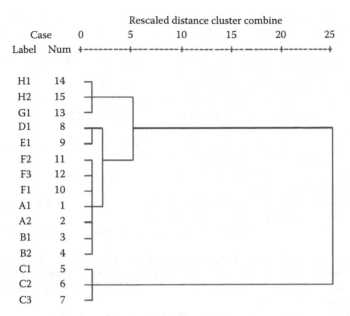

FIGURE 1.24 An example dendrogram. (From Xu, L. et al., *Anal. Chim. Acta*, 633, 136, 2009.)

analytical chemist. Because of this, it is important to note that often the chemometrician processing the data is not the person responsible for collecting the data, care needs to be taken to generate useful data, as the saying goes in chemometric circles garbage in equals garbage out. However, chemometrics is a valuable tool for mass spectrometry and used in a suitable manner can aid many areas of mass spectrometry-coupled chromatographic research.

REFERENCES

1. deHoffmann, E.; Stroobant, V.; *Mass Spectrometry Principles and Applications,* John Wiley & Sons Ltd., Chichester, U.K., 2007.
2. McLafferty, F.W.; Turecek, F.; *Interpretation of Mass Spectra*, University Science Books, Mill Valley, CA, 1993.
3. Thomson, J.J.; *Philosophical Magazine*, 1897, 44, 293.
4. Goldstein, E.; *Berlin Ber.*, 1886, 39, 691.
5. Wein, W.; *Verhanal. Phys. Ges.*, 1898, 17.
6. Thomson, J.J.; *Rays of Positive Electricity and Their Application to Chemical Analysis*, Longmans, Green and Co., London, U.K., 1913.
7. Dempster, A.J.; *Physical Reviews*, 1918, 11, 316.
8. Aston, F.W.; *Philosophical Magazine*, 1919, 38, 707.
9. Cameron, A.E.; Eggers, D.F.; *Review of Scientific Instruments*, 1948, 19, 605.
10. Stephens, W.; *Physical Reviews*, 1946, 69, 691.
11. McLaffery, F.W.; *Applied Spectroscopy*, 1957, 11, 145.
12. Gohlke, R.S.; *Analytical Chemistry*, 1959, 31, 535.
13. Munson, M.S.B.; Field, F.H.; *Journal of the American Chemical Society*, 1966, 20, 359.
14. Karataev, V.I.; Shmikk, D.V.; Mamyrin, B.A.; *Soviet Physics Technical Physics-USSR*, 1972, 16, 1177.

15. Arpino, P.; Baldwin, M.A.; McLaffer, F.W.; *Biomedical Mass Spectrometry*, 1974, 1, 80.
16. Olivares, J.A.; Nguyen, N.T.; Yonker, C.R.; Smith, R.D.; *Analytical Chemistry*, 1987, 59, 1230.
17. Fenn, J.B.; Mann, M.; Meng, C.K.; Wong, S.F.; Whitehouse, C.M.; *Science*, 1989, 246, 64.
18. Gelpi, E.; *Journal of Mass Spectrometry*, 2002, 37, 241.
19. Gaskell, S.; *Journal of Mass Spectrometry*, 1997, 32, 1378.
20. Abian, J.; *Journal of Mass Spectrometry*, 1999, 35, 157.
21. Karas, M.; Hillenkamp, F.; *Analytical Chemistry*, 1988, 60, 2299.
22. Aebersold, R.; Mann, M.; *Nature*, 2003, 442, 198.
23. Barber, M.; Bordoli, R.S.; Sedgwick, R.D.; Tyler, A.N.; *Journal of the Chemical Society—Chemical Communications*, 1981, 325.
24. Surman, D.J.; Vickerman, J.C.; *Journal of the Chemical Society—Chemical Communications*, 1981, 324.
25. Dobberstein, P.; Korte, E.; Meyerhoff, G.; Pesch, R.; *International Journal of Mass Spectrometry and Ion Processes*, 1983, 46, 185.
26. Oka, H.; Ito, Y.; Ikai, Y.; Kagami, T.; Harada, K.; *Journal of Chromatography A*, 1998, 812, 309.
27. Oka, H.; Ikai, Y.; Hayakawa, J.; Harada, K.; Asukabe, H.; Suzuki, M.; Himei, R.; Horie, M.; Nakazawa, H.; Macneil, J.D.; *Journal of Agricultural and Food Chemistry*, 1994, 42, 2215.
28. Watson, D.G.; Davidson, A.G.; Knight, B.I.; *Rapid Communications in Mass Spectrometry*, 1997, 11, 415.
29. Michalke, B.; *Trac-Trends in Analytical Chemistry*, 2002, 21, 142.
30. Szpunar, J.; Lobinski, R.; *Hyphenated Techniques in Speciation Analysis*, The Royal Society of Chemistry, Cambridge, U.K., 2003.
31. Nic, M.; Jirat, J.; Kosata, B.; in *IUPAC*, 2008.
32. Henzel, W.J.; Stults, J.T.; Watanabe, C.; in *Third Symposium of the Protein Society*, Seattle, WA, 1989.
33. Henzel, W.J.; Billeci, T.M.; Stults, J.T.; Wong, S.C.; Grimley, C.; Watanabe, C.; *Proceedings of the National Academy of Sciences of the United States of America*, 1993, 90, 5011.
34. Mann, M.; Hojrup, P.; Roepstorff, P.; *Biological Mass Spectrometry*, 1993, 22, 338.
35. Pappin, D.J.C.; Hojrup, P.; Bleasby, A.J.; *Current Biology*, 1993, 3, 327.
36. James, P.; Quadroni, M.; Carafoli, E.; Gonnet, G.; *Biochemical and Biophysical Research Communications*, 1993, 195, 58.
37. Haroudin, J.; *Mass Spectrometry Reviews*, 2007, 26, 672.
38. Wiley, W.C.; MacLaren, I.H.; *Review of Scientific Instruments*, 1955, 26, 1150.
39. Ioanoviciu, D.; *International Journal of Mass Spectrometry*, 2001, 206, 211.
40. Paul, W.; Steinwedel, H.; *Zeitschrift Fur Naturforschung Section A—A Journal of Physical Sciences*, 1953, 8, 448.
41. Fernández, L.E.M.; *Carbohydrate Polymers*, 2007, 68, 797.
42. Yost, R.A.; Enke, C.G.; *Journal of the American Chemical Society*, 1978, 100, 2274.
43. Yost, R.A.; Enke, C.G.; *Analytical Chemistry*, 1979, 51, 1251.
44. Cody, R.B.; in B.N. Pramanik, A.K. Ganguly, M.L. Gross (eds.), *Applied Electrospray Mass Spectrometry*, Marcel Dekker Inc., New York, 2002, p. 434.
45. DeCarlo, P.F.; Kimmel, J.R.; Trimborn, A.; Northway, M.J.; Jayne, J.T.; Aiken, A.C.; Gonin, M.; Fuhrer, K.; Horvath, T.; Docherty, K.S.; Worsnop, D.R.; Jimenez, J.L.; *Analytical Chemistry*, 2006, 78, 8281.
46. Shi, S.D.H.; Hendrickson, C.L.; Marshall, A.G.; *Proceedings of the National Academy of Sciences of the United States of America*, 1998, 95, 11532.
47. Bristow, A.W.T.; Webb, K.S.; *Journal of the American Society for Mass Spectrometry*, 2003, 14, 1086.
48. Gao, S.; Zhang, Z.-P.; Karnes, H.T.; *Journal of Chromatography B*, 2005, 825, 98.

49. Hayati, I.; Bailey, A.I.; Tadros, T.F.; *Journal of Colloid and Interface Science*, 1987, 117, 205.
50. Bruins, A.P.; *Journal of Chromatography A*, 1998, 794, 345.
51. Kostiainen, R.; Kauppila, T.J.; *Journal of Chromatography A*, Editors' Choice III, 2009, 1216, 685.
52. Kostiainen, R.; Bruins, A.P.; *Rapid Communications in Mass Spectrometry*, 1996, 10, 1393.
53. Mihailova, A.; Lundanes, E.; Greibrokk, T.; *Journal of Separation Science*, 2006, 29, 576.
54. Goodacre, R.; York, E.V.; Heald, J.K.; Scott, I.M.; *Phytochemistry*, 2003, 62, 859.
55. Daszykowski, M.; Wu, W.; Nicholls, A.W.; Ball, R.J.; Czekaj, T.; Walczak, B.; *Journal of Chemometrics*, 2007, 21, 292.
56. Sorensen, H.A.; Petersen, M.K.; Jacobsen, S.; Sondergaard, I.; *Journal of Mass Spectrometry*, 2004, 39, 607.
57. Powell, D.W.; Weaver, C.M.; Jennings, J.L.; McAfee, K.J.; He, Y.; Weil, P.A.; Link, A.J.; *Molecular and Cellular Biology*, 2004, 24, 7249.
58. Adams, M.A.; *Chemometrics in Analytical Spectroscopy*, The Royal Society of Chemistry, Cambridge, U.K., 1995.
59. Xu, L.; Han, X.; Qi, Y.; Xu, Y.; Yin, L.; Peng, J.; Liu, K.; Sun, C.; *Analytica Chimica Acta*, 2009, 633, 136.

2 Analytical Aspects of Modern Gas Chromatography: Mass Spectrometry

Shin Miin Song and Philip J. Marriott

CONTENTS

2.1 Introduction 31
    2.1.1 Scope of the Chapter..33
2.2 GC-MS Setup..33
2.3 Ionization Techniques...35
    2.3.1 Electron Ionization 36
        2.3.1.1 Full-Scan and Selected-Ion Monitoring Modes..................36
        2.3.1.2 Electron Impact: A "Hard" Ionization Technique 37
    2.3.2 Chemical Ionization...39
2.4 Mass Analyzers...42
    2.4.1 Quadrupole Mass Analyzers 43
    2.4.2 Magnetic Sector Mass Analyzers 45
    2.4.3 Ion Trap Mass Analyzer 46
    2.4.4 Tandem Mass Spectrometry 47
    2.4.5 Time-of-Flight Mass Analyzer..48
2.5 Data Evaluation...50
2.6 New Developments and Emerging Trends 58
References..59

2.1 INTRODUCTION

It is now apparent that the coupling of gas chromatography (GC) with mass spectrometry (MS), following development of GC in the 1950s has revolutionized the way analytical problems for volatile chemical analytes are approached today, although it is entirely logical that this coupling has had the impact that it has. This is the most successful of all hyphenation to produce a synergy offering much more than either technique alone.

Modern GC-MS is the progeny of technological advancements in chromatography, derivatization methods, high performance electronics, computational software, material science, robotic automation, and instrumental designs. The different state-of-the-art GC-MS configurations available today are manifestations of these advancements as their improved sensitivity, versatility, and specificity push detection limits even lower to enable access to new sample materials that would not have been possible decades before. These innovations are the primary reasons that keep GC-MS relevant through almost five decades of technological evolution.

As GC and MS are essentially two fundamentally different, independent analytical techniques, the hyphenation of GC and MS exploits the best features of each technique to give the most information within a single chromatographic run time: GC to provide high separation efficiency of volatile components in a sample mixture to give retention time information; MS to provide a unique fragmentation pattern of the unknown component to facilitate the deduction of its identity. This is elegantly supported by the compatibility of the two dimensions, each employing gaseous sample phase in the implementation.

A competent GC-MS procedure requires an adequate knowledge of the capabilities and limitations of the various GC-MS configurations and operational modes, as well as a sufficient understanding of their underlying principles. Be it the development of a GC-MS procedure or the purchase of a new GC-MS equipment, it is helpful to begin the decision making process by answering these few questions:

- What are the short-term and long-term analytical goals to be achieved?
- What are the requirements for the application(s) in terms of specificity, sensitivity, and/or speed of data acquisition?
- What is the MS resolution required of the application(s)?
- What are the chemical characteristics of the analytes, e.g., mass range, sample complexity?
- What is the laboratory budget?
- What are the costs that will be involved, e.g., purchase cost, setup cost, maintenance cost, consumables cost, training cost and personnel cost (salary), etc.?
- What is the estimated sample throughput of the analytical procedure?
- Are there opportunities for optional system upgrade features for keeping up-to-date with future analytical needs?

While cost will invariably be the major deciding factor in the purchase of a new GC-MS equipment, it is just as important to appreciate the performance attributes and capabilities of different GC-MS instrumentations offered by different manufacturers, as well as considerations of reliability and maintenance.

While liquid chromatography coupled to mass spectrometry (LC-MS) is becoming increasingly popular for the analysis of low volatility compounds, innovations in derivatization techniques to confer volatility and stability extend both GC and GC-MS to a much wider suite of analytes [1]. The Phenomenex EZ-faast kit is an example of a product that was developed to meet the demands of modern analysis. This product is specifically designed for amino acids analyses by significantly reducing sample preparation time with novel derivatization solutions.

Modern GC-MS systems are now replete with attractive features aimed to significantly decrease laboratory turnaround times and to push the limits of analytical boundaries even further. Convenience and speed are the new buzzwords as manufacturers launch different types of upgrades in the form of innovative computational software and hardware designs to boost the capabilities of a seemingly mature analytical technique. Keeping abreast of new developments in this area therefore empowers the analyst to make more informed decisions for solving new analytical problems.

2.1.1 SCOPE OF THE CHAPTER

The intent of this chapter is to discuss the conceptual basics for understanding the rationale behind the vital steps of GC-MS operations before taking on any GC-MS related tasks such as method development or routine procedures. Emphasis is placed on the analytical aspects of modern GC-MS instrumentation that will provide users with the essential skills to get the most value out of a contemporary GC-MS system. A section will be dedicated to emerging GC-MS technologies, which include new upgrades and multidimensional solutions to meet the growing demands of today's and tomorrow's analytical needs. An exhaustive treatise of GC-MS theories and in-depth mass spectral interpretative guide will not be attempted in this chapter as they have been more than comprehensively covered in other available GC-MS texts, but rather the authors strive to capture the essence of modern GC-MS for readers to have an appreciation of the technique as a powerful and versatile tool in an analyst's repertoire.

2.2 GC-MS SETUP

Figure 2.1a is a general schematic description and overview of a typical GC-MS procedure. Sample mixture is first introduced through the injector in which it is vaporized. Separation of the vaporized components in the sample mixture then occurs in the capillary GC column according to their varying degrees of interaction (distribution constant, K) with the liquid stationary phase that is coated on the inside of the GC column. The separated components from the GC are then introduced into the mass spectrometer through a heated transfer line. Ionization of the sample components then takes place, followed by the separation of the ionized fragments in the mass analyzer. The signals generated are then amplified by the electron multiplier or other ion detector system, to be detected, and the acquired data are recorded into the computer system for storage and further processing. The computer also doubles up as a control system that communicates experimental parameters (e.g., transfer line temperature, GC temperature program, mass range, etc.) to the GC-MS when they are loaded onto the fully integrated software program that is unique to the GC-MS manufacturer. At the end of a GC-MS analytical run, a total ion current chromatogram (TIC) is generated of the sample mixture (see Figure 2.1b) together with the mass spectra of the sample components (see Figure 2.1c).

The TIC is analogous to a universal detection gas chromatogram, in which the sum of the abundance of each ion detected over the specified mass range is plotted against the chromatographic run time. This mirrors the GC trace of the sample

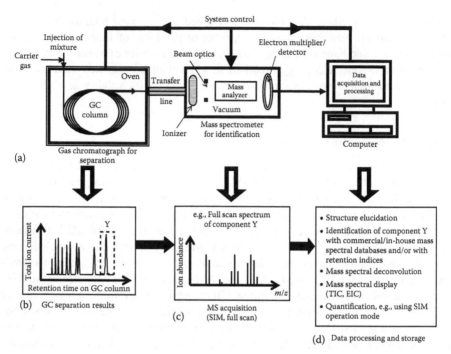

FIGURE 2.1 A schematic description of a general GC-MS procedure. (a) instrument schematic, (b) typical GC output, (c) MS spectral acquisition, (d) data interpretation.

obtained with a traditional GC flame ionization detector (considering the specific response factors for the sample components). The mass spectrum is a display of the fragmentation pattern of a sample component according to the abundance of each mass-to-charge ratio (m/z) of the detected ions and this display is often used as a diagnostic fingerprint of the compound to facilitate the elucidation of its structural identity. Unless otherwise stated, the charge on each ion fragment is usually unity. Hence the m/z values shown on the mass spectra directly represent the masses of the fragmented ion species of a sample component.

The operation of the MS system must be kept at low-pressure (i.e., high vacuum) conditions to maximize the ionization efficiency and ion detection. For example, under electron ionization (EI), low-pressure conditions are maintained to minimize the chances of ion–molecule collisions in the ionization chamber. Ion–molecule collisions in EI reduce the amount of ions (ion flux) reaching the detector, and will complicate the structure elucidation process of an unknown compound.

Reliable interfacing technology for the GC-MS coupling has overcome the teething problems of its early days (largely related to the larger flows associated with packed column GC operation) and has now been successfully integrated into all commercial GC-MS systems. One of the most straightforward coupling strategies is to subject the GC column to vacuum outlet conditions by directly placing the end of the capillary column into the MS ion source. As a result of the vacuum outlet conditions in the aforementioned arrangement, retention times shown on the GC-MS total ion current chromatogram will be shorter than that obtained from a similar translated GC-non MS detection method operated under atmospheric pressure conditions. Certainly, pressure difference (i.e., vacuum vs. atmospheric conditions) is not the only factor that influences the retention data on GC-MS and GC-non MS operations. Extra variability can be expected as a result of carrier gas leaks, exhausted filters, poor column installation, incorrect detector flows [2] on either of the GC system will inevitably contribute to shifts in retention data as well. An awareness of these is useful when trying to reference GC-MS retention data with that of a similar GC-non MS analysis. Perhaps one of the other major differences between routine GC-MS and GC operation is the choice of carrier gas. While many GC users have moved to H_2 carrier gas, He is still preferred in GC-MS due to the (perceived) problems of ion or molecule reactivity with H_2 gas in the ion source.

Once the GC-MS data are acquired and stored on the computer, they can be subsequently displayed in several ways to facilitate qualitative and quantitative determinations (see Figure 2.1d). Besides TIC chromatograms, data of the same analytical run may also be retrieved in the form of an extracted ion chromatogram (EIC) that selectively displays the signal of ions with a specific m/z value. The various ways of data manipulation and processing are further discussed in Section 2.5.

2.3 IONIZATION TECHNIQUES

The purpose of the ionization process is to supply energy to an analyte molecule so that the molecular structure will break apart in a characteristic manner. The fragment ions are analogous to the pieces of a jigsaw puzzle that will provide the analyst with patterns or clues to elucidate the structure of the previously intact molecule, i.e., identification of an unknown analyte or confirmation of a suspected analyte. Interpretation of the mass spectrum is today both well known, but is still an acquired art. Ionization patterns arise according to established rules, but recognizing and application of the rules of MS normally require an experienced eye. Detailed discussion of these rules and conventions is beyond the scope of this chapter, but the technical aspects of ionization will be briefly described.

The choice of ionization technique to be used for a GC-MS analysis is determined by the type of analytical problem to be solved and the chemical characteristics of the analytes involved. Two different ionization techniques, such as electron ionization (EI) and chemical ionization (CI), are sometimes necessarily used together in a GC-MS procedure to successfully identify the analyte so that mass spectral information derived from each ionization technique complement each other: EI to obtain structural information; CI (a softer, less energetic ionization method) to give molecular mass information.

The option to perform either EI or CI in the same equipment is now a common feature in many modern GC-MS apparatus. Today, harnessing the information from both EI and CI is made even easier through the use of a common ion source for EI and CI operations to further reduce apparatus downtime. Other commercial GC-MS systems use a method known as ACE™, which has the ability to alternate between EI and CI on successive scans to generate both EI and CI mass spectra with a single injection. These recent developments certainly open up more opportunities for improving productivity and simplifying procedures in GC-MS analyses.

2.3.1 Electron Ionization

Electron ionization is the most common ionization mode used in GC-MS, and for most users, the only other method consideration is to operate EI ion detection in the full-scan or selected ion-monitoring (SIM) mode.

Unless otherwise stated, EI mass spectra are acquired at 70 eV, which is the accepted standardized energy provided to the electrons used to knock out a valence electron from the gaseous molecule so that ionization of the molecule occurs. The pressure in the ion source during EI is approximately 10^{-5} mbar, maintained by a vacuum, usually via a turbo molecular pump.

The simplified ionization process is described in the following chemical equation, showing that an electron (e^-) in the bond between X and Y parts of the molecule is ionized:

$$X-Y \quad + \quad e^- \quad \rightarrow \quad [X-Y]^{+\bullet} \quad + \quad 2e^-$$

Neutral molecule   Energized electron   Ionized molecule   Electrons

The internal energy of the ionized species is then dissipated via a number of processes (e.g., vibration) one of which may be fragmentation of the molecule.

2.3.1.1 Full-Scan and Selected-Ion Monitoring Modes

Full-scan mode is implemented by acquiring the mass spectrum over a specified mass range that is appropriate for the application of interest from an upper mass that includes the molecular ion (or highest mass ion), and this is typically over the range from 40 to 500 u (u = unified atomic mass unit). The consistency and uniqueness of the mass spectra acquired by full-scan EI mode at 70 eV enable comparable GC-MS spectra on all relevant GC-MS instruments. As a result, identification of unknown compounds may be further simplified by comparing the similarity of their full-scan EI mass spectra with large mass spectral libraries (or database compilations) of known compounds which (hopefully) contains that of the unknown. The use of mass spectral libraries for identification will be further elaborated in Section 2.5. Full-scan mode in (EI) GC-MS finds one of its greatest uses in systematic toxicological analyses (STA) in forensic toxicology for detecting all substances of toxicological relevance in physiological samples such as blood. Full-scan is the method of choice for comprehensive screening procedures.

By instructing the MS to monitor only a few preselected ions (i.e., 4–8 ions) rather than all 460 ions in a typical full-scan mode (e.g., 40–500 u mass range), the SIM operation mode is in effect. Depending on the number of ions selected for monitoring, sensitivity of the analyses can be 100 times greater than the full-scan mode. The SIM operation mode can be used as a screening tool to flag the presence of any substance included in a specified list, as well as for the quantification of targeted substances with the incorporation of a labeled isotopic internal standard. The SIM mode is particularly useful in detecting low concentrations of target analytes in moderate-to-highly complex matrices when full-scan MS mode fails. Hence, much lower limits of detection can be realized with SIM than the full-scan mode, albeit with specificity trade-offs for the former.

In practice, about 4–6 ions are selected for SIM. In ensuring sufficient specificity and sensitivity in the SIM mode, the ions selected should ideally be unique and abundant. The specificity of an ion is closely governed by its origin. The molecular ion is the most specific ion in a mass spectrum although its use in SIM may not be as common due to its limited abundance in many molecules, or because not many compounds will have this mass in their spectra. It is more common to employ a common daughter ion that will reveal the presence of many compounds (often of the same compound class). High mass ions are almost invariably considered more specific than the lower mass ones.

With less than 10 ions for identification in SIM, the reliability and accuracy of deductive findings hinge on the choice of ions, and the probability of false-positive interpretative results are much greater for SIM than for the full-scan mode. The choice of less specific ions may also be turned to the analysts' advantage by having common qualifier ions to screen for structurally and pharmacologically similar substances (i.e., same drug class) as a means to keep up-to-date with trends such as new designer drugs in a "zero-tolerance" doping regulatory environment. Subsequent confirmation of the positive screening results with a more specific chemical analytical procedure that is capable of providing greater informing power [3] is highly recommended for unequivocal identification of the substance. It may be necessary to choose more than one ion per molecule to improve identification specificity. The use of groups of selected ions covering different retention time windows allows a larger number of ions to be employed in the analysis.

Having to decide between full-scan and SIM mode for a given analysis may well be a thing of the past. With benchtop quadrupole MS, the attainment of faster data acquisition has been realized in recent times with the advent of improved electronics that enable both SIM and full-scan data to be obtained in a single run. This feature will certainly bridge the gap between the sensitivity and specificity trade-offs in the full-scan and SIM mode, respectively, to give the analyst greater versatility and options to approach an analytical problem.

The SIM mode is not available with time-of-flight (TOF) mass analyzers.

2.3.1.2   Electron Impact: A "Hard" Ionization Technique

(EI) GC-MS is considered a "hard" ionization technique as the ionization energy of 70 eV exceeds the ionization potential of most organic molecules (typically 10–20 eV). The surplus energy results in further characteristic fragmentation of the molecule so that the EI mass spectrum will also show ion signals that arise from these secondary fragmentation pathways and together with the presence of the molecular ion, deduction of the molecule identity can be made.

The excess energy from the EI ionization treatment at 70 eV may result in the absence of the molecular ion peak of some molecules in their mass spectra due to the complete fragmentation of their molecular ions in the ion source. The presence of the molecular ion (giving the molecular mass) in the mass spectrum is one of the most important pieces of information needed to correctly deduce the structural identity of the unknown molecule, especially when distinguishing between analytes from the same homologous series. In order to obtain this vital clue, i.e., molecular ion, other strategies must be sought, namely, by chemical derivatization or by "softer" ionization techniques.

The clever implementation of different chemical derivatization strategies can be used to selectively manipulate the fragmentation pathway(s) of a molecule such that additional and specific information of an unknown molecular structure can be obtained in GC-MS analyses. This is in addition to the simultaneous conferment of much improved volatility and thermal stability to the target analytes for overall enhanced chromatographic performance.

In general, molecular ion abundance in (EI) GC-MS can be enhanced by the introduction of a functional moiety with a very low ionization potential. Figure 2.2 shows how the EI mass spectrum of methamphetamine can be selectively altered through different chemical derivatization methods.

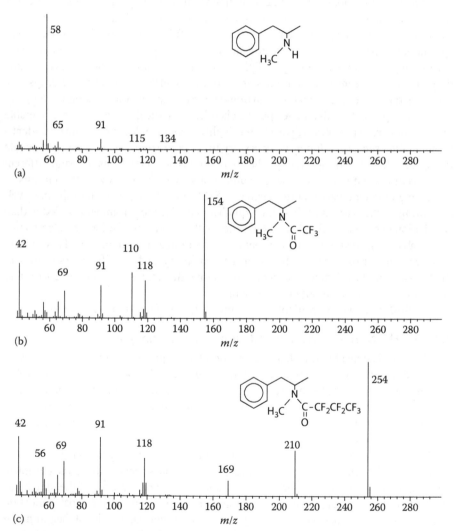

FIGURE 2.2   Electron ionization mass spectra of methamphetamine (a) underivatized, (b) trifluoroacetyl, and (c) heptafluorobutyryl derivatives. (Reprinted from Halket, J.M., in: Blau, K. and Halket, J.M. (eds.), *Handbook of Derivatives for Chromatography*, 2nd edn., John Wiley & Sons Ltd., Chichester, U.K., 1993, pp. 297–324. With permission from copyright owner.)

Besides the ability to increase the molecular ion abundance, chemical derivatization can also be used to specifically determine the position of double bonds, as in the case of unsaturated fatty acids by converting them into pyrrolidide derivatives (this stabilizes the terminal acid end of the molecule to give preferential ionization and fragmentation neutral loss of the hydrocarbon chain end). In addition, chemical derivatization significantly improves the specificity of the mass spectral information in (EI) GC-MS by directing the fragmentation pathways to higher mass region since chemical noise interferences generally occur in the lower mass region. However, derivatization can also mask molecular identity by reducing diagnostic fragmentation patterns through major fragment losses arising from the derivative group rather than the remainder of the molecule.

The choice of derivatization strategy may also be dictated by the laboratory access to reference mass spectral library(s) used for identification. Trimethylsilylation (TMS) is one of the most common derivatization method used for GC and is just as equally accepted for GC-MS analyses due to the moderately abundant $[M-15]^+$ ion generated, as well as the availability of TMS-derivative mass spectra in many established commercial reference databases.

More in-depth treatment on the topic of chemical derivatization for GC-MS and mass spectrometry are available in references [4,5] while other "soft" ionization techniques in GC-MS are elaborated in the subsequent sections.

2.3.2 CHEMICAL IONIZATION

Since its first introduction by Munson and Field [6] in 1966, CI in mass spectrometry has secured a place alongside the classical mode of EI to augment the capabilities of GC-MS for analyses. In CI, the ionization of gaseous sample molecules occurs in two steps. First, a suitable reagent gas is introduced into the ion source and is ionized by EI into reagent ions. The ionization of the sample molecules then occurs upon their collision with the reagent ions, usually through the transfer of a reactive group to give pseudo-molecular ions. As the amount of energy transferred to the sample molecules is much less in CI than in EI, CI is sometimes termed a "soft" ionization technique.

Following is a conceptual scheme for the formation of positive ions via proton transfer reaction by CI.

Formation of reagent ions by EI:

$$R-H \quad + \quad e^- \quad \rightarrow \quad [R-H]^+$$

Reagent gas   Energized electron   Reagent ion

CI via proton transfer reaction to M:

$$[R-H]^+ \quad + \quad M \quad \rightarrow \quad [M-H]^+$$

Reagent ion   Sample molecule   Quasi-molecular ion

Note that the line here refers to the bond between the species.

The different operational modes and the wide selection of reagent gases give (CI) GC-MS the flexibility and control over the sites as well as the degree of fragmentation

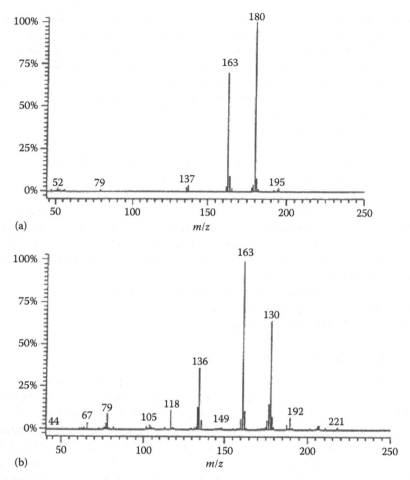

FIGURE 2.3 Chemical ionization mass spectra of MDA. Reactant gas (a) methanol, (b) acetonitrile. (From Pellegrini, M. et al., *J. Chromatogr. B*, 769, 243, 2002.)

of a sample molecule (see Figure 2.3) and consequently on the type of information to be extracted from the CI mass spectrum. Methane, ammonia, and isobutane are examples of reagent gases commonly used in CI.

In CI, the efficiency of the ionization process and the complexity of the mass spectrum are intimately related to the level of exothermicity in the reaction between the reagent gas ions and the sample molecules. If the reaction were too exothermic, appreciable fragmentation of the ionized molecule would be observed in the mass spectrum and vice versa. The complexity of the mass spectrum may be customized to a particular type of application through the choice of reagent gas used.

CI in GC-MS is not exclusively reserved for molecular weight determination only. With the wide range of reagent gases available, different selectivity in the ionization sites can be achieved during CI to enable isomeric forms of a compound to be distinguished [7], thus facilitating structure elucidation and identification as well. This is not possible in EI without employing some form of derivatization or chromatographic strategies. Similar to EI, quantitative determinations can also be made with CI upon the addition of an appropriate internal standard.

Under the conditions of positive ion CI mode based on proton transfer reaction, the mass spectrum is characterized by a prominent quasi-molecular ion peak [MH]$^+$ that would enable the molecular weight of the unknown compound to be derived. The use of positive ion CI in the SIM mode as a rapid screening method for amphetamines and related drugs in urine has been reported [8].

For negative ion CI, electron capture is by far the most widely used reaction as it can provide up to 1000 times greater sensitivity over its positive ion counterpart and EI. For this reason, negative ion CI by electron capture reaction is a valuable tool for trace analyses of environmental pollutants (e.g., polychlorinated biphenyls, PCBs) and innocuous potent drugs (e.g., benzodiazepines) that are present in complex samples. In general, a [M–H]$^-$ ion peak, i.e., corresponding to the molecular mass minus the proton of one mass unit, is featured prominently in many negative ion CI mass spectra.

For analytes with poor electron affinity, chemical derivatization can be employed to extend negative ion CI (electron capture) to a wider range of applications in which high levels of sensitivity are essential. For special applications where even greater sensitivity is desired, negative ion CI (electron capture) operated in the SIM mode may also be used albeit with specificity trade-offs. Specificity in (CI) GC-MS may be further improved by coupling an additional resolution element to the system, either in the chromatographic (e.g., comprehensive two-dimensional [2D] GC or GC × GC) [9] or mass spectrometric domain [10] (tandem mass spectrometry or MS/MS).

Unlike EI that produces a unique and uniform fragmentation pattern at 70 eV, problems with the reproducibility of CI mass spectra are sometimes encountered in CI as the ionization process is much more sensitive to delicate changes in the ionization conditions, e.g., ion source temperature and pressure, purity of reagent gases, etc. More frequent maintenance is also required of a dedicated (CI) GC-MS system, such as regular cleaning of the ion source as the greater exposure to different contaminants makes the ion source more susceptible to fouling. Due to the relatively simple CI mass spectra and reproducibility considerations, the construction of large mass spectral libraries for identification purposes has limited utility in CI. Overall, the operation of CI and the interpretation of CI mass spectra demand greater technical competence and skills.

The considerable advances in GC-MS instrumental designs over the past decade have certainly contributed to the growing use of CI for analyses. Systems are now available that can be operated in positive-ion and negative-ion CI mode such that both their mass spectra may be acquired simultaneously (or more correctly, in rapid sequential scanning) with one injection. Such solutions make it even easier and more economical to harness information obtained from CI and EI to give greater certainty to identification.

Further reading on chemical ionization is available [11].

2.4   MASS ANALYZERS

The key role of the mass analyzer in a GC-MS system is to separate ions (of a defined mass range) according to their mass-to-charge ratios (*m/z*). Separation of these ions occurs by virtue of their differences in time or in space depending on the principles that govern this stage, as they travel through the mass analyzer.

Mass range, data acquisition rate, and resolving power are the key attributes that characterize a mass analyzer. Together, they contribute toward the strengths and limitations of a GC-MS system.

In full-scan operation, the mass range used must primarily take into consideration the general fragmentation pattern of the target analytes. For instance, volatile samples such as essential oils contain mainly small molecules, hence their molecular masses generally do not exceed 300 u. Therefore, using a mass range of 30–260 u along with good chromatography should provide adequate specificity to identify the analytes of interest in the essential oil sample. A suitable lower mass may be chosen (e.g., 50 u) such that interferents from the solvent (e.g., ethanol, M_R 46 g/mol) and/or air components (e.g., carbon dioxide M_R 44 g/mol) are selectively excluded from the mass spectrum. For library matching purposes, the user must be aware that differences in the mass range used between the reference spectrum and the experimental spectrum may adversely contribute to the quality of the mass spectral matches obtained [12]. Where chemical derivatization is compulsory, the derivatization technique chosen should not form derivatives with molecular masses that fall beyond the mass range of the mass analyzer. A mass range of 40–550 u is typically used for a general screening of samples containing volatile and semi-volatile analytes.

In choosing the data acquisition rate, the following rule of thumb may be established: the chromatographic peak should be sampled no less than six times (and often, much more) for good definition of the peak. The signal-to-noise ratio (S/N) is, however, inversely proportional to the data acquisition rate. Thus, the final data acquisition rate should be sufficiently high without substantial loss in S/N.

The ability of a mass analyzer to differentiate between two adjacent ions is given by its resolving power (*R*). A high-resolution mass spectrometer (HRMS) has a resolving power of at least 5–10,000 such that the *m/z* values can be given to several decimal places for accurate mass measurements. Hence, HRMS enables the elemental composition of an unknown to be determined and may also be used as a tool to study or verify fragmentation mechanisms. GC-HRMS systems are classically based on double-focusing magnetic sector mass analyzers, although similar systems based on TOF mass analyzers have also recently become available (e.g., the Micromass GCT system). Conversely, a low-resolution mass spectrometer (LRMS) has a resolving power of approximately 1000 and below, to give *m/z* values as integer numbers. HRMS instruments are more expensive than LRMS options.

The quadrupole, magnetic sector, and ion-trap mass analyzers are the most commonly encountered in a routine GC-MS laboratory, and their use is still highly relevant in solving today's analytical problems. The renaissance of TOF mass analyzers is largely due to the advances in microelectronics that provide the required computational power to keep up with the phenomenal amount of the TOF data and the speed at which they are generated. TOFMS analyzers are becoming increasingly more common.

TABLE 2.1

Comparing the Performance Characteristics of Magnetic Sector, Quadrupole, Ion Trap, and Time-of-Flight Mass Analyzers

Property	Magnetic Sector	Quadrupole	Ion Trap	Time-of-Flight
Resolution	10^2–10^5	10^2–10^5	10^2–10^5	10^2–10^5
Mass accuracy (ppm)	1–5	100	50–100	5–50
Upper mass range	10^4	10^4	10^5	$>10^5$
Linear dynamic range	10^9	10^7	10^2–10^5	10^2–10^6
Precision (%)	0.001–1	0.1–5	0.2–5	0.1–1.0
Abundance sensitivity	10^6–10^9	10^4–10^5	10^3	Upto 10^6
Efficiency (%)	<1	<1–95	<1–95	1–100
Data acquisition rate	Slow	Moderate	Moderate	Very fast
Cost	Moderate	Low	Low	High
Size	Floor standing	Benchtop	Benchtop	Benchtop[a]

Source: Adapted from *The Essence of Chromatography* p. 750, Table 9.4 [24].

[a] Both benchtop and floor standing instruments have been described.

The principles behind the main types of mass analyzers are briefly described later and a comparison of the different mass analyzer features are summarized in Table 2.1.

2.4.1 QUADRUPOLE MASS ANALYZERS

As the name implies, the quadrupole ("quad-rupole") mass analyzer is made up of four cylindrical or hyperbolic rods of highly precise geometry, with each rod occupying a corner of a square arrangement (see Figure 2.4). A voltage comprising of a direct current voltage (V_{dc}) and a radio frequency voltage (V_{rf}), is exerted across opposite pairs of the poles to create an electric field in which the ions would oscillate and traverse. The adjacent pairs of rods are of opposite polarities, as denoted by the *plus* and *minus* signs in Figure 2.4. By varying both the V_{dc} and V_{rf} together while keeping their ratios constant, the predetermined mass range may be repetitively scanned for ions in increasing or decreasing m/z values. Hence, the separation of the ions can be said to occur "electronically" as their trajectories are controlled by the voltage settings, which influence the type of ions that ultimately reach the detector.

Quadrupole mass analyzers generally have a resolving power of about 1000 and are capable of unit mass resolution to give m/z values to the nearest integer numbers. The inherent medium-level mass resolution from quadrupole mass analyzers can easily be compensated for with good chromatography or additional sample work-up steps.

Undoubtedly, benchtop GC-MS systems based on quadrupole mass analyzers (GC-qMS) are the workhorse in *Analytical Chemistry* as they are relatively inexpensive, robust, and versatile in the various operational modes available. As the

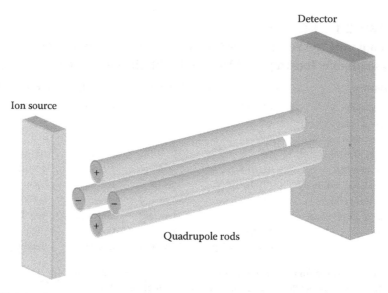

FIGURE 2.4   Quadrupole mass analyzer.

merits of fast GC and multidimensional GC solutions are fast gaining recognition and acceptance across the analytical community, these GC applications that can generate chromatographic peak widths as narrow as 80 ms, challenge the practical limits of the quadrupole scan rate.

An effective approach to increase the quadrupole scan rate is to trim the mass scan range and minimize ion dwell times to the extent that they are practically possible (note that this is application dependent). This creative approach enables improvements in scanning rates (albeit with some sensitivity and specificity trade-offs) before other expensive instrumental solutions are sought. This approach has been successfully adopted to improve the quadrupole scan rate in several multidimensional GC studies, which include essential oils [13] and perfume analyses [14]. In the essential oils study using the Agilent HP5973 quadrupole mass analyzer, a scan rate of approximately 20 Hz was obtained with a mass range of 188 u and minimum ion dwell times.

The need for faster quadrupole scan rates have been acknowledged by various GC-MS manufacturers as they recently launched quadrupole mass spectrometers that offer scan rates up to 10–20,000 u/s. While this translates into faster data acquisition, the calculation of duty cycle, and number of scans/s, depends very much on the interscan delay time (or reset time) where the scan rate might be represented as

$$\text{Scan rate} = \frac{10,000}{[(\text{scan time for the mass range}) + (\text{interscan delay})]}$$

It can be appreciated that the upper effective scan speed is limited by the interscan delay time.

2.4.2 MAGNETIC SECTOR MASS ANALYZERS

Under the influence of a magnetic field, the ions follow a curved flight path as they enter into the mass analyzer region. The correlation between the m/z value of the ion and the magnetic field may be expressed mathematically according to Equation 2.1:

$$\frac{m}{z} = \frac{(B^2 r^2)}{2V} \tag{2.1}$$

where
  m/z is the mass-to-charge ratio of the ion
  B is the magnetic field strength
  r is the radius of the circular path that the ion traverses
  V is the voltage that is applied to accelerate the ion

Based on the aforementioned working principle, a mass analyzer may be constructed using an electromagnet. By varying the magnetic field strength (B) while keeping the accelerating voltage constant (V), the masses in a defined mass range are scanned. Separation of the masses occurs in *space*, as each ion follows a different trajectory (as defined by the radius, i.e., r) as it traverses the curved flight tube to the magnetic sector. A schematic diagram of a single-focusing magnetic sector mass analyzer is shown in Figure 2.5.

By placing an electric field before or after the magnetic field, systems with high mass resolution are realized and are also referred to as double-focusing instruments.

The electric field acts as an electrostatic analyzer to correct for the differential initial velocities (momentum) of the ions. Double-focusing systems have a resolving power of up to 40,000 (i.e., four times more than the single-focusing systems), so that the mass of an ion can be given to four decimal places.

While the mass resolution in magnetic sector instruments is certainly superior to its quadrupole counterparts, this advantage, however, also comes with a higher price. Magnetic sector instruments are generally more complex and their inherently slow scan rates preclude their use for fast chromatography applications. They are, however, becoming mandatory for some studies, such as dioxin analysis where multiple labeled internal and surrogate standards and accurate mass detection of fragment ions may be critical in identification of required components.

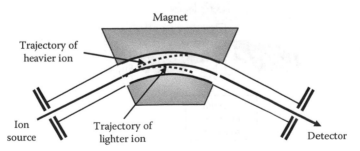

FIGURE 2.5 Single-focusing magnetic sector mass analyzer.

2.4.3 ION TRAP MASS ANALYZER

The most distinguishing feature of the ion trap is its ability to concentrate the ions formed by storing them over a defined period of time, in addition to its mass analyzing function. Ionization can take place within (internal ionization) or outside (external ionization) the ion trap in a pulsed fashion. Therefore, the ion trap is really an ion source, an ion-focusing device and a mass analyzer combined into one compact assembly.

The heart of the ion trap is a donut-shaped ring electrode that is positioned between a pair of end-cap electrodes (see Figure 2.6). A radio frequency voltage is applied to the ring electrode. If external ionization is used, sample ions are repetitively pulsed into the mass analyzer (i.e., ring cavity) region through an electrostatic ion gate at the top end-cap electrode. For internal ionization, the electrostatic ion gate is replaced by an electron gate, which regulates the entry of the electrons (from the filament) into the ion trap for ionization of the sample molecules. The primary manipulation of V_{rf} on the ring electrode controls the stability of the ions in the ion trap. Stable ions are held in the ion trap while unstable ions are ejected out of the ion trap to be detected. By varying V_{rf} through a range of values, storage of ions and mass scanning are performed sequentially.

There is a limit to the number of ions that can be contained within the ion trap. When this limit is exceeded, the number of ions in the trap (ring electrode) region is large enough to generate an undesirable electric field that could potentially change the ion trajectories such that the mass analysis step would be adversely affected. This could arise through ion–ion interactions, or space-charge effects. For this reason, ion trap mass analyzers may not be particularly suited in analyses where highly complex and concentrated matrices are involved. The automatic gain control (AGC) feature is incorporated in all modern ion-trap equipment, which takes the guesswork out of determining the optimum ionization time within the ion trap.

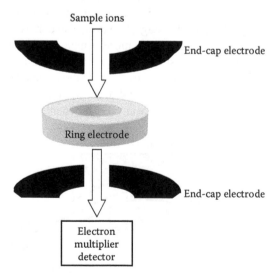

FIGURE 2.6   Ion trap mass analyzer using external ionization.

Some notable differences in the analytical data obtained from ion trap and quadrupole mass analyzers have been reported. Excessive GC peak tailing of some late eluting benzodiazepines [15] and greater variability in ion ratio measurements have been observed on ion trap instruments [16] vs. its quadrupole counterparts.

Whenever possible, the experimental and reference mass spectra for library matching should be synchronized on the same type of instruments to ensure that the possibility of generating false-negative and/or -positive results are minimized during the identification process. Thus the similarity—or difference—of ion trap vs. quadrupole vs. magnetic sector spectra may need to be confirmed before a library match quality is accepted.

An in-depth treatment of ion trap mass analyzers is available in Ref. [17].

2.4.4 TANDEM MASS SPECTROMETRY

For trace analysis, GC coupled to tandem mass spectrometry (GC-MS/MS) is one of the techniques of choice as it is capable of exquisite specificity with the incorporation of an additional mass analysis step. GC-MS/MS can be performed on various mass analyzer configurations in association with different ionization and operational modes.

GC-MS/MS systems based on quadrupole mass analyzers comprise of three sets of quadrupole stages (i.e., triple quadrupole MS), which are conceptually represented in Figure 2.7. Mass analysis is carried out only in the first (MS^1) and the last quadrupole set (MS^2) of triple quadrupole arrangement. The middle quadrupole functions solely as a collision cell, where the selected ions (from MS^1) are subjected to further fragmentation by an inert gas (e.g., helium, argon) to generate a characteristic mass spectrum of the daughter ions in MS^2. Under this system configuration, there are several ways to implement the MS/MS experiment:

1. *Parent-ion scan*
   Masses of a defined mass range are scanned in MS^1, in which all the parent ions undergo further fragmentation via collision-induced dissociations (CID) with the inert gas. Only daughter ions with particular m/z values produced from CID are selectively monitored in MS^2. This mode allows analytes with related structural features to be identified through their common fragment ions. This could be, for instance, the $m/z = 91$ ion for certain aromatic compounds.

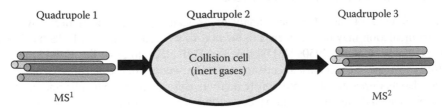

FIGURE 2.7 Conceptual representation of tandem mass spectrometry (MS/MS) based on a triple quadrupole instrument.

2. *Daughter-ion scan*

Ions of a particular m/z value are selectively isolated and monitored in MS^1. The selected ion undergoes CID fragmentation, to generate a mass spectrum of the daughter-ion products as MS^2 is scanned. The mass spectrum obtained from the second mass analysis stage provides highly specific structural information about the analyte, from which unequivocal identification may be made with much greater confidence.

3. *Neutral-loss scan*

In this operational mode, scanning is conducted in both MS^1 and MS^2 at a predetermined mass difference ΔX between these two mass analyzers. ΔX corresponds to the mass of the neutral fragment ion that is indicative of the presence of a characteristic functional group, e.g., loss of water (18 u) from hydroxy (–OH) group of alcohols. Here, only parent ions that lose the neutral fragment can be detected in MS^2 through their daughter ions. This could be, e.g., the M-31 loss for methyl esters of fatty acids, with MS^1 and MS_2 scanned at a mass difference of 31.

4. *Selected reaction monitoring (SRM) or multiple reaction monitoring (MRM)*

This mode offers the greatest sensitivity by having both MS^1 and MS^2 operating in the SIM mode so the ion dwell times in the mass analyzers are maximized. A limited number of characteristic and abundant qualifier ions are first isolated in MS^1. Daughter ions (with predetermined m/z values) that are derived from the highly structure-specific fragmentation pathway(s) via CID, are detected in MS^2. This mode integrates superior sensitivity and specificity for the confirmation and quantification of trace levels of analytes in very complicated matrices. By associating a time element into the SRM, MRM is implemented so that the monitoring of several specific reactions can occur sequentially as the GC analysis progresses.

The advantage of ion trap MS/MS systems rests in the fact that no hardware modifications to the original ion trap mass analyzer need to be made in order to perform MS/MS experiments. All of the ion selection and ion-decomposition processes occur in the ion trap, with expulsion of unwanted ions from the trap as necessary. This benefit is offset somewhat by the unavailability of tandem operations (1) and (3), due to the limitations imposed by the inherent principles governing ion trap mass analyzers.

2.4.5 TIME-OF-FLIGHT MASS ANALYZER

The reference of the TOF mass analyzer as state-of-the-art in modern GC-MS instrumentation may be considered to be a bit of a paradox since TOFMS has been in existence since the 1950s [18]. Today, the TOF claim to fame lies in its inherent ability to acquire a full mass spectrum on a microsecond timescale, which ironically was also its impediment, as previously no tools were available to handle and process the phenomenal amount of data generated in such a short time. It was through the recent advances made in the field of microelectronics that transforms the storage and processing of TOFMS data into a feasible reality.

The mass analysis process in the TOFMS begins with the formation of ions through pulsed ionization, after which the ions are accelerated into a field-free drift tube for separation. The ions are then resolved by the differences in their arrival times (after traversing the entire length of the drift tube) at the detector. Heavier ions with larger m/z values will take a longer time to traverse the entire length of the drift tube, while lighter ions with smaller m/z values will take a shorter time. This may be mathematically represented in Equation 2.2:

$$t = L \cdot \sqrt{\frac{m}{2zV}} \tag{2.2}$$

where
  t is the time the ion takes to traverse the entire length of the drift tube (or flight time)
  L is the length of the drift tube
  m is the mass of the ion
  z is the charge of the ion
  V is the acceleration potential

Theoretically, ions with the same m/z values should arrive at the detector simultaneously, provided that they possess the same kinetic energy upon acceleration. In practice, this is complicated by the dispersion of their kinetic energies prior to their acceleration. This creates a series of ion arrival times when there should only be one single value for the ions of a specific m/z value. By placing an electrostatic field that impedes the motion of the energetic ions, energy focusing is realized and this is accomplished by a device known as a reflectron (or an ion mirror). An illustration of the described TOF mass analyzer is given in Figure 2.8.

Under the employed ionization mode, which purges ions into the drift tube every 200 µs (i.e., 5000 times/s), a complete mass spectrum is generated with each purge event. An approach to improve S/N ratio in some commercial TOF systems is to sum the spectra from several purge events (i.e., transient spectra) prior to data transfer, to give a net data acquisition rate that is calculated according to

$$\text{TOFMS acquisition rate} = \frac{\text{Total transient spectra (5000 spectra/s)}}{\text{Number of summed transient spectra}} \tag{2.3}$$

Based on the aforementioned expression, a net data rate of 50 Hz is acquired by summing (or bunching) 100 transient spectra for data transfer.

Currently, the TOFMS is the only mass spectrometer that can acquire data up to 500 Hz that makes it ideal for ultra-fast GC studies, which include very fast multidimensional GC (MDGC) applications. In recent years, there has been an explosion of studies dedicated to MDGC, in particular comprehensive 2D GC (GC × GC). Coupling of GC × GC to TOFMS (GC × GC-TOFMS) is a powerful analytical tool for the characterization of highly complex samples such as cigarette smoke [19] and natural products [20].

With much higher data acquisition rates than its scanning counterparts, there can now be greater data density to provide better definition of the chromatographic peak profiles. This potentially improves quantitative results and allows closely eluting chromatographic peaks to be detected.

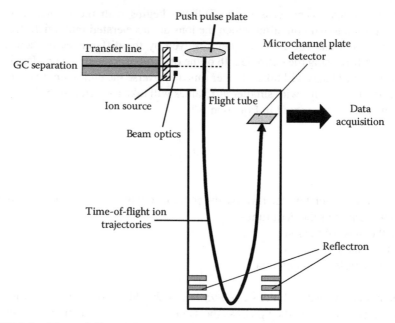

FIGURE 2.8　Time-of-flight (TOF) mass analyzer.

The absence of mass spectral skewing is a notable feature of TOFMS data. Since the ratios between the ions are constant across a single-component GC peak, algorithms may be formulated to mathematically pull apart the mass spectra of co-eluting analytes. This mass spectral deconvolution feature is built into the software that comes with some GC-TOFMS systems.

At present, the few commercial GC-TOFMS systems in operation means it carries a high price range among the unit mass resolution mass analyzer systems. The mass range of most commercial GC-TOFMS systems is currently set at 1–1000 u with comparable resolving power (unit mass resolution) as the quadrupole mass analyzer. A high-resolution TOFMS system has also recently been launched, that is capable of providing accurate mass measurements. With up to 10 times faster data acquisition rate (2 vs. 20 Hz) over its HRMS counterpart based on magnetic sector technology, the high-resolution TOFMS is expected to be a serious competitor to magnetic sector mass analyzers in the future. Further new developments include GC-Q-TOFMS systems, based on successful LC counterpart instruments. This brings added capability to the MS/MS experiment.

2.5　DATA EVALUATION

The growing sophistication of computer technology has dramatically transformed the management and processing of data in modern GC-MS analyses. Today, virtually all GC-MS data are stored on electronic media for easy retrieval and may be shared by a collegiate of scientists via the Internet/Intranet. Large print copies of mass spectral compilations are now replaced by their electronic versions with built-in algorithms and macros to greater facilitate qualitative and quantitative determinations of an analyte.

An effective strategy for evaluating GC-MS data begins with an understanding of the functions behind the various processing features. These will be illustrated with selective application studies that will demonstrate the important aspects of data presentation and data analysis.

Case Study 1: Class Identification in a Kerosene Sample Using Extracted Ion Chromatogram

Data acquired by GC-MS are essentially three-dimensional, and provide mass (*m/z* value), ion abundance and retention time information. Once GC-MS data are acquired and stored on the computer disk drive, the analyst has at his/her disposal, a variety of ways to visualize and evaluate the results.

Figure 2.9a is a TIC chromatogram of a kerosene sample acquired by (EI) GC-MS using a quadrupole mass analyzer. Clearly, this TIC chromatogram is a two-dimensional display of the total abundance of the ions detected plotted against the retention time of the sample components. Without any specific mass information, the function of the TIC presentation is to primarily provide a facile assessment

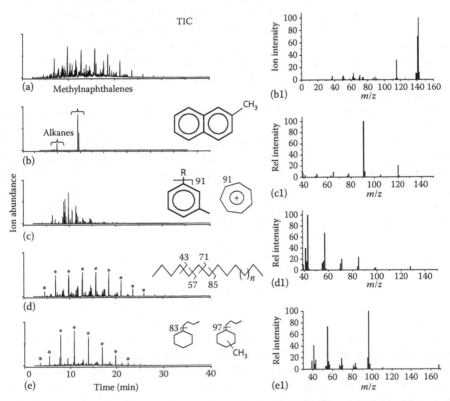

FIGURE 2.9 TIC of kerosene as well as extracted ion chromatograms (EIC), chemical structures and representative mass spectra of naphthalenes, monoaromatics, n-alkanes and cyclic alkanes. (a) TIC chromatogram of kerosene, (b) EIC of 142 u (methyl naphthalenes), (c) summed EIC of (77 u + 91 u + 105 u) (monoaromatics), (d) EIC of 85 u (n-alkanes), (e) EIC of 83 (cyclic alkanes). (b1)–(e1) are representative mass methylalkylcyclohexane, respectively.

of the GC-MS data so as to have an idea of the sample complexity and the quality of the chromatographic separations. Provided that there is sufficient chromatographic resolution, each sample component can be identified by averaging the mass spectra near the peak apex, where the ion abundances are the most constant and representative.

From Figure 2.9a, a few characteristics of the sample can be established. First, the multitude of peaks observed in the TIC chromatogram evidently indicates the great complexity of the kerosene sample, while the "hump-like" baseline suggests the presence of many unresolved components buried amid the total ion current signals. In order to get reliable mass spectra for overlapping peaks, background subtraction methods will be required. The more abundant the component, the more reliable will be its spectral purity. Conversely, for smaller components, their spectra will be less reliable, and background subtraction less able to correct the spectrum of the target compound.

Now, suppose it is of interest to detect a specific class of aromatic compounds, e.g., the C1-naphthalenes in the kerosene sample. By changing the display from a TIC chromatogram to an EIC or mass fragmentogram, the signal of specific ions can be selectively retrieved from the same data file, and plotted against retention time. The EIC based on the characteristic ion (142 m/z) of the C1-naphthalenes is shown in Figure 2.9b. In this case, the molecular ion carries much of the ion signal, and so can be used as a relatively sensitive EIC display of these diaromatics. The $[M-15]^+$ ion is also significant in the mass spectrum of these compounds, so could likewise be used for the EIC plot.

By doing the same for characteristic ions of monoaromatics (71 + 91 + 105 m/z), linear (85 m/z) and cyclic (83 m/z) alkanes, their EICs are displayed in Figure 2.9c through e, respectively.

Certainly, the success of the EIC in providing class identification of a sample lies in the judicious choice of ions. As seen in Figure 2.9d and e, the use of ions with m/z 85 and 83 are not exclusively unique for linear and cyclic alkanes, respectively, and there are interferences (or other compound classes) that also produce the selected ions. Nonetheless, the use of EIC is able to significantly narrow down the number of peaks (from Figure 2.9a) for further investigation. Having obtained the retention times of the component peaks in Figure 2.9d and e, closer examination of their mass spectra confirms the presence of linear and cyclic alkane homologous series, which are marked with asterisks in Figure 2.9d and e.

Figure 2.9b1–e1 indicates typical mass spectra of the aforementioned components, which justifies the selection of the ions in each case. The common ion arising for the common fragmentation reaction in each case is also indicated. It is easy to recognize that 83 m/z will be selective toward the monosubstituted cyclic alkane (cyclohexane), and by inference, 97 m/z will be selective toward methyl-alkyl disubstituted cyclic alkanes where fragmentation occurs at the bond between the cyclic group and the longer alkyl chain substituent. The mono-alkyl substituted cyclic pentane series will likewise therefore be expected to be selected by use of an EIC of 69 m/z. That the cyclic compounds have a diagnostic ion two mass units lower than saturated alkanes is evident when one considers that cyclization of the ring results in loss of two protons, hence 83 m/z vs. 85 m/z for the ions with six carbon atoms.

Besides providing class identification, the EIC can also be used to estimate the spectral contribution of two partially resolved chromatographic peaks, provided that each of these peaks contains characteristic fragment ions of their own.

Case Study 2: Characterization of Lavender (*Lavendula angustifolia* L.) Essential Oil with Mass Spectral Libraries and Retention Index Data

Characterization of a sample is perhaps the most classic use of GC-MS analyses. Lavender essential oil has commercial relevance as it is a key ingredient in industrial perfumes and fragrance materials. Consumer perception of such products is strongly influenced by their distinct aroma. Thus, the ability to unravel the constituents of raw materials such as Lavender essential oil is central to the development of an effective quality control protocol during the manufacturing process of a product such as hand cream, or a soap.

Assuming that each organic molecule fragments in a characteristic manner at 70 eV, the construction of large mass spectral libraries greatly simplifies the identification procedure of a compound by comparing its spectral similarity with reference mass spectra in the database. Mass spectral libraries based on EI at 70 eV provide the most comprehensive collection of references for such purposes, which include several databases in specialized fields of applications. The mass spectral library compiled by Robert P. Adams [21] is an example of such specialized libraries, which contain the mass spectra of approximately 1500 essential oil components.

Depending on the search algorithm incorporated into the computer software, the search results are generally ranked according to the degree of similarity (expressed in %) between the unknown mass spectrum and the reference mass spectra in the database. Note that the best library match does not necessarily identify the compound unequivocally. One reason for this is that many different compounds (in this case, terpenes) have very similar mass spectra. Fortunately, many of such compounds have different physical characteristics such as boiling point, which enable their identification to be made based on their differences in retention times, after separation on a suitable GC stationary phase. Hence, whenever possible, retention time data should always be used with mass spectral data to support compound identification. Sometimes, it is necessary to use two different column phases for identification, either as parallel or sequential analysis. It is often still considered only a tentative identification, unless an authentic compound is subsequently co-injected with the sample, or is run under identical conditions to confirm correspondence of retention time and mass spectrum.

The TIC chromatogram of *Lavendula angustifolia* essential oil acquired by (EI) GC-MS on a quadrupole instrument is presented in Figure 2.10. Due to the spectral similarity of many terpene components in essential oils, both retention index (RI) data and library match results will be used to assist compound identification. In this example, RIs are acquired under linear temperature program condition with the *n*-alkane series used as reference retention compounds. Here, only the compounds labeled peaks 1 and 2 are tentatively identified and discussed in detail.

For peak #1, its mass spectrum was searched against the Adams Library with its search results (hit list ranking, match quality) and retention indices given in Table 2.2.

The probability-based matching (PBM) algorithm [22] incorporated into the computer software found linalool to have the greatest mass spectral similarity to the sample mass spectrum with high match quality of 90%, followed by linalool butanoate (52%) and linalyl isobutanoate (50%) in the second and third position, respectively. Next, the RI of peak #1 is checked against the reference RI (provided by Adams Library) to confirm the tentative identification. Since temperature program

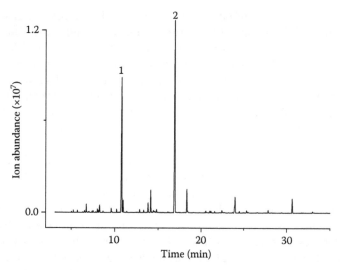

FIGURE 2.10   TIC chromatogram of lavender essential oil acquired by (EI) GC-MS.

TABLE 2.2

Identification of Peaks 1 and 2, Figure 2.10

Peak No.	t_R (min)	Calculated Retention Index[a]	Hit List Rankings	Match Quality (%)	Ref. Index[c]	Retention
				Library Matches[b]		
1	10.81	1100	Linalool (154.1 g/mol)	90	1097	
			Linalool butanoate (224.2 g/mol)	52	1423	
			Linalyl isobutanoate (224.2 g/mol)	50	1375	
2	16.94	1255	β-Pinene (136.1 g/mol)	80	979	
			α-Fenchene (136.1 g/mol)	59	953	
			Sabinene (136.1 g/mol)	59	975	

[a] Calculated experimental linear programmed temperature retention index (van den Dool & Kratz procedure).

[b] Retrieved library reference mass spectral matches to the experimental spectrum.

[c] Reference retention index values provided by the Adams library for the library match spectra.

conditions were applied for the analysis, the RI for peak #1 was obtained using the van den Dool and Kratz formula.

The RI of peak #1 is calculated to be 1100. Out of the library search results obtained, the reference RI of linalool (i.e., 1097) comes closest to the experimental RI. A difference of three RI units is acceptable having taken into consideration possible slight variations in retention time (attributed to slight overloading of peak #1) and system configurations (e.g., different column manufacturer, vacuum, source design). With an

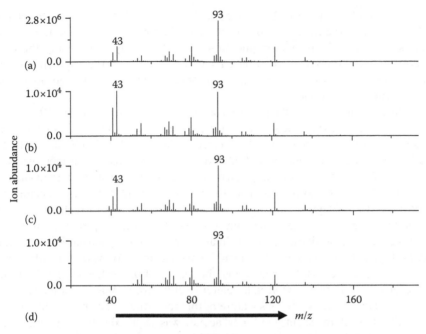

FIGURE 2.11   (a) Experimental mass spectrum of peak #2 (Figure 2.10). (b) Mass spectrum of linalyl acetate in Adams Library. (c) Mass spectrum of linalyl acetate in Wiley6 Library with reference entry number of 64,027. (d) Mass spectrum of linalyl acetate in Wiley6 Library with reference entry number of 64,036.

excellent library match of 90% and a good correlation in RI, deductive findings are made with much greater confidence, which identifies peak #1 as linalool.

Not all cases are so straightforward. Let us now consider the mass spectrum of peak #2.

Applying the same strategy, the mass spectrum of peak #2 (Figure 2.11a) is searched against the entries of Adams Library with the computer software. The hit list rankings and match qualities are presented in Table 2.2.

The "best" match is given by β-pinene, which has a mass spectral match of 80%. This is then followed by α-fenchene (59%) and sabinene (59%), which are ranked in the second and third place, respectively. There are a few discrepancies with these assignments.

With a common nominal molecular mass of just 136.1 g/mol, β-pinene, α-fenchene, and sabinene are expected to elute quite early in the GC analysis (at a relatively low temperature). This reasoning is supported by the apparent disparity between the RIs obtained of these three compounds (i.e., at least 250 RI units less) with respect to that of peak #2.

Let us consider the statistics of the library match again. According to the PBM search algorithm, the mass spectrum of β-pinene does indeed bear good resemblance to that of peak #2, which is indicated by a library match of 80%. Taking the RI discrepancies and the match statistics into account, the possibility that the identity of peak #2 could be a structural homologue or derivative of β-pinene but with a longer side chain is acceptable.

So is it possible to use the RI value of peak #2 (RI = 1255) as a guide to retrieve and further examine the mass spectra of possible pinene-type compounds in the Adams Library with similar fragmentation ions? By only considering the mass spectra of compounds with reference RI within 5 units of the experimental RI (i.e., 1255 ± 5), the number of possible compounds selected for further investigation is significantly reduced.

Using this strategy, one of the possible compounds is linalyl acetate (M_R 195.1 g/mol; reference RI 1257) and its mass spectrum in the Adams, and the two Wiley6 libraries are shown in Figure 2.11b through d respectively. As exemplified in Figure 2.11b, the relative abundances between ions of 43 and 93 m/z clearly has a considerable influence on the PBM search procedure, which as expected, did not suggest linalyl acetate to be a possibility (i.e., no correct match). On the other hand, the linalyl acetate mass spectrum in the Wiley6 library (see Figure 2.11c) provided an excellent match of 91% to the experimental mass spectrum (of peak #2) with a top ranking in the hit list. Note the great similarity between linalyl acetate mass spectrum in the Wiley6 library (Figure 2.11c) and the experimental mass spectrum (of peak #2) with respect to mass range as well as the relative ion abundance of m/z value 43 and 93. In the final example, another linalyl acetate reference entry from Wiley6 library is shown in Figure 2.11d. The most obvious feature of this spectrum is the absence of ion with m/z of 43 as this reference entry of linalyl acetate was acquired with a mass range that did not include this ion. Nevertheless, the PBM search procedure rated this reference mass spectrum of 91% similarity match to the experimental mass spectrum.

So is peak #2 linalyl acetate? This speculation may be confirmed by the analysis of a pure linalyl acetate standard to check against the appearance of its mass spectrum and retention time. It is apparent from this exercise that mass spectral variability of the same compound exists, not only between different libraries, but also within the same library. Factors that contribute to these mass spectral variabilities (e.g., relative ion abundances) include system pressure, nature of inlet, ion source and temperature, as well as mass range selection during their acquisition.

The message here is clear. Despite the availability of more features in the modern GC-MS system for simplifying the data analysis task, such as through sophisticated software packages, these features are only tools, not a substitute for the analyst's knowledge. Hence, the experience and expertise of the analyst are in fact the most important link to a successful analytical solution.

Case Study 3: Isotope Dilution Mass Spectrometry for Quantitative Analysis of Naphthalene in Petroleum-Derived Sample

Isotope dilution mass spectrometry is essentially an internal standard quantitative method, which involves the spiking of the sample with an isotope derivative (enriched analogue) of the target analyte. As the chemical and physical properties of the isotope derivative and the target analyte are virtually identical, highly accurate and precise quantitative measurement of the target analyte can be attained by the incorporation of this standard into the sample prior to any sample preparative treatment. In doing so, sample recovery, variations, and errors in all stages of experiments (including instrumental fluctuations) can be corrected for in the final quantitative measurement.

In this application, a deuterium-labeled analogue of naphthalene (i.e., d^8-naphthalene) is added to a petroleum-derived sample, in which its naphthalene concentration is to be determined. Naphthalene ($C_{10}H_{58}$) has a mass of 128 m/z while d^8-naphthalene has a mass of 136 m/z (i.e., 8 mass units more). Note that an isotopically enriched standard is not as likely to have interferences from naturally occurring organic compounds, and so should be well identified and isolated mass spectrometrically from the matrix.

The isotope derivative may be added in three ways:

1. A small volume of reasonably concentrated labeled standard is added into the sample such that the total volume is not significantly altered.
2. Calibration standards and sample can be prepared by having the same concentration of labeled standard spiked into them. The concentration of the naphthalene can then be determined from the graphical plot of peak-area ratios of m/z 128/136 vs. naphthalene concentration, with their relative response factors taken into consideration.
3. The two solutions (i.e., labeled standard and kerosene sample) of similar expected concentrations are mixed in similar proportions.

Here, method (3) is employed where equal volumes (500 μL each) of the sample and the isotope derivative (50 mg/L d^8-naphthalene in methanol) are mixed. The resultant mixture is then injected (0.2 μL splitless) into the GC-MS quadrupole instrument for analysis.

Preferably, the two compounds should have the same elution time, but with highly efficient capillary GC columns, these two closely related analogues can often be chromatographically resolved (the greater degree of enrichment, the larger the retention difference), as shown in Figure 2.12. It is usually found that deuterium-labeled

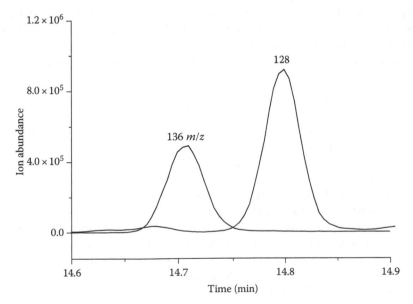

FIGURE 2.12 Extracted ion chromatogram of naphthalene (m/z 128) and d^8-naphthalene (m/z 136). Refer to text for experimental details.

compounds elute prior to their unlabeled analogue. In contrast, isotopically enriched standards such as ^{13}C, ^{14}C, and ^{18}O labeled standards elute much more closely to their non-labeled analogue. To be useful, it is usually necessary to have sufficient number of labeled atoms to shift the mass of the labeled compound away from the mass of the normal compound, considering its natural isotopic distribution.

By direct comparison of the response intensities of these two respective ions (i.e., 128 and 136 m/z), the naphthalene concentration in the kerosene sample can be calculated by the following equation:

$$\frac{A(\text{naphthalene})}{A(\text{internal standard})} \cdot \text{Relative response factor} = \frac{C(\text{naphthalene})}{C(\text{internal standard})} \quad (2.4)$$

where
  A(naphthalene) is the peak area of naphthalene given by the signal of 128 m/z ion
  A(internal standard) is the peak area of isotope derivative, d^8-naphthalene of 136 m/z ion
  C(naphthalene) is the concentration of naphthalene in petroleum-derived sample
  C(internal standard) is the concentration of labeled standard, d^8-naphthalene spiked into the petroleum-derived sample

Owing to the precision and accuracy of isotope dilution mass spectrometry, this quantitative technique (based on method 2) is widely used in areas such as environmental and clinical studies. Suitable isotope derivatives of the target analytes are also generally costly and may not always be commercially available. In this case, they would have to be synthesized at the expense of time and labor. The routine application of this technique would also require meticulous and regular calibration of apparatus to ensure integrity of the results obtained. The general benefit of isotope standards is that they can be added to the original sample and be processed through the sample extraction procedure to correct for extraction losses. Here, the recovery of the incurred compound and the isotope standard are assumed equivalent.

2.6   NEW DEVELOPMENTS AND EMERGING TRENDS

With increasing emphasis on ultra-trace analysis and faster sample throughput, future developments in GC-MS will be steered by the quest for enhanced selectivity, sensitivity, and speed. Strategies to improve analytical sensitivity in GC are manifold [23] and need not be confined to a drastic switch to expensive mass spectrometric equipment. Sensitivity enhancements are available in the form of gold (inert) ionization sources that can potentially improve S/N for active analytes. Speed, with reference to laboratory sample throughput, is expected, as laboratories are moving toward automating sample preparation. New GC-MS accessories are available that provide opportunities to halve equipment downtimes such as during column changeover. Manufacturers are also likely to capitalize on the growing market of multidimensional and fast GC to push the limits of MS data acquisition capability even further as they launch different equipment upgrades and designs to boost scan rates. As the area of microelectronics and computer operation continue to advance, the future outlook for the non-scanning TOFMS technology is bright and they are the most likely MS instruments to replace conventional quadrupole and double-focusing magnetic sector instruments for routine GC-MS analysis.

REFERENCES

1. Wells, R.J.; *Journal of Chromatography A*, 1999, 843, 1.
2. Hinshaw, J.V.; *LC-GC Europe*, 2004, 18, 22.
3. Fetterolf, D.D.; Yost, R.A.; *International Journal of Mass Spectroscopy and Ion Processes*, 1984, 62, 33.
4. Halket, J.M.; In *Handbook of Derivatives for Chromatography*, 2nd edn., Blau, K.; Halket, J.M.; Eds.; John Wiley & Sons Ltd., Chichester, U.K., 1993, pp. 297–324.
5. Anderegg, R.; *Journal of Mass Spectrometry Reviews*, 1988, 7, 395.
6. Munson, M.S.B.; Field, F.H.; *Journal of the American Chemical Society*, 1966, 88, 2621.
7. Slater, G.P.; Manville, J.F.; *Journal of Chromatography*, 1993, 648, 433.
8. Pellegrini, M.; Rosati, F.; Pacifici, R.; Zuccaro, P.; Romolo, F.S.; Lopez, A.; *Journal of Chromatography B*, 2002, 769, 243.
9. Korytár, P.; Parera, J.; Leonards, P.E.G.; De Boer, J.; Brinkman, U.A.Th.; *Journal of Chromatography A*, 2005, 1067, 255.
10. Frison, G.; Favretto, D.; Tedeschi, L.; Ferrara, S.D.; *Forensic Science International*, 2003, 133, 171.
11. Harrison, A.G.; *Chemical Ionization Mass Spectrometry*, 2nd edn., CRC Press, Inc., Boca Raton, FL, 1992.
12. Aebi, B.; Bernhard, W.L.; *Journal of Analytical Toxicology*, 2002, 26, 149.
13. Shellie, R.A.; Marriott, P.J.; *Analyst*, 2003, 128, 879.
14. Mondello, L.; Casilli, A.; Tranchida, P.Q.; Dugo, G.; Dugo, P.; *Journal of Chromatography A*, 2005, 1067, 235.
15. Borrey, D.; Meyer, E.; Lambert, W.; De Leenheer, A.P.; *Journal of Chromatography A*, 1998, 819, 125.
16. Fitzgerald, R.L.; O'neal, C.L.; Hart, B.J.; Poklis, A.; Herold, D.A.; *Journal of Analytical Toxicology*, 1997, 21, 445.
17. March, R.E.; Todd, J.F.J.; Eds.; *Practical Aspects of Ion Trap Mass Spectrometry*, CRC Press, Boca Raton, FL, 1995.
18. Gohlke, R.S.; *Analytical Chemistry*, 1959, 31, 535.
19. Dallüge, J.; Van Stee, L.L.P.; Xu, X.; Williams, J.; Beens, J.; Vreuls, R.J.J.; Brinkman, U.A.Th.; *Journal of Chromatography A*, 2002, 974, 169.
20. Shellie, R.; Marriott, P.; *Journal of High Resolution Chromatography*, 2000, 23, 554.
21. Adams, R.P.; *Identification of Essential Oil Components by Gas Chromatography/Quadrupole Mass Spectrometry*, Allured Publishing Corp., Carol Stream, IL, 2001.
22. Stauffer, D.B.; McLafferty, F.W.; Ellis, R.D.; Peterson, D.W.; *Analytical Chemistry*, 1985, 57, 1056.
23. Song, S.M.; Marriott, P.; Ryan, D.; Wynne, P.; *LC-GC North America*, 2006, 24, 1012.
24. Poole, C.F.; *The Essence of Chromatography*, Elsevier Science B.V., Amsterdam, the Netherlands, 2003.

3 Coupling Liquid Chromatography and Other Separation Techniques to Nuclear Magnetic Resonance Spectroscopy

Cristina Daolio and Bernd Schneider

CONTENTS

3.1 Introduction 62
3.2 Practical Aspects 63
    3.2.1 Sample Preparation 63
    3.2.2 Liquid Chromatography 64
    3.2.3 Nuclear Magnetic Resonance Spectroscopy 66
3.3 Modes of Operation 68
    3.3.1 Continuous-Flow LC-NMR Mode 68
    3.3.2 Stopped-Flow LC-NMR Mode 70
    3.3.3 Loop Storage LC-NMR Mode 71
    3.3.4 LC-SPE-NMR Mode 73
3.4 Miniaturization 76
3.5 Applications 79
    3.5.1 LC-NMR for Identification, Structure Elucidation,
        and De-Replication of Natural Products 79
    3.5.2 Using LC-NMR to Analyze the Tissue-Specific Occurrence
        of Natural Products 80
    3.5.3 LC-NMR in Biosynthetic Studies 81
    3.5.4 LC-NMR for Studies on Xenobiotic Metabolism and Degradation 83
    3.5.5 LC-NMR for Studying Environmental Samples 83
    3.5.6 Using LC-NMR for Studies on Drug Metabolism, Degradation,
        and Impurities 86
    3.5.7 LC-NMR in Structure Elucidation and Analysis of Isomers 88

      3.5.8   LC-NMR for Reaction Monitoring and Chemical
              Product Analysis 89
      3.5.9   LC-NMR in Polymer Research 91
3.6   Conclusions and Outlook 92
Acknowledgments 93
References 93

3.1   INTRODUCTION

Many well-established analytical methods are available to identify organic compounds of synthetic and natural origin and to elucidate unknown structures. Nuclear magnetic resonance (NMR) spectroscopy and mass spectrometry (MS) are by far the most important. Often a single technique such as NMR or MS is sufficient to assign a compound to a particular structural class or even to identify it. However, in most cases, a combination of two or more methods is required to make identification unambiguous. The use of NMR spectroscopy is usually necessary if a new structure has to be elucidated. Traditionally, NMR has been applied to single compounds that had to be separated from mixtures prior to analysis. Although in an increasing number of cases, identification has been achieved from a reaction mixture or an extract obtained from natural sources, separation is still beneficial, especially for new structures and for complex matrices. So-called hyphenated analytical systems comprising a separation technique, such as gas chromatography (GC) or liquid chromatography (LC), combined with one or more spectroscopic detectors are nowadays widely used in analysis of all kinds of mixtures and provide reliable information on the samples under study [1,2].

One of the oldest hyphenated techniques, LC-UV, has been used successfully to rapidly screen complex mixtures. The absorbance data obtained are insufficient for *de novo* structure elucidation. Hyphenating LC with spectroscopic techniques such as MS or NMR provides a more useful tool for rapid data collection and structure elucidation [3]. In fact, due to its higher sensitivity, LC-MS has been more extensively applied than NMR. However, if we compare the two techniques, the data obtained with MS allow molecular mass and certain functional groups to be determined but often do not permit the structure of unknown compounds to be unequivocally elucidated, especially the structures of regio- or stereoisomers [4,5]. NMR provides detailed structural information and is particularly suited for differentiating isomers and analysis of stereochemical features, even from 1D ^{1}H NMR spectra.

Combining liquid chromatography with nuclear magnetic resonance spectroscopy (LC-NMR) was first proposed in the late 1970s [6–8]. At that time, the method was limited by the inherently poor sensitivity of NMR spectroscopy and therefore did not attract the attention of researchers that would have been necessary to establish it as a routine analytical approach [9]. However, subsequent technical and methodological improvements such as pulse sequences for efficient suppression of solvent signals [10–12], the use of high-field magnets, the development of cryogenically cooled probes [13], and miniaturization [14,15] have dramatically improved NMR sensitivity and contributed to LC-NMR's development into a well-established analytical tool. More recently, the introduction of post-column solid-phase extraction (SPE)

[16,17], which enables compounds of interest to be concentrated prior to NMR, has led to the extended hyphenated technique, LC-DAD-SPE-NMR.

Additional progress in hyphenation techniques has been achieved by implementing MS. The coupling of the two most important spectroscopic techniques, NMR and MS, to LC has expanded the structure-solving capabilities by making it possible to obtain MS, NMR, and UV/Vis data simultaneously from the same chromatographic peak [18,19]. Hyphenated LC-MS-NMR supplements conventional methods of separation and identification and makes them more effective, but is not always able to completely replace them in identifying or structurally elucidating new compounds. There will still be cases that require more steps to purify the target compound, e.g., when the separation problem is too complex or the resolution of the chromatographic peak is insufficient for online spectroscopy to be applied. Since each analytical method has its own advantages and limitations, analytical chemists have available a repertoire of techniques which are appropriate to solve a particular problem in the most efficient way. Here we discuss the capability of the different variants of coupling LC and other separation techniques to NMR and how the analytical chemist can select an approach to solve specific analytical issues.

3.2   PRACTICAL ASPECTS

3.2.1   SAMPLE PREPARATION

The procedures for sample preparation for LC-NMR are usually the same as for analytical high-performance liquid chromatography (HPLC): a homogeneous and clear solution without any non-dissolved particles is required and the solvents used for injection should not disturb the chromatographic process or the detection of the analyte. In many cases, the samples for HPLC analysis require pretreatment, including operations such as weighing and/or dilution before injection [20]. Considering the circumstance that extracts in general might be complex mixtures of various substances covering a broad range of polarity, including both lipophilic and hydrophilic components, pre-purifying or fractionating the crude extract could improve the chromatographic resolution. Enriching the desired product fraction could prevent the column from being overloaded by unwanted components and thus could enhance the concentration of the analytes above the detection limit [21,22].

For successful LC-NMR analysis, sample preparation is dictated according to the nature of the sample and should be carefully optimized. A wide range of techniques can be applied to samples prior to their injection in the HPLC system. The most common step is filtration, but centrifugation, pre-column SPE, liquid–liquid extraction, sonication, lyophilization, supercritical fluid extraction or any other appropriate extraction or purification procedure may also be applied. Removing strongly lipophilic fractions (de-fatting) is one of the steps that is crucial to improve separation performance and enhancing concentration of analytes in the volume to be injected to HPLC. Preferably, the sample should be injected in the starting eluent mixture [23]. This requires proper solubility of the sample in small solvent volume. However, some matrices may be insoluble in common solvents, so the analytes must be fractionated before HPLC. On the other hand, concentrated samples must often be diluted to avoid overloading the column or saturating the detector. A suitable fractionation and sample enrichment procedure is online pre-column SPE [24]. An alternative approach to any pretreatment of the sample is to employ multidimensional column chromatography (or column switching). In this approach, a portion of the chromatogram from an initial column is transferred selectively to a second column for further separation. This procedure maximizes the injection of the desired analyte band onto the second column while minimizing the amount of interfering compounds injected.

Developing an adequate sample pretreatment procedure can be more challenging then selecting a useful solvent system and gradient for HPLC separation [21]. Finally, the mode of LC-NMR hyphenation determines to some extent the sample preparation. For example, it is advantageous to dissolve samples dedicated to online and stopped-flow LC-NMR in deuterated solvents; non-deuterated solvents can be used for LC-SPE-NMR without negatively affecting NMR measurement.

3.2.2 LIQUID CHROMATOGRAPHY

There are several requirements for stationary and mobile phase, running conditions, and equipment, when HPLC is coupled with NMR. In early applications of LC-NMR, normal-phase chromatographic columns were employed as stationary phases and halogenated solvents such carbon tetrachloride [6,7] or trichlorotrifluoroethane [8] as mobile phases. For example, Bayer et al. [7] carried out hyphenated experiments on standard compounds, using normal-phase columns and D_2O-saturated carbon tetrachloride as mobile phase. The non-deuterated solvent was feasible for HPLC because an ampoule filled with D_2O was placed near the NMR flow cell or in the center of the cell to adjust for magnetic field inhomogeneity. Despite some problems in LC-NMR associated with the use of binary solvent mixtures, reversed-phase HPLC with MeCN-water or MeOH-water became the standard separation method in hyphenated systems for most applications. Due to the large signals of non-deuterated solvent components and the resulting difficulties in detecting tiny signals of the analytes in such ^1H NMR spectra, D_2O had to substitute for H_2O as an eluent. Reducing the intensities of the residual water signal and the signal of the second solvent component, if also used in deuterated form, resulted in improved signal-to-noise ratios (S/N) of the NMR spectra, but even suppressed signals may be large compared to the signals of the analytes. Techniques for suppressing solvents have been established, which has helped to overcome many solvent-related compatibility problems of LC-NMR and essentially contributed to further improving the quality of the NMR spectra. The benefit of solvent suppression for LC-NMR coupling will be discussed in the following (see Section 3.2.3).

The two components of a binary HPLC solvent system may be applied in a constant ratio (isocratic conditions) or as a gradient. Elution under isocratic conditions has the drawback that chromatographic peak broadening increases significantly with longer retention times. Hence, due to the given volume of the NMR flow cell, the fraction of the HPLC peak accessible to NMR measurement will be reduced for late-eluting components, thereby decreasing the S/N ratio of the obtained NMR spectrum. HPLC peak broadening is less pronounced during gradient elution. Therefore, no significant decrease in sensitivity occurs for components eluting late from the

HPLC column. However, the two components of HPLC solvents possess different magnetic susceptibilities and the change in the solvent composition during the gradient course of the chromatographic run will very likely interfere with the homogeneity of the magnetic field. As a result, broadening of NMR signals will occur and negatively affect the S/N ratio and resolution of the NMR spectrum. Additionally, fluctuations in the solvent chemical shifts complicate the interpretation of NMR spectra especially under continuous-flow conditions (also referred to as on-flow). The negative effects of solvent gradients could be eliminated by stopping the chromatographic solvent flow so that static solvent conditions are obtained in the NMR flow cell. Interestingly, static detection conditions have been achieved using solvent gradients and a two-pump system [25]. When an equal but reverse gradient is created on the second column and the two solvent gradients are combined before the NMR detector, the composition of solvent reaching the NMR flow cell is kept constant. High-quality NMR spectra have been obtained using this approach.

Ideally, the chromatographic peak width should be adjusted so that it is less than or equal to the active volume of the NMR flow cell [26]. Hence, not only should HPLC conditions be optimized as carefully as possible but also size and type of the NMR flow cell has to be taken into account in hyphenation (see Section 3.2.3). One of the conventional ways to reduce HPLC peak width is to add a small proportion of an acid component such as formic acid, acetic acid, phosphoric acid, or trifluoro acetic acid to the eluent. The two last acids are given the preference in LC-NMR because they do not give rise to additional signals in the ^{1}H NMR spectrum or otherwise disturb the LC-NMR measurement [19].

In conventional LC-NMR coupling, standard reversed-phase columns of 4 or 4.6 mm i.d. are currently being used. Due to the intrinsically low sensitivity of NMR spectroscopy, the columns need to be loaded with much higher sample amounts (often up to several 100 µg) than are used in analytical HPLC with UV or DAD detection. In order to deliver an appropriate amount of minor components of a mixture to the NMR flow cell, the column has to be overloaded. Under such conditions, chromatographic separation is frequently suboptimal and normally well-resolved peaks may (partially) overlap. To overcome problems arising from partially co-eluting HPLC peaks, peak segments may be transferred to the NMR flow cell; in addition, so-called time-slice experiments can be carried out, in which the HPLC flow is stopped at regular time intervals to measure NMR spectra of each fraction. If these fractions are mixtures of incompletely separated components, the NMR spectra show signals of both of them at different intensities. NMR difference spectra of the individual compounds then can be generated using processing software. It should be noted that high column loading can be better tolerated if a relatively small peak of interest appears before a partially co-eluting peak rather than after [26].

Because, with the exception of D_2O, completely deuterated eluents are too expensive for routine analysis on standard HPLC columns, a need to reduce solvent consumption was recognized soon after LC-NMR became a routine analytical method [27,28]. Coupling miniaturized chromatographic separation techniques such as capillary HPLC (capLC) and capillary electrophoresis (CE) to NMR offers several advantages compared to the conventional hyphenated LC-NMR. Fully deuterated solvents can be used because the required flow rates are low and the need for solvent suppression is reduced. Moreover, using capillary columns instead of standard columns increases separation efficiency significantly.

In general, in addition to C18 stationary phases, a variety of other reversed-phase materials such as C30 [29] are suitable for LC-NMR, but there are also reports demonstrating the feasibility of alternative sorbents. Examples are ion exchange resins, which have been used to separate protein mixtures [30] and size exclusion gels for the separation of polymers (see Section 3.5.9). Thus, the introduction of alternative stationary phases to LC-NMR has considerably extended the range of applications. This is also the case for recently reported use of hydrophilic interaction chromatography coupled to LC-NMR for the separation and identification of extremely polar analytes in body fluids [31]. Due to the problems caused by protonated solvents in NMR detection, nonprotonated mobile phases were used in the early days of LC-NMR coupling [6,7]. In addition, supercritical CO_2 was demonstrated to be a feasible mobile phase not only in fluid chromatography but also in supercritical fluid chromatography-NMR coupling [32].

Apart from problems regarding the choice of the solvents and stationary phase, only a low pulsation LC pump with a short mixing cycle can provide proper gradient formation. In addition, the UV or DAD detector cell should be as small as possible to keep back mixing to a minimum. It is important to note that a high-quality chromatographic separation contributes essentially to the success of the NMR measurement. Thus, the separation should be optimized and executed as carefully as possible [33].

3.2.3   Nuclear Magnetic Resonance Spectroscopy

The field strength of NMR magnets used for hyphenation with LC is mostly between 9.4 T, corresponding to 400 MHz ^1H resonance frequency, and 16.4 T (700 MHz). This field strength corresponds to the spectrometer types usually applied to NMR analysis of natural products and other small molecules. LC-NMR spectroscopy uses probe heads that, in contrast to conventional NMR probes, do not make use of exchangeable NMR tubes but contain a flow-through cell. The flow cell is connected to the HPLC via a transfer capillary (inlet) and, via an outlet capillary, to a fraction collector or waste container. These permanent connections make spinning the sample impossible. However, the requirement for spinning is reduced in modern spectrometers due to the excellent field homogeneity of magnets of the newest generation, computer-controlled methods for further optimizing field homogeneity, and the trend to miniaturize flow cells. The capillary connection between the UV or DAD detector and the NMR magnet should be as short as possible, but interference of the stray field of the NMR magnet has to be considered. As a compromise, the HPLC is usually positioned slightly outside the 5 mT line (corresponding, e.g., to about 1.5 m for a non-shielded 11.7 T magnet) of the stray field [1]. Nowadays, actively shielded NMR magnets are frequently used, as their 5 mT lines are only slightly outside or even within the body of the magnet, reducing the distance needed between the HPLC system and the magnet. As a result, a shorter capillary connection with smaller volume and less diffusion of the sample is possible.

The active volumes of conventional flow cells are between 30 and 120 μL. These volumes provide convenient conditions with regard to the intensity and line width of the resulting NMR signals. However, if the active volume of the flow cell exceeds the sample volume, the sample may be diluted and sensitivity and resolution drop [34]. The design of the flow cell is also important for the performance of the NMR probe [9]. Since the sample does not rotate, the radio frequency coils can be mounted directly on the outer surface of the glass body of the flow cell. This arrangement guarantees an optimal filling factor and, consequently, improves the sensitivity of the LC-NMR probes compared to conventional NMR measurements in tubes [35,36]. Recently, capillary flow NMR cells of a detection volume in the order of 1.5 µL have been used for coupling with capLC [37] and even detection volumes in the submicroliter range have been reported. Additionally, the small size of the solenoid coils of such micro- and nano-flow cells improves the filling factor and enhances mass sensitivity [38]. Moreover, these microcoil probes with effective volume in the microliter range have significantly reduced the amount of mobile phase needed to perform LC-NMR analysis, thus making fully deuterated solvents economically feasible.

Given the specific requirements of LC-NMR discussed earlier, it is obvious that the overall construction of LC-NMR probes is dictated by the architecture of the flow cell and the capillary connections. Hence, it is trivial that a LC-NMR probe is not useful for measuring samples in tubes and vice versa, i.e., different probe types are needed. Using two or more alternating probes on one spectrometer is not a problem with room temperature probes, which can be rapidly removed from the NMR magnet and replaced by a different probe. However, exchanging cryogenically cooled probes includes warm up/cool down periods which last several hours. Cryoprobes that can be operated both with tubes and in the flow mode have been developed. Inserting a so-called Cryofit™ (www.bruker-biospin.com) into a 5 mm cryoprobe converts a normal probe into a flow probe within a few minutes without interrupting the cold time. Cryoprobes, both of the convertible and non-convertible types, have improved the sensitivity of NMR dramatically by cooling the receiver coil and pre-amplifier to cryogenic temperatures (~20 K) while the sample remains at ambient temperature [13]. The use of a cryogenic instead of a conventional probe improves S/N ratio by about a factor of 4 or reduces measuring time of a spectrum of the same S/N by a factor of 16. In the low nanogram range, samples can be detected, and in the high nanogram range, structural analysis can be carried out. The cryogenic NMR technology, which also allows measurements under coupling conditions [39,40], represents one of the major improvements in the sensitivity of NMR spectroscopy during the last decade. Room temperature LC-NMR probes are built as inverse-detection probes and therefore are able to sensitively measure ^{1}H NMR spectra and ^{1}H-detected two-dimensional (2D) experiments but are hardly suitable for acquiring ^{13}C spectra. This situation has changed thanks to the availability of cryogenically cooled flow probes which are available also in the ^{13}C{^{1}H} coil geometry. Remarkably, sensitivity gains due to the introduction of cryogenic NMR technology and the post-column SPE (see Section 3.3.4) have enabled the direct ^{13}C NMR spectra of unlabeled samples with an inverse detection probe to be measured.

As mentioned earlier, using reversed-phase HPLC conditions requires binary solvent mixtures such as MeCN–water or MeOH–water. This is not the preferred situation for NMR spectroscopy because each of the two solvent components gives rise to disturbing signals in the ^{1}H NMR spectrum. In addition to the ^{1}H NMR signal of [2-^{12}C]MeCN and [^{12}C]MeOH, respectively, the ^{13}C coupling satellites from the 1.1% naturally abundant ^{13}C isotopomers appear in the spectra. The signals of MeCN and its satellites, e.g., cover a part of the spectrum as broad as 0.3–0.4 ppm and may obscure signals of the analyte. In order to avoid such broad solvent signals and to adjust the receiver gain properly, the intensity of the solvent signals should be reduced to the height of the analyte signals. This can be achieved using deuterated solvents, by applying solvent suppression techniques or by a combination of both. Reduced intensities of the residual water (HDO) signal and the signal of the second solvent component, if also used in deuterated form, enable enhanced receiver gain and result in improved S/N ratios of the NMR spectra but may be still large compared to the signals of the analytes. Solvent suppression techniques have been established which have helped overcome many solvent-related compatibility problems of LC-NMR and essentially contributed to further improving the quality of the NMR spectra. Solvent suppression can be accomplished, e.g., by NOESY-type presaturation [10,11] or by the water suppression enhanced through T_1 effects' (WET) sequence [12]. The WET sequence is more efficient in continuous-flow measurement because it requires shorter delays than does the presaturation technique. In most LC-NMR applications, resonances of HDO, MeCN, and MeOH can be efficiently suppressed. However, the recorded spectra will still show the ^{13}C satellite peaks of solvent molecules. In samples containing low concentrations of the analyte, the ^{13}C satellites can still overlap or obscure NMR signals. In this case, ^{13}C decoupling collapses the satellites peaks and further improves the solvent suppression [34].

However, even most efficiently suppressed solvent signals still cover a certain part of the spectrum and may hide some of the resonances of the analyte of interest. This general drawback of direct LC-NMR coupling can be reduced by running the same sample once more in another eluent system having different chemical shift values. Using two different organic modifiers, e.g., MeCN (δ1.8–2.2)-water in one run and MeOH (δ3.1–3.4)-water in the second run, can help detect obscured signals [41].

Further compatibility issues between HPLC and NMR are connected to the different timing requirements of the two techniques. Different approaches to overcoming these issues are implemented in the interfaces linking HPLC with NMR. Time management in LC-NMR, in addition to solvent flow handling and other points, is an important part of the following paragraph.

3.3 MODES OF OPERATION

3.3.1 Continuous-Flow LC-NMR Mode

In continuous-flow mode, the chromatographic system is directly connected to an NMR detection cell which is inside the magnet (Figure 3.1). The NMR spectra are acquired continuously while the chromatographically separated peaks are flowing through the detection cell. The result is a series of one-dimensional (1D) NMR spectra which cover the whole chromatogram and are typically displayed as a 2D matrix (NMR chromatogram, pseudo 2D NMR spectrum) showing NMR spectra against retention time in the form of a stacked plot or contour plot. Overlapping

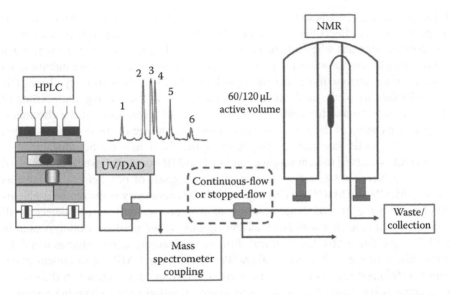

FIGURE 3.1 Scheme of an LC-NMR hardware setup for continuous-flow and stopped-flow mode.

peaks in the timescale (LC dimension) can be separated in the ^{1}H NMR spectrum (chemical shift dimension) [10].

With actively shielded 11.7 T cryomagnets, the LC system can be as close as 30–50 cm vs. a distance of 1.5–2 m for conventional magnets of the same field strength. Normally, a UV or DAD detector is used in the LC system to monitor the chromatographic run and to record absorption spectra, supplementing analytical data obtained from NMR. However, in continuous-flow mode, a HPLC detector is not indispensable because there is no peak selection and the NMR is not triggered by means of a detector signal.

In continuous-flow mode, NMR signal intensity and line width are significantly affected by both the flow rate and the detection volume of the flow cell [10]. The ratio of flow cell volume to flow rate defines the residence time of an analyte in the flow cell. After a radio frequency pulse has been applied to the sample in the active volume, the number of excited nuclei in the detection cell decreases not only due to spin-lattice (T_1) and spin–spin relaxation (T_2) but also because excited nuclei leave the cell and are replaced by unexcited nuclei. Hence, for a given flow cell, the S/N and line width can be improved by optimizing the residence time, i.e., the flow rate, of HPLC. In practice, flow rates of 0.6–1.0 mL min^{-1} are the exception rather than the rule in continuous-flow LC-NMR. Often flow rates, in comparison with the stopped-flow or loop storage mode, are significantly reduced, e.g., to 0.3 mL min^{-1} [42] or 0.1 mL min^{-1} [43]. Extremely slow elution rates, as low as 0.006 mL min^{-1} [44] have been used to increase residence time and thus compensate for insufficient sensitivity.

In addition to optimizing residence time through flow rate, increasing the sample concentration as much as possible in the active volume of the flow cell is a way to improve the S/N ratio. One of the possibilities is to use highly loadable semi-preparative columns of 8 mm diameter [43] in combination with low flow rates.

Injecting samples in amounts of several milligrams into standard columns of 4 or 4.6 mm diameter affects the chromatographic resolution and separation, since resolution often degrades when sample injection is scaled up to a certain level [45]. Such conditions also compromise the lifetime of the columns due to a more intense mass transfer of the analyte molecules through the particles of the stationary phase. The limited solubility of the analytes in the mobile phase at the starting composition of the eluent may be another issue compromising HPLC separation. If the chromatographic separation of the extract starts at a low percentage of the organic modifier, the solubility of the late-eluting compounds in the eluent may not be satisfactory.

Several aspects of solvent suppression in LC-NMR coupling have been discussed earlier (see Section 3.2.3). Suppression of large signals of non-deuterated solvents (mostly MeCN and MeOH) and of residual HDO is necessary to overcome problems connected to the limited dynamic range of the receiver unit, i.e., to detect small analyte signals together with large solvent signals. When gradient elution is used in HPLC coupled to NMR, the chemical shifts of the solvents signals change while the composition of the mobile phase alters. Performing the NMR measurement under flow conditions complicates the issue because the frequency with which the solvent resonance is irradiated has to be continuously adjusted to the changing chemical shifts. A one-scan-one-pulse experiment ("scout scan") is used prior to the WET sequence [12] for this purpose, while the transmitter automatically is adjusted to keep the largest peaks (non-deuterated MeCN) at a constant frequency [41].

The continuous-flow mode has been frequently used to analyze the major components of the mixture, investigate reaction processes almost in real time, study chemotaxonomy, de-replicate known compounds and identify drug impurities, among other purposes (see Section 3.5). However, often LC-NMR on continuous-flow mode has provided only preliminary information about the samples and has been followed, e.g., by stopped-flow experiments. More accurate data, such as those from 2D NMR experiments which are necessary for unambiguous structural assignments, have rarely been obtained under such conditions [46]. The usual solution of this problem is to acquire spectra under stopped-flow, loop storage or LC-SPE-NMR conditions.

3.3.2   STOPPED-FLOW LC-NMR MODE

The stopped-flow mode is probably the most versatile and widely employed technique among the direct LC-NMR coupling methods because the spectra are measured under static conditions, the NMR measurement time is not limited by the chromatographic run, and users can control the quality of the NMR spectra during the measurement. Hence, NMR measurements are much more sensitive in the stopped-flow mode compared to the continuous-flow mode. HPLC peaks are selected based on the UV/DAD chromatogram. Alternatively, fluorescence or radioactivity detectors can also be used to detect signals of interest. HPLC flow is stopped when a selected peak has arrived at the center of the flow cell. Therefore, the delay time between UV detector and NMR flow cell must be carefully determined [10]. Selection of HPLC peaks can be performed manually or under automatic computer control according to preselected retention time windows or peak shape criteria such as height or slope steepness. In the "time-slicing" mode, which is a variant of the stopped-flow mode, the HPLC pump is stopped at preselected intervals during the flow of the eluent through the NMR detector cell. Time slicing can be used instead of triggering the NMR by signals received from the UV or DAD detector. Hence, time slicing is very useful to detect non-chromophores [47] or other hard-to-detect substances. It is also useful to check for purity of peaks. In any case, in the stopped-flow mode, NMR spectra are measured in the order of elution from the LC column. Moreover, the LC-NMR interface enables the NMR probe to be bypassed but fractions of the HPLC eluent to be sent to the waste (Figure 3.1). Due to the static NMR conditions of the stopped-flow mode, measurement times can be prolonged to meet the NMR spectroscopic requirements for recording 1D spectra of optimal S/N ratio and time-consuming 2D NMR spectra [48]. Moreover, acquisition parameters, including the homogenization of the magnetic field and the parameters for solvent suppression can be optimized. The NMR measurement can be automatically stopped after the S/N ratio reaches a predefined level. After the NMR measurement is completed, the LC flow will be restored until the next LC peak reaches the center of the NMR flow cell. Thus, the whole stopped-flow LC-NMR run can be automated. However, in fully automatic mode, HPLC and NMR must be synchronized so that several peaks can be measured without further action by the operator [49].

While one or multiple NMR experiments for a certain peak are running, additional LC peaks remain in the chromatographic system. Depending on the chromatographic conditions and the length of the stop period, diffusion of the already separated analytes may occur in the column or the transfer capillary. As a result, these peaks will be broadened and finally the concentration of the analytes will be decreased [50]. However, this effect is significantly reduced if a gradient elution is employed. If this is the case, the analytes are confined to narrow regions within the stationary phase, and sharp peaks may be eluted despite prolonged periods without flow. In other cases, during long-lasting NMR measurements, the separation of close peaks previously achieved by LC can be negatively affected [1,50]. Alternative LC-NMR techniques such as a loop storage or LC-SPE-NMR set up may be recommended in such cases.

Numerous examples (see Section 3.5) have demonstrated the excellent performance of the stopped-flow LC-NMR mode, especially for analyzing extracts from plants and other natural sources. However, such extracts often contain a limited number of major components and a tremendous number of minor compounds. Analysis of the latter often is more difficult not only because of their low concentrations, which are connected to column-loading limitations, but also because partial co-elution with major compounds may occur. For example, an analyte eluting as a large peak may contaminate subsequent peaks, especially the minor ones, through diffusion and the peak broadening effect of the flow cell. Even when there is little peak tailing, washing out a large peak from the flow cell may take more time than is available before the next peak approaches. Hence, it is problematic to measure small peaks eluting immediately or shortly after large ones. The opposite case usually does not cause any problems.

3.3.3 LOOP STORAGE LC-NMR MODE

Loop storage is another of the direct LC-NMR approaches that can be used to analyze samples containing more than one component of interest. In this mode,

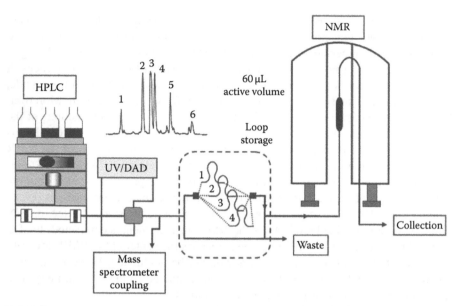

FIGURE 3.2 Scheme of LC-NMR hardware setup for the loop storage mode.

chromatographically separated peaks are stored in hoses or capillaries, so-called loops for later NMR measurements. After separation, which is monitored by an UV, DAD or other detector, a series of individual chromatographic peaks are allocated to a corresponding number of loops (Figure 3.2). Later, the stored peaks can be transferred to the NMR probe and analyzed in an order different from that of the chromatographic run. This is different from continuous-flow and stopped-flow modes, in which peaks always are measured in the order of elution from the column. Another advantage of loop storage mode is that the chromatographic eluent flow is not interrupted and, hence, peak broadening due to diffusion on the column can be largely excluded. If sufficient material has been captured in the loop, all kinds of 1D and 2D NMR experiments can be carried out. A prerequisite for storing separated peaks in loops is, of course, that the analytes must be stable in the eluent mixture, which often contains acid components (see Section 3.2.3) during the extended period of analysis.

In early LC-NMR interfaces, the loop storage system was made of hoses and their volume was bigger than that of the flow cell, leading to a dilution effect and finally decreasing the S/N ratio of NMR spectra. Modern loop collection devices are made of capillaries with a storage volume adjusted to that of the flow cell, so only negligible peak broadening is observed. However, since broadness of chromatographic peaks depends on specific LC conditions, the elution volume of the chromatographic peak is independent of that of the loop volume and that of the flow cell, either in the loop storage mode or with other direct LC-NMR approaches. Another feature of modern loop storage devices is that they allow for direct transfer HPLC → storage loop → NMR probe; alternatively, the HPLC and NMR devices of the LC-NMR platform can be operated completely independently of each other. Separated HPLC peaks can be stored in the loops while the NMR spectrometer is used for other measurements. Otherwise it is possible to automatically transfer analytes from the loop storage interface to the NMR probe and measure spectra. In addition, it is even possible to perform the chromatographic runs in a separate HPLC system or to transfer the cassette containing the loops to another NMR or MS instrument for analysis.

Cross-contamination can be avoided in the loop storage mode because the loops, in addition to the flow cell, are extensively washed after transfer. Also, the loop storage mode is successfully employed in the analysis of polar or high hydrophilic compounds that cannot be trapped on commonly used cartridges of the LC-SPE-NMR system [51]. It should be noted that NMR spectra of peaks collected in storage loops will be measured in the eluent from HPLC. Therefore, the solvents or at least the water component has to be used in deuterated form and solvent suppression has to be applied.

3.3.4 LC-SPE-NMR Mode

In addition to the established direct LC-NMR hyphenation techniques described earlier, an innovative approach, namely, LC-SPE-NMR coupling, has been introduced to the analysis of compounds in mixture [16,17]. LC-SPE-NMR connects the chromatographic device with the spectrometer via a SPE interface (Figure 3.3) operating at HPLC pressure. The fundamental difference between direct coupling techniques and LC-SPE-NMR is solvent handling. Direct LC-NMR coupling (continuous-flow, stopped-flow, loop storage) transfers the analyte from the LC to the NMR probe as a solution in the chromatographic solvent, which is then also used for NMR measurement. In contrast, the post-column SPE module mediates an exchange from HPLC solvents to the deuterated NMR solvents. While problems connected to solvent incompatibility between HPLC and NMR were inherent to direct LC-NMR coupling (see Sections 3.3.1 through 3.3.3), this is no longer the case for LC-SPE-NMR and both chromatographic separation and NMR analysis can be performed under

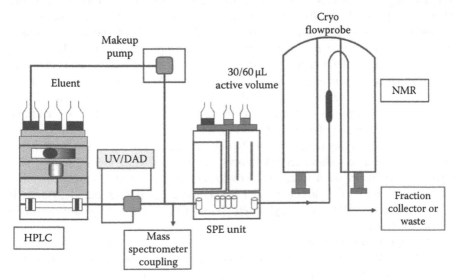

FIGURE 3.3 Scheme of LC-SPE-NMR hardware setup.

optimal conditions. Chromatographic separations do not need to be optimized for NMR-specific conditions, but typical reversed-phase solvent systems routinely used for offline-HPLC are applicable. In particular, there is no need for expensive deuterated solvents to be used for chromatographic separation. The use of any buffer additive is possible, including formic acid and acetic acid, both of which give rise to NMR signals and therefore have to be avoided in direct LC-NMR coupling.

In fact, the concept of SPE is closely related to a pre-cleaning or pre-concentrating step prior to injecting samples in the HPLC system. The combination of pre-column SPE with LC-NMR [24] and LC-SPE-NMR [52] has been reported. However, in the post-column LC-SPE-NMR approach, SPE is not used to enrich unseparated analytes from a diluted sample prior to LC but to capture individual peaks or compounds after separation. The introduction of post-column analyte trapping has eliminated not only problems of solvent incompatibility but also some of the other drawbacks of direct LC-NMR discussed earlier. For example, diffusion-mediated peak broadening, which in the stopped-flow mode can occur during extended periods of solvent flow interruption, is no longer a problem in LC-SPE-NMR because LC is not stopped during NMR measurement. Moreover, even broad LC peaks or peaks from semi-preparative columns [51] can be completely trapped on a SPE cartridge. In an LC-SPE-NMR setup, each analyte detected in the LC run can be selected for automatic trapping on SPE cartridges. Analyte selection can be achieved by considering the intensity of the chromatographic peaks at a certain wavelength, the retention time, or in an unspecific time-sliced way. After being eluted from the HPLC column and passing the UV or DAD detector, some water must be added (typically in a ratio between 1:3 and 1:4) to the LC eluent, by means of a make-up pump, to lower its eluotropic strength.

Retention at the stationary phase of the SPE cartridge and subsequent elution are crucial steps in LC-SPE-NMR coupling. The partition factor between the stationary and mobile phase is affected by the proportion of the organic modifier in the mobile phase after post-column dilution with water, the volume of the mobile phase, and the analyte concentration in the mobile phase [53]. In addition, the analyte amount in the stationary phase retained from preceding trappings (in multiple trapping mode) has to be taken into account. Thus, well-optimized SPE parameters can drastically enhance the amount of analyte reaching the NMR probe [51,54–56]. These parameters include the choice of the SPE sorbent, which due to limitations of post-column dilution ratios, is extremely important. Among the commercially available sorbents, Resin GP, a poly(divinylbenzene), and Resin SH, a strongly hydrophobic polystyrene-poly(divinylbenzene) copolymer, are highly efficient in retaining most analytes including a wide variety of natural products [53]. Cartridges filled with CN (cyanopropyldimethylsilyl silica), C2 (ethyldimethylsilyl silica), C8 (octyldimethylsilyl silica), C8(EC) (octyldimethylsilyl silica, end-capped), C18 (octadecyldimethylsilyl silica), C18(HD) (octadecyldimethylsilyl silica, high-density) are useful for special purposes.

Although impressive results have been obtained with various types of analytes, a recent LC-SPE-NMR study has shown that polarity of a compound, which was estimated based on calculated octanol/water partition coefficients (log K_{OW}; log P), is not a reliable parameter for predicting SPE retention [51].

Hence, predicting SPE retention properties and trapping of highly polar analytes is still problematic. Trapping of highly hydrophilic compounds can be improved when the water content of the eluent is maximized by post-column dilution. However, direct LC-NMR (stopped-flow, loop storage) seems to be the method of choice in such cases.

In direct LC-NMR, regardless of the specific mode, the sample amount from only one injection to the chromatographic column is available for NMR measurement. This is different in LC-SPE-NMR because a particular chromatographic peak from two, three, or more HPLC runs can be automatically trapped on a single SPE cartridge ("multiple trapping"). The efficiency of multiple trappings, i.e., the number of trappings which are possible before breakthrough occurs, depends on the properties of the analyte, the partition factor on the cartridge, etc. [57]. Multiple trapping, in addition to solvent change, is one of the most important advantages of LC-SPE-NMR.

The next step after trapping can be flushing the loaded cartridges with water to remove remaining acidic components of the elution buffer and protect sensitive analytes from decomposition, followed by drying the cartridge with a soft stream of nitrogen gas to remove the residual amounts of non-deuterated chromatographic solvent. After drying, the analyte is eluted into the NMR flow cell using a fully deuterated solvent. Often MeCN-d_3, MeOH-d_4, acetone-d_6, or CDCl$_3$ are employed to desorb the analyte from SPE cartridge and transfer it to the NMR probe. It is important that the eluotropic capacity of the deuterated solvent and the solubility of the analyte in that solvent are sufficient to elute the analyte with the solvent front. If the volume of the deuterated solvent exceeds the volume of the flow cell, which is usually 30 or 60 μL, the S/N ratio of the resulting NMR spectrum drops. In such a case, a different eluent should be used for desorption. Many deuterated solvents are appropriate for eluting the SPE cartridge into the NMR probe, another important advantage of LC-SPE-NMR. Thus, a solvent can be used for which chemical shift data are reported or spectra are available in a database, allowing for a direct comparison with measured spectra. Desorption with fully deuterated solvents also reduces the need for solvent suppression and results in NMR spectra comparable to those obtained in spinning tubes [51].

Because of possibility for capturing an entire chromatographic peak from multiple runs, reducing the volume of the peak during desorption from the SPE cartridge to that of the flow cell volume, and enabling a fully deuterated solvent of choice for NMR analysis to be used, LC-SPE-NMR is much more sensitive than conventional direct LC-NMR coupling techniques. Furthermore, LC-SPE-NMR takes advantage of cryogenically cooled probe heads to further enhance sensitivity. Taken together, these benefits enable routine measurements to be made, including of all kinds of ^{1}H-detected 2D homo- and heterocorrelation NMR experiments such as ^{1}H,^{1}H COSY, TOCSY, HSQC, and HMBC and, in addition, of direct ^{13}C spectra. Since the 2D NMR methods are highly important for structural elucidation of novel compounds, why LC-SPE-NMR has been successfully employed in natural product analysis becomes clear. In summary, fully automated measurements of a whole set of 1D and 2D NMR spectra of multiple peaks from a single LC run are routine and feasible.

3.4 MINIATURIZATION

Miniaturization is not a new trend in LC-NMR coupling. In the middle of the 1990s, microliter and sub-microliter NMR flow cells were developed [58] and coupled to capillary electrophoresis (CE) [59] and capillary HPLC (capLC) [60]. Miniaturized saddle coils were used [28] or a copper wire was wrapped directly around the surface of a fused silica separation capillary to form a solenoid microcoil [59]. The resulting flow cell had a volume as small as 5 nL and limits of detection in the nanogram range were achieved for NMR spectra which were recorded in a few minutes [58]. Selenoidal microcoils are horizontally oriented, perpendicular to the static magnetic field (B_0) (Figure 3.4), and immersed in a magnetic susceptibility matching fluid to improve resolution.

However, although methods continued to be developed, many early publications dealing with various capillary separation methods such as capLC [60], CE [59], and capillary electrochromatography (CEC-NMR) [28], coupled to capillary NMR

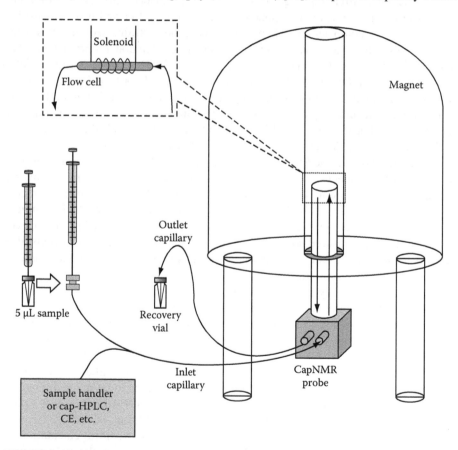

FIGURE 3.4 Scheme of the setup and probe design for capNMR spectroscopy. (Courtesy of Schroeder, F.C. and Gronquist, M.: Extending the scope of NMR spectroscopy with microcoil probes. *Angew. Chem. Int. Ed.* 2006, 45, 7122–7131. Copyright Wiley-VCH Verlag GmbH & Co. KGaA. Reproduced with permission.)

(capNMR) were carried out to prove their feasibility. The number of reported practical applications using "real-life samples" (see Section 3.5) increased significantly as miniaturized separation modules and NMR probes became commercially available. Coupling capillary isotachophoresis (cITP) to nanoliter microcoil NMR probes is another promising innovation [61]. In cITP, the sample matrix containing the dilute, charged analytes is sandwiched between two buffers, a leading electrolyte with a larger and trailing electrolyte with a smaller electrophoretic mobility. Applying an electric field not only separates the components of the sample based on their electrophoretic mobility, but also dramatically concentrates the analytes by two to three orders of magnitude, making the sample volume compatible to that of the micro flow cell [62,63].

Thus, a fundamental problem of NMR coupling methods, namely, how to match the volume of the sample to the active volume of the NMR micro-flow cell, does not exist in cITP. In contrast to HPLC, in which variable column dimensions can be used, the inner diameter of electrophoresis capillaries is generally restricted to less than about 100 µm [64]. Due to the limited buffer amount used for electrophoretic separation, the analyte volumes are close to those of the capNMR probe; therefore, the eluent volumes are not a problem of CE. Instead, the size of the NMR detector has to be reduced to correspond to that of the separation peak.

In contrast, volume incompatibility remains a challenge when HPLC is coupled with capNMR. In this case, the larger chromatographic elution volume needs to be adapted to the smaller volume of the flow cell. As discussed earlier, post-column SPE can be used for this purpose. In a proof-of-concept study, a standard-diameter (4.6 mm) HPLC column was combined with capNMR via post-column SPE. Trapping was performed on a 10 × 1 mm cartridge instead of the 10 × 2 mm cartridge that is normally used and (partially) eluted into a 5 µL capillary microcoil NMR probe with an active volume of 1.5 µL [37] (see also Section 3.5.1). 1D and 2D NMR spectra were recorded from separated peaks (Figure 3.5). Another approach to minimizing the elution peak volume directly in the chromatographic run prior to NMR acquisition is capLC [65]. Technical improvements have made capLC-NMR nowadays an established method with many reported examples [15,66] (see Section 3.5). As with conventional LC-NMR, continuous-flow and stopped-flow modes are feasible with capLC–capNMR coupling as well.

Coupling capLC or UPLC to NMR requires some technical development of the chromatographic equipment. For example, improved stationary phase materials such as reversed-phase sorbents with reduced particle size and chromatographic instrumentation with reduced dead volume in junctions and inner-surface-coated transfer capillaries enhance the overall performance of the separation process. Miniaturized separation techniques are more efficient than standard methods and are able to elevate concentrations at the eluting peak maximum, which is ideal for adjusting the analyte precisely to the volume of the capillary flow cell. The commonly used flow rate in capLC is two orders of magnitude smaller than that of classical HPLC [67]. As a consequence, the consumption of deuterated water is dramatically reduced and makes the use of fully deuterated solvents, including organic modifiers, economically feasible. Moreover, due to the increased analyte/solvent ratio, NMR solvent suppression is easier and may no longer be necessary.

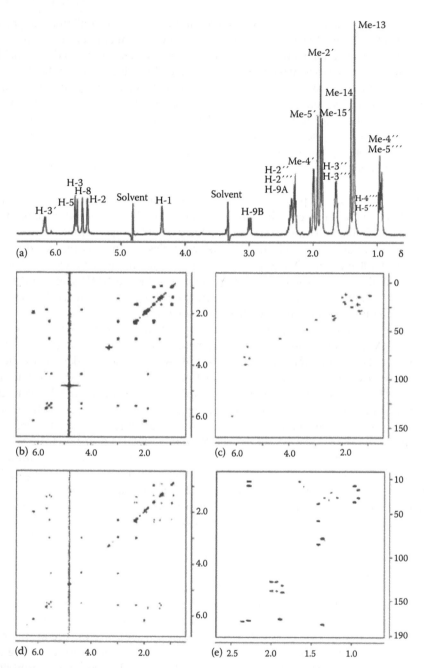

FIGURE 3.5 LC-SPE-capNMR spectra of thapsigargicin (a, b) and thapsigargin (c, d, e): (a) ¹H NMR spectrum, (b) COSY spectrum, (c) HSQC spectrum, (d) NOESY spectrum, (e) aliphatic region of an HMBC spectrum. (Reprinted with permission from Lambert, M., Wolfender, J.-L., Stærk, D., Christensen, S.B., Hostettmann, K., and Jaroszewski, J.W., Identification of natural products using HPLC-SPE combined with capNMR, *Anal. Chem.*, 79, 727–735, 2007. Copyright 2008 American Chemical Society.)

Excellent sensitivity of the NMR measurement is essential for identifying and structurally elucidating mass-limited samples [68]. For a fixed length-to-diameter ratio, mass sensitivity (S/N ratio obtained per mole of the analyte) depends on the diameter of the radio frequency coil; a miniaturized coil design increases sensitivity. Moreover, in terms of sensitivity, solenoid coils are superior to saddle coils of a comparable active volume [69], especially when immersed in susceptibility-matched perfluorinated fluids. Therefore, although both saddle coils and solenoid coils have been used in capNMR, only probes equipped with solenoidal microcoils are already commercially available. Overall, miniaturization enhances the sensitivity of NMR spectroscopy and of various NMR coupling methods. In other words, the demand for better sensitivity is a major driving force behind miniaturization. However, despite the excellent performance of microcoils for mass-limited samples, their use is limited by the solubility of the analyte. As an alternative to microprobes with relatively poor concentration sensitivity, cryogenically cooled probes (see Section 3.2.3) with a relatively large active flow cell volume can be used for sensitive analysis of analytes with limited solubility.

3.5 APPLICATIONS

3.5.1 LC-NMR for Identification, Structure Elucidation, and De-Replication of Natural Products

The structure determination of natural products has been a major focus of conventional LC-NMR. Examples from plant secondary metabolites and its use as a de-replication strategy have been reported [3,70–74] and reviewed several times [75,76]. The introduction of the post-column SPE interface provides further sensitivity gains through repeated SPE steps [16,17,77]. The advantages of the SPE interface compared to loop storage and stopped-flow HPLC have been reported [78,79] and discussed earlier. Since post-column SPE was introduced to LC-NMR hyphenation (see Section 3.3.4), this approach has rapidly been accepted by analytical chemists and an increasing number of applications have been reported in natural product chemistry [80–84] and other disciplines (see following). Multiple trapping and elution with small volumes of deuterated solvent concentrates the analytes prior to NMR, significantly improving the S/N ratio, and facilitates acquisition of NMR data including heteronuclear 2D NMR experiments.

These advantages were also shown in studies on *Hypericum perforatum*, which were carried out both by conventional LC-NMR at 500 MHz [85] and LC-DAD-SPE-NMR at 400 MHz [81]. In the latter investigation, multiple trapping on SPE cartridges was used. In addition to identifying flavonoids, phenolic acids, and naphtodianthrones, the authors found two new phloroglucinols, hyperfirin and adhyperfirin, which had previously been described as hypothetical biosynthetic intermediates of hyperforin and adhyperforin. In contrast, direct LC-NMR was able to characterize only some flavonoids and hypericin [85]. Because hyperforin and analogous natural products are instable, they easily undergo oxidation and photochemical conversion [86]. Therefore, they are difficult to isolate by preparative separation and it is difficult to measure a full set of their spectra in the conventional NMR tube.

In neither online chromatography nor the SPE cartridge are they exposed to light or oxygenation because a stream of nitrogen is used for drying. Hence, LC-SPE-NMR has proved a useful tool for rapidly identifying unstable compounds. Products of hyperforin oxidation have been also analyzed by LC-NMR [42]. The LC-DAD-SPE-NMR technique (600 MHz; 30 µL inverse ^{1}H$\{^{13}$C$\}$ flow probe) was used, e.g., for elucidating structures and assigning ^{1}H and ^{13}C NMR signals of a variety of complex flavonol glucosides and cardenolides from *Kanahia laniflora* [87]. The authors emphasized the potential of LC-SPE-NMR to distinguish between positional isomers and stereoisomers.

Conventional sample preparation methods such as pre-column SPE fractionation are frequently used before a sample is subjected to LC-SPE-NMR. For example, a fraction prepared by off-line SPE from an extract of a miniscule single tuber of *Corydalis solida* (Papaveraceae) was subjected to LC-SPE-NMR hyphenation. Chromatographic peaks were trapped in the SPE device and transferred to a 600 MHz NMR spectrometer equipped with a 30 µL cryofit insert fitted into a 3 mm cryoprobe. Measurements of homo- and heteronuclear 1D and 2D NMR data enabled three minor benzyl isoquinoline-related alkaloids to be identified [56].

Another trend in LC-NMR hyphenation is the use of miniaturized NMR flow cells [14,15,60,88] (see Section 3.4). The off-line combination of standard HPLC columns (4.6 mm i.d.) with a solenoidal microcoil NMR probe (5 µL total volume; 1.5 µL active volume) was recently used to rapidly identify complex sesquiterpene lactones and esterified phenylpropanoids from a toluene-soluble fraction of an ethanolic extract of *Thapsia garganica* fruits [37]. The 10 × 2 mm i.d. SPE cartridges normally used were replaced by 10 × 1 mm i.d. cartridges, which were partially eluted into the capNMR probe (Figure 3.5).

3.5.2 USING LC-NMR TO ANALYZE THE TISSUE-SPECIFIC OCCURRENCE OF NATURAL PRODUCTS

Ruta graveolens is a medicinal plant which in its roots and aerial parts contains more than 120 natural products of different classes, including acridone alkaloids, coumarins, essential oils, flavonoids, and furoquinolines. Among these compounds, acridone alkaloids, which are likely involved in ecological interaction between the plant and its environment, are considered root-specific natural products. This suggestion is supported by the finding that acridone alkaloids accumulate not only in the root of intact plants but also in genetically transformed root cultures. Such "hairy roots" have been used to study tissue- and cell-specific distribution in more detail, employing LC-^{1}H NMR coupling at 500 MHz [89]. A 4 mm flow probe (120 µL active volume) was used in the stopped-flow mode to analyze extracts of dissected root tips and the zones of cell elongation and differentiation. The latter zones contained acridone glucosides, and large amounts of acridone alkaloids, mainly rutacridone. Gravacridondiol glucosides were identified as the dominant natural product of the root tips.

A similar approach was used to study the occurrence of phenylphenalenone-related compounds in root segments of whole plants and different *in vitro* culture lines of *Xiphidium caeruleum* [90]. As little as 10 mg lyophilized root material was extracted with DMSO-d_6 and the extract was subjected to LC-^{1}H NMR. The major peaks were identified as two phenylbenzoisochromenones, which showed remarkable concentration gradients in the opposite longitudinal direction of the root. Furthermore, LC-NMR was used to discriminate phenylphenalenone glycosides, which show very similar retention times and UV spectra. The position of an unusual allophanyl moiety at the secondary hydroxyl group of the glycosidically bound glucose and its orientation between the glucosyl moiety and the lateral phenyl were established [90]. Spring and coworkers [91] proposed LC-NMR be used to study the chemical composition of plant glandular trichomes and the obtained information used for chemotaxonomic assessments.

3.5.3 LC-NMR IN BIOSYNTHETIC STUDIES

Stable isotope labeling, e.g., ^{13}C and/or ^{2}H, is frequently used in biosynthetic studies. Enhanced levels of ^{13}C are indirectly detectable from the occurrence of ^{1}H–^{13}C-coupling satellites in the ^{1}H NMR spectrum [92]. In other words, the multiplicity and the integral ratio of central and satellite lines of a ^{1}H NMR signal can encode information on the extent of ^{13}C enrichment above natural abundance in a specific position of the carbon skeleton. This information is useful for detecting the incorporation of ^{13}C from labeled precursors in biosynthetic studies, and for quantitatively determining the isotopologue composition of labeled biosynthetic products [93]. Hence, the readily detectable effect of ^{13}C enrichment on the multiplicity of ^{1}H NMR signals enables 1D ^{1}H NMR spectroscopy to be used in biosynthetic studies, regardless of whether the spectrum has been measured after off-line separation or under hyphenated conditions.

The stable isotope labeling method was used to investigate the phenylpropanoid metabolism of root cultures of *Anigozanthos preissii* [94]. In this study, a 4 mm NMR flow probe (120 µL), which was connected to a 500 MHz magnet via a Bruker peak sampling unit (BPSU 12) as an interface, was used in the stopped-flow mode. In addition to ^{13}C enrichment, labeling with deuterium, i.e., substitution of non-exchangeable ^{1}H by ^{2}H, is detectable by LC-^{1}H NMR spectroscopy because of missing or reduced intensity of specific ^{1}H NMR signals and/or the collapse of multiplets due to the absence of ^{1}H–^{1}H spin–spin couplings. Biosynthetic products formed from ^{13}C and ^{13}C/^{2}H-labeled precursors, respectively, were detected (Figure 3.6). The signal of H-2 (δ 6.37), in addition to the ^{1}H–^{1}H coupling ($J_{\text{H-2-H-3}}$ = 16 Hz), displays the ^{1}H–^{13}C coupling ($J_{\text{H-2-C-2}}$ = 162 Hz), indicating the ^{13}C in position 2 (Figure 3.6, panel a). The large ^{1}H–^{13}C coupling is still visible in the spectrum of deuterated ferulic acid (Figure 3.6, panel b), but the satellites collapsed to (broad) singlets because ^{1}H-3 has been replaced by deuterium. After the spectra were analyzed and signal lines were integrated, various isotopologues could be identified and quantified. Further details of a multiple hydrogenation/dehydrogenation mechanism in phenylpropanoid interconversion were also inferred from such experiments [94,95].

The introduction of SPE as an interface and the availability and adoption of cryogenically cooled probes for LC-NMR coupling have made it possible to measure ^{13}C NMR spectra directly even from unlabeled samples containing ^{13}C at natural abundance. Hence, the analysis of ^{13}C-enriched samples in biosynthetic studies by detecting the ^{13}C nucleus directly is no longer a problem.

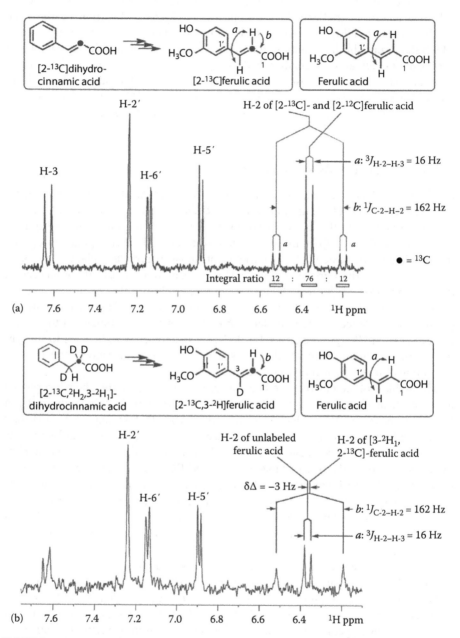

FIGURE 3.6 Biosynthetic formation of ferulic acid from labeled dihydrocinnamic acid observed by LC-^{1}H NMR spectroscopy (500 MHz). (a) Detection of ^{13}C in [2-^{13}C]ferulic acid. (b) Detection of ^{13}C and ^{2}H in [2-^{13}C,3-^{2}H]ferulic acid. (Reprinted from *Prog. Nucl. Magn. Reson. Spectrosc.*, 51, Schneider, B., Nuclear magnetic resonance spectroscopy in biosynthetic studies, 155–198, Copyright 2007, with permission from Elsevier.)

3.5.4 LC-NMR FOR STUDIES ON XENOBIOTIC METABOLISM AND DEGRADATION

Xenobiotic metabolism in crop plants must be investigated before agrochemicals and pesticides can be legally registered and applied in agriculture and horticulture. Conventional analytical methods such as radiolabeling followed by isolation and structure identification of the metabolites can be extremely laborious. In order to check whether hyphenated techniques are useful tools with which to study agrochemical metabolism, Bailey et al. [96] hydroponically applied a model compound, 5-nitropyridone, to maize plants and analyzed the extracts using LC-NMR-MS. The authors succeeded in directly identifying three carbohydrate conjugates of the parent substance, namely, the *N*-glucoside, *N*-malonylglucoside, and *O*-malonylglucoside [96].

Another model compound, the ecotoxin 3-trifluoromethylaniline (3-TFMA), was used to explore the possibility of studying the metabolic fate of xenobiotic compounds by LC-^{19}F NMR spectroscopy [97]. Earthworms of *Eisenia veneta* spp. were subjected to 3-TFMA and the extracts of individual earthworms analyzed by LC-^{19}F NMR spectroscopy in the continuous-flow mode to establish HPLC chromatographic retention times of fluorinated metabolites in a single analytical step. Directly coupled LC-^{19}F/^{1}H NMR spectroscopy was then used to identify the parent compounds and α- and β-glucoside conjugates of 3-TFMA. Concentration prior to analysis was a possible way to identify minor metabolites which escaped direct identification from crude extracts.

Stopped-flow LC-^{1}H NMR and LC-ESIMS methods were developed and applied to separate and characterize the by-products arising from the TiO_2-catalyzed photodegradation of the herbicide iodosulfuron methyl ester in aqueous solution under UV irradiation [98]. ^{1}H NMR data permitted the identities of major products and the differentiation of several positional isomers to be unequivocally confirmed, in particular, the hydroxylation isomers. Cyanuric acid was identified as the final organic product. In this study, a [1]H{[13]C} inverse flow probe (3 mm, active volume 60 μL) was used for LC-^{1}H NMR measurements.

3.5.5 LC-NMR FOR STUDYING ENVIRONMENTAL SAMPLES

Soluble environmental pollutants and their metabolites, ground water contaminated with ammunition residues, and leachate samples from landfills are further targets of identification using LC-NMR methods. The structures of nitroaromatic compounds occurring in traces down to the microgram-per-liter level in groundwater collected near a former ammunition plant have been confirmed using continuous-flow LC-NMR of SPE-enriched samples [44]. HPLC was performed on an RP-18 column with a reduced flow rate of 0.05 mL min^{-1} or even slower under isocratic conditions. Such a slow flow rate resulted in increased peak widths by factor of 10 (for early eluting peaks) and 25 (for a late eluting compound) without decreasing the separation efficiency but with increasing the number of scans that could be made and improving the S/N ratio of the 600 MHz NMR spectrum (4 mm flow cell/detection volume 120 μL). The best sensitivity was obtained at a flow rate of 0.006 mL min^{-1}.

Benfenati et al. [99] have shown that on-flow LC-NMR analysis of leachate samples from an industrial landfill was possible using a 4 mm/120 μL room temperature probe at 600 MHz. Extracting rows of the pseudo-2D NMR spectrum resulted in well-resolved partial spectra of the aromatic regions; however, the high-field part of the spectra were crowded from solvent signals (MeCN as the LC solvent, propionitrile as an impurity in the LC solvent) and overlapping signals of aliphatic components. Hence, identification was restricted to major aromatic components for which standards were available.

The effluent of a textile company was investigated by LC-NMR, and in parallel by LC-MS, in order to characterize the complex mixture of polar, nonvolatile organic constituents in industrial wastewater and to check the suitability of combining the two hyphenated techniques for nontarget analysis [100]. Various compounds such as anthraquinone-type dyes and their by-products, a fluorescent brightener, a by-product from polyester production, and anionic and nonionic surfactants and their degradation products were identified. A mixture of MeCN and D_2O was used as a mobile phase for HPLC. ^1H NMR spectra (600 MHz) were recorded in an inverse flow probe (4 mm; detection volume 120 μL) in the stopped-flow mode.

Nonliving natural organic matter is a complex mixture of mostly uncharacterized components occurring in the environment. More detailed identification of this mixture, the components of which belong to the most abundant organic carbon reserves on earth, is of considerable interest because of its role in multiple environmental processes. Conventional techniques are insufficient to accomplish the separation and structural characterization of such heterogeneous mixtures. LC-NMR was applied to natural organic matter collected from freshwater environments, and both LC-NMR and LC-SPE-NMR were applied to an alkaline soil extract [101]. A Bruker peak sampling unit (BPSU) was used to couple the reversed-phase C_{18} HPLC column to the SPE unit. NMR experiments (600 MHz) were carried out in a ^1H–^{13}C inverse-detection 3 mm flow probe (active volume 60 μL). Time-slice ^1H NMR experiments (see Section 3.3.2) were carried out in order to measure the NMR spectra of chromatographically unresolved components. This approach was only moderately successful for samples of nonliving natural organic matter. However, although the UV profiles appeared to improve only slightly, retention times were enhanced and NMR data in the region of aromatic signals were dramatically improved after the ion pair reagent tetrabutylammonium hydrogen sulfate was added to the mobile phase and the analyte. A drawback of using the ion pair reagent is its strong signals in the high field region of the ^1H NMR spectrum between 0.4 and 3.5 ppm; these completely mask weaker signals from the sample and significantly reduce the information that can be extracted from this region of the spectrum. Application of LC-SPE-NMR to complex natural samples was successfully tested in the same study [10] with soil fulvic acid. Development of SPE cartridges that selectively retain specific components, thereby providing an additional separation step post-LC, has also been recommended. In addition, coupling SPE-NMR to multidimensional LC approaches with a series of orthogonal columns to separate mixtures, which are as complex as natural organic matter, has been proposed.

A continuous-flow LC-^1H NMR method has been developed to analyze polycyclic aromatic hydrocarbons in soil samples [102]. The results were compared with those obtained by LC-DAD, LC-fluorescence detection, and GC-MS. The best agreement was found between the LC-NMR results and GC-MS. Thus, LC-NMR has been shown to be a valuable reference method for the verification of other techniques used in the analysis of polycyclic aromatic hydrocarbons and to uncover methodical errors of established methods (Figure 3.7). The suitability of LC-NMR for characterizing environmental contaminants and their transformation products and future developments has been discussed by Cardoza et al. [103].

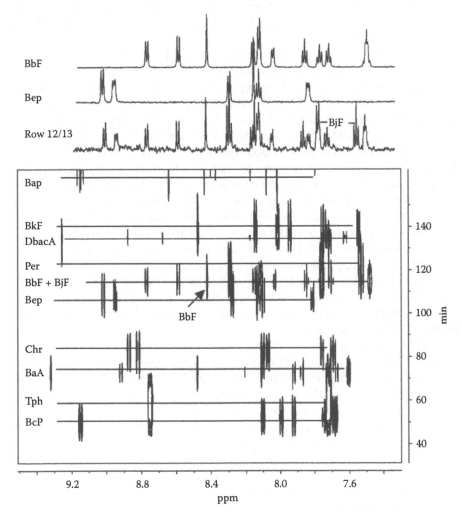

FIGURE 3.7 NMR chromatogram of a mixture of polycyclic aromatic hydrocarbons and the addition of rows 12 and 13. MeCN/D$_2$O (90/10 v/v), flow 0.05 mL min^{-1}, 56 scans row^{-1}. Abbreviations: BcP, benzo[c]phenanthrene; Tph, triphenylene; BaA, benz[a]anthracene; Chr, chrysene; BbF, benzo[b]fluoranthene; BjF, benzo[j]fluoranthene; Bep, benzo[e]pyrene; Per, perylene; DbacA, dibenz[a,c]anthracene; BkF, benzo[k]fluoranthene; Bap, benzo[a] pyrene. (With kind permission from Springer Science+Business Media: *Anal. Bioanal. Chem.*, Development of an HPLC-NMR method for the determination of PAHs in soil samples—A comparison with conventional methods, 373, 2002, 810–819, Weißhoff, H., Preiß, A., Nehls, I., and Mügge, T.W.C., figure number 5, original copyright.)

3.5.6 USING LC-NMR FOR STUDIES ON DRUG METABOLISM, DEGRADATION, AND IMPURITIES

The identification of impurities and degradation products is crucial to the development of medicinal drugs. Drug impurities at >0.1% levels have to be identified for risk assessment at the drug discovery–development interface [34,54,104]. LC-NMR has become an important tool in pharmaceutical research to analyze drug metabolism [105] and impurities originating from production, formulation, or degradation processes [106,107].

The degradation of the synthetic estrogenic steroid ethynylestradiol, which was used in oral contraceptives, was investigated using LC-NMR [108]. A variety of autoxidation products, namely, 9-hydroperoxide and dimers formed via intermolecular ortho–ortho C–C and ortho C–O bond formation, were identified. In most experiments, D_2O/MeCN was used as a mobile phase but some HPLC runs were conducted with D_2O/MeCN-d_4. ^{1}H NMR, COSY, and NOESY spectra were measured in the stopped-flow mode at 500 MHz using the WET sequence to suppress MeCN and D_2O signals. A flow probe with a flow cell volume of 60 μL was used. In order to maximize the sensitivity and quality of the NMR data, the amount of material injected for LC-NMR analyses was up to 300-fold more than the amount used for the HPLC-UV analyses. This increased column loading slightly deteriorated chromatographic separation but did not interfere with the collection of NMR data.

An unknown impurity (N1) in the drug 5-aminosalicylic acid was identified by LC-NMR in the stopped-flow mode using an inverse 4 mm detection ^{1}H/^{13}C flow probe (cell volume 120 μL) and LC-MS [109]. After the structure of N1 was proposed from LC-NMR and LC-MS data to be a sulfated derivative 3-phenyl-5-aminosalicylic acid, the compound was isolated and its identity confirmed by homo- and heteronuclear 2D NMR experiments.

Various hyphenated analytical techniques such as GC-MS, LC-MS, and LC-SPE-NMR have been employed to obtain complementary information about the structure of a degradation product of TCH346 (dibenzo[b,f]oxepin-10-yl-methyl-methyl-prop-2-ynyl-amine hydrogen maleate), whose use in the treatment of neurodegenerative disorders is under investigation [110]. The peak of the unknown compound was separated by reversed-phase HPLC using MeCN-H_2O, trapped on a SPE cartridge and eluted with $CDCl_3$ into the NMR flow cell (30 μL active volume) of a 3 mm cryofit dual inverse probe. ^{1}H NMR spectra were measured at a resonance frequency of 600 MHz. The analytical data including ^{1}H-, ^{13}C-, and 2D NMR spectra characterized the structure of the analyte as dibenzo[b,f]oxepin-10-carbaldehyde.

Xanthohumol, a prenylated chalcone, from hops (*Humulus lupulus*) has recently attracted attention because of its interesting antiproliferative, antimutagenic, and antioxidative properties. In order to investigate its metabolism in rats, xanthohumol (1 g kg^{-1} body weight) was suspended in starch solution and fed to female rats [111]. Feces were collected and, after extraction and fractionation using pre-column SPE, subjected to LC-^{1}H NMR spectroscopic analysis (500 MHz) in the stopped-flow mode. Products resulting from hydroxylation, O-methylation, O-acetylation, epoxidation, cyclization involving the prenyl side chain, and glucuronidation were identified.

Besides the known sulfate and glucuronide metabolites, previously undetected metabolites of acetaminophen were directly observable in a 15 min LC-NMR on-flow experiment using a cryogenic flow probe at 500 MHz. The cryoflow probe used for these experiments was built in a dual inverse $^1H\{^{13}C\}$ configuration with a glass flow cell (active volume $\sim$40 µL) in a vertical geometry [39].

In a study of trace degradants in a pharmaceutical formulation, cITP-NMR was used to demonstrate separation and detection of 0.1% 4-aminophenol as a spiked impurity in the presence of a 1000-fold excess of acetaminophen and tablet matrix components [112]. Analysis of an acetaminophen thermal degradation sample revealed resonances of several cationic impurities in addition to 4-aminophenol.

Since studies of the metabolic fate of active drugs depends on highly sensitive methods to detect and discriminate metabolites from matrix components, radiolabeling has been used in many such investigations and is still a method of choice. For LC-NMR, a radiodetector instead of or in addition to the UV of DAD detectors is suitable for detecting radioactive metabolites before NMR analysis. Lenz et al. [113] determined the major urinary and biliary metabolites of [^{14}C]-ZD6126, a phosphorylated prodrug, following its administration to rats and dogs, using HPLC coupled to radioactivity detector (RAD) and NMR spectroscopy (LC-RAD-NMR). In this study, a HPLC radioactivity monitor was additionally interfaced prior to the DAD in order to detect the ^{14}C radiochemical profiles of the biofluid samples. The radiolabeled components were subsequently transferred online to a 500 MHz NMR spectrometer equipped with a dual tunable ^{1}H/^{19}F LC flow probe (3 mm, 60 µL active volume) and measured in the stopped-flow mode. Despite severe LC column overloading, the major metabolites were identified from their ^{1}H NMR spectra to be glucuronides, a sulfate conjugate and products of demethylation. The structures were confirmed by LC-MS.

The use of ^{19}F NMR in metabolic studies has the advantages that its excellent sensitivity nearly equals ^{1}H NMR and that the spectrum is free of matrix signals. ^{19}F NMR can also be used instead of radiolabeling when the labeled carbon or proton is lost during metabolism. The in vivo human metabolism of the antiviral compound 5-chloro-1-(29,39-dideoxy-39-fluoro-erythro-pentofuranosyl)uracil has been studied by hyphenated methods, including continuous-flow LC-^{19}F NMR and stopped-flow LC-^{1}H NMR [114]. The use of the LC-^{19}F NMR experiment yielded chromatographic retention times and ^{19}F chemical shifts for the parent drug, the glucuronide conjugate of the parent and an early eluting polar metabolite. The parent drug and its glucuronide conjugate were easily characterized by directly coupled LC-^{1}H NMR spectroscopy and 2D TOCSY experiments. Continuous-flow LC-^{19}F NMR spectra (470 MHz) were acquired using a 3 mm dual ^{19}F/^{1}H flow probe (60 µL active volume). Stopped-flow LC-^{1}H NMR spectra (500 MHz) were measured at retention times obtained from the continuous flow LC-^{19}F NMR experiment using a 4 mm dual ^{1}H/^{13}C flow probe (120 µL active volume) [114]. A similar approach, namely, continuous-flow LC-^{19}F NMR and stopped-flow LC-^{1}H NMR, was employed, among other methods, to study the metabolism 2-trifluoromethylaniline and its acetanilide in the rat [115] (Figure 3.8). Several additional practical applications of LC-SPE-NMR based on the identification of both drug metabolites and drug impurities in addition to methodological aspects have been reported [54,116,117].

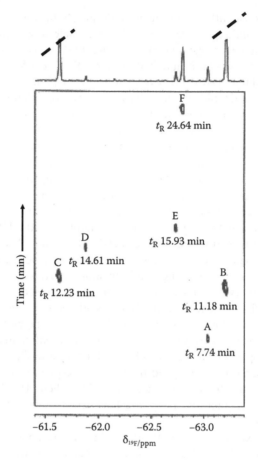

FIGURE 3.8   Pseudo-2D contour plot of a continuous-flow ^{19}F NMR (470 MHz) detected HPLC separation of the metabolites from urine of rats treated with 2-trifluoromethylaniline. (Reprinted from *J. Pharm. Biomed. Anal.*, 30, Tugnait, M., Lenz, E.M., Hofmann, M., Spraul, M., Wilson, I.D., Lindon, J.C., and Nicholson, J.K., The metabolism of 2-trifluormethylaniline and its acetanilide in the rat by ^{19}F NMR monitored enzyme hydrolysis and $^{1}H/^{19}F$ HPLC-NMR spectroscopy, 1561–1574, Copyright 2003, with permission from Elsevier.)

3.5.7   LC-NMR in Structure Elucidation and Analysis of Isomers

One of the major advantages of NMR spectroscopy is its unique ability to distinguish between isomeric chemical structures. These include positional isomers and account for various stereoisomeric forms. Separation and identification of all *trans-* and various *cis-*isomers of carotinoids are important because these compounds are all active in different ways. LC-^{1}H NMR has been used successfully to separate and identify various *cis/trans* isomers of retinoic acid [5]. One of the advantages of coupled LC-NMR experiments is that separation and detection are performed in a closed system. Therefore, light-induced re-isomerization of separated *cis/trans* isomers during analysis, or decomposition caused by contact with oxygen can be excluded.

Structure elucidation of a tryptic peptide fragment of ZAP70, a protein kinase, has been achieved with only minute amounts of the sample using a capillary HPLC connected to a $^1H\{^{13}C\}$ microprobe NMR (600 MHz) [118]. The solenoid copper microcoil was surrounded by a susceptibility-matched fluid in order to decrease field inhomogeneities. After the ^1H and ^{13}C signals of the peptide were assigned, the study focused on determining the phosphorylation state of two adjacent tyrosine units (Y_1 and Y_2), i.e., distinguishing between the four possible peptide species containing 0, 1, or 2 phosphoric acid moieties (p) at the p-hydroxyl group of the tyrosine residues. First, the four synthetic peptides (Y_1Y_2, Y_1pY_2, pY_1Y_2, pY_1pY_2) were injected via a syringe in order to analyze the effect of the phosphorylation patterns on the chemical shifts of the aromatic tyrosine signals. Then a mixture of approximately 300 pmol of each peptide was separated on a C_{18} capLC column and ^1H NMR spectra of each component were recorded in the solenoid microprobe (1.5 µL active volume) using the stopped-flow mode. Although the S/N ratio of the spectra was poor compared to the ratios obtained from larger amounts of syringe-injected individual peptides, the suitability of online capLC/microprobe NMR for determining peptide phosphorylation states has been demonstrated in this study [118].

3.5.8 LC-NMR FOR REACTION MONITORING
AND CHEMICAL PRODUCT ANALYSIS

The absolute configurations of the secondary alcohol function of polyacetylenes isolated from root tuber material of various Apiaceae species were determined by analyzing the Mosher esters by LC-SPE-NMR (500 MHz ^1H resonance frequency; 3 mm LC-SEI $^{13}C\{^1H\}$ probe; 60 µL active volume of the flow cell). The chiral polyacetylenes were reacted with an excess of either (2R)- or (2S)-2-methoxy-2-trifluoromethyl-phenylacetylchloride (MTPA-Cl), and the resulting mixture subjected to LC-^1H- and LC-^{19}F NMR-DAD-NMR. The ^1H-NMR spectrum of aethusanol A and the ^1H- and ^{19}F NMR spectra (to detect the trifluoromethyl signal, the proton coil of the probe was manually tuned to 470 MHz) of its (R)- and (S)-Mosher esters are shown in Figure 3.9 as an example. ^1H NMR peak assignment of the two Mosher ester derivatives was facilitated by comparing their shift values and coupling patterns to those of the respective spectra of the parent secondary alcohol. Subsequently, $\delta(S)-(R)$ values were calculated and used as input into an empirical model [119]. Using this approach, aethusanol A, e.g., was found to be (R)-configured. The advantage of using LC-SPE-NMR in this study was that isolating the analytes on a microscale, which in the case of unstable polyacetylenes may result in decomposition, was avoided. Another advantage of LC-SPE-NMR is that solvent change on the SPE cartridges allows to measure NMR spectra of different samples under identical conditions. Measurements under identical conditions are important in chemical-shift-sensitive experiments such as Mosher ester analysis. In a study by Seger et al. [120], elution of the lipophilic analytes from the SPE cartridges was achieved by $CDCl_3$, an approach rarely used in LC-SPE-NMR.

LC-^1H NMR has been used to identify products of isolated or recombinant biosynthetic enzymes. For example, products of a peroxidase-catalyzed dimerization of naphthylisoquinoline alkaloids have been analyzed [121]. Korupensamine B, the

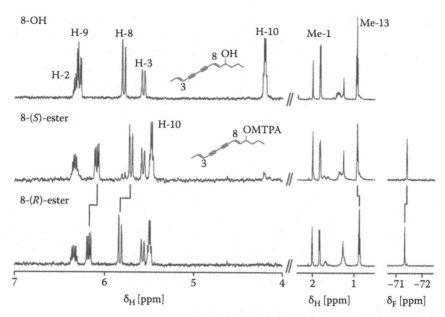

FIGURE 3.9 LC-SPE-^{1}H NMR (500 MHz) and LC-SPE-^{19}F NMR (470 MHz) spectra of a polyacetylene (top trace) and its Mosher ester derivatives. Bars indicate shift differences $\Delta\delta(S)$–(R) between the (S)- and (R)-ester derivatives. (Reprinted from *J. Chromatogr. A*, 1136, Seger, C., Godejohann, M., Spraul, M., Stuppner, H., Hadacek, F., Reaction product analysis by high-performance liquid chromatography-solid-phase extraction-nuclear magnetic resonance: Application to the absolute configuration determination of naturally occurring polyyne alcohols, 82–88, Copyright 2006, with permission from Elsevier.)

monomeric substrate, and michellamine C, the dimeric product of phenolic coupling, were identified by comparing their ^{1}H NMR spectra (600 MHz) obtained by HPLC of the enzyme assay mixture under coupling conditions, with spectra of authentic compounds.

Keto and enol tautomers of some aliphatic and alicyclic α-dicarbonyl compounds can be separated using HPLC at a low temperature. Solutions of ethylbutyryl acetate contain a relatively large percentage of the enol, and the conversion to the keto form is slow enough to be monitored by ^{1}H NMR spectroscopy. In order to study the ketonization rate of the enol form, a solution of ethylbutyryl acetate was subjected to LC-NMR (500 MHz) in the stopped-flow mode [122]. The HPLC was stopped when the enol peak was in the detection cell (70 μL active volume interchangeable flow cell) of a cryogenic probe and the kinetics of ketonization was analyzed. Using this approach, the impact of various factors such as acid effect, deuterium isotope effect, and solvent effects on the ketonization reaction have been investigated.

Chemical substance libraries obtained from combinatorial synthesis are subject to analysis using coupling methods such as LC-NMR because simultaneous separation and identification is demanded. On-flow LC-NMR spectroscopy at 600 MHz on a 3 mm flow probe was used to separate and characterize a synthetic library of tripeptide combinations of alanine, methionine, and tyrosine as the C-terminal amides [123]. Most of the peptides were identified on the basis of ¹H NMR chemical shifts and coupling constants, including the use of diagnostic values. The effects of pH and varying proportions of MeCN and water that occur during the gradient elution HPLC run on ¹H NMR chemical shifts were evaluated. Further examples of analyzed combichemical libraries include a mixture of four-ring positional (dimethoxybenzoyl)glycine isomers obtained from a split-mix synthesis, and a five-component pentapeptide mixture containing two isomolecular weight peptides. These samples were analyzed by on-flow and stopped-flow LC-NMR at 500 MHz [124].

3.5.9 LC-NMR IN POLYMER RESEARCH

Conventional reversed-phase HPLC under standard conditions is capable of separating small molecules from natural sources or synthesis, but is not generally suitable for polymers. Instead, special methods such as size exclusion chromatography (SEC) or liquid chromatography at critical conditions (LCCC, also referred to as liquid chromatography at the critical point of adsorption) are required for separation [125]. SEC, which is also known as gel permeation chromatography (GPC), separates mixtures according to the different molar masses of their components. In contrast, LCCC separates polymeric mixtures according to their chemical heterogeneity but eliminates effects due to entropic interactions, i.e., molar mass distribution is no longer a separation criterion. For example, polymers containing certain structural groups elute together regardless of their molecular masses. Therefore, LCCC helps eliminate some of the problems inherent to copolymer analysis, which is complicated because of the simultaneous distribution in molar mass and chemical distribution. Not only do special LC procedures have to be used, but also the high molecular mass of the analytes and their broad signals limit the sensitivity of polymer analysis by LC-NMR coupling.

Nevertheless, a number of interesting examples of LC-NMR hyphenation employed in polymer research have been reported. GPC-NMR has been used to determine molar masses and mass distribution of macromolecules [126–128], the molar mass dependence of tacticity [129], and copolymer composition [130,131] of polymer samples. Recently, GPC-NMR analysis of the molar mass and the chemical structures of complex polyolefins have been reported [132]. Different GPC columns were run with 1,2,4-trichlorobenzene as a mobile phase at an operating temperature of 130°C. ¹H NMR spectra were measured with a high temperature NMR flow probe containing a 120 μL flow cell and operating at 120°C in the on-flow or stopped-flow mode.

Kitayama et al. [133] analyzed the tacticity distribution of stereoregular poly(ethyl methacrylate)s using LCCC on a column packed with silica gel bonded with aminopropyl groups. An evaporating light-scattering detector was used to monitor the column effluent at slow flow rates of 0.2 mL min⁻¹. The eluent was composed of acetone, acetone-d_6, and cyclohexane and showed only two solvent peaks in the NMR spectrum (750 MHz). The WET sequence [12] was used for solvent suppression. Functionality-type distribution and molar mass distribution of technical polyethylene oxides were studied by means of LC-NMR coupling [134]. The structure of the end groups and distribution of the chain length of oligostyrenes have been investigated using on-flow LC-NMR and various off-line NMR experiments [135]. Furthermore, LC-NMR and GPC-NMR provided information about the chemical composition distribution of poly(styrene-co-ethyl acrylate) and similar copolymers, including quantitative data [136,137].

Online coupling of LC to NMR was used to characterize fatty alcohol ethoxylate-based surfactants [138]. Different surfactant mixtures were separated by a combined exclusion-adsorption method of LC and simultaneously characterized by ^{1}H NMR spectroscopy at 400 MHz using a flow probe (120 µL). Information about the degree of oligomerization of the ethylene oxide chain and the chemical structure of the terminal groups was obtained in these experiments.

A mixture of fatty alcohol ethoxylates, prepared by polymerization of ethylene oxide with fatty alcohols, was recently analyzed by LCCC-NMR to gain information about the chain length of their end groups [139]. The authors used MeOH/D$_2$O as a mobile phase on a C18 column, monitored the chromatography by an evaporating light-scattering detector, and measured ^{1}H NMR spectra (500 MHz) in the on-flow mode. Results obtained from a triple-resonance room temperature probe equipped with an interchangeable 60 µL flow cell were compared with those from a cryogenic flow probe with an active volume of 90 µL. Although it was not possible to fully detect all signals of the end groups using the room temperature probe, complete assignment was achieved with the cryogenic flow probe. The chain lengths of the alkyl end groups were determined from the integral ratios of the methylene and methyl group signals.

Hiller et al. [140] used LCCC-NMR to analyze polystyrene-*block*-poly(methyl methacrylate) (PS-*b*-PMMA) samples. Non-deuterated MeCN and tetrahydrofurane were used as the mobile phase on RP18 columns. The on-flow NMR measurements (400 MHz) were performed in the unlocked mode on a ^{1}H{^{13}C} inverse detection flow probe (60 µL flow cell). The materials were found to be chemically heterogeneous and to contain fractions of polystyrene homopolymer as well as copolymer fractions of different composition, i.e., the true chemical composition of the block copolymers can significantly differ from the average composition.

3.6 CONCLUSIONS AND OUTLOOK

LC-NMR has been accepted as an attractive and valuable tool for studying the chemical composition of complex mixtures such as plant extracts, animal biofluids, environmental samples or synthetic reaction libraries, and even polymers. Frequently, a compromise has to be made between the optimum conditions for chromatography and those for NMR. In this context, a number of issues and examples concerning in particular the application of LC-NMR to low-level pharmaceutical impurities have been discussed in the literature [45]. In special cases, separation by LC followed by conventional off-line NMR analysis in a tube may yield results superior to those achieved with the flow probe. However, the advantages of automated LC-SPE-NMR and further recent developments have already overcome most of the previously expressed concerns.

Applying LC-NMR to the analysis of minor components in mixtures remained limited to microgram-per-component amounts as long as NMR spectroscopy was not considered a trace analytical method. This situation has changed dramatically since the field strength of NMR magnets, which are currently installed in LC-NMR laboratories, increased to the range between 9.4 and 14.1 T or even higher. Even more than the field strength of the magnets, probe technology has progressed as a result of improvements in miniaturized flow cells [14,141] and cryogenic probes [13] in flow NMR and LC-NMR hyphenation [39]. In addition, as discussed earlier, post-column SPE interfaces [53,56] are available, which allow a substantial increase in the amount of the analyte to be transferred from the HPLC to the NMR flow probe. The possibility for solvent exchange is another important advantage of the SPE interface.

In the early days of LC-NMR, stopped-flow mode was considered the most sensitive technique available because the residence time in the flow NMR cell was not limited and measurements could be extended in order to obtain better S/N ratio. Continuous flow was and still is useful to obtain a quick overview, e.g., for de-replication purposes, but this method did not usually allow 2D NMR measurements to be made. Recently developed ultrafast multidimensional NMR methods, such as the Hadamard spectroscopy [142], which significantly shorten NMR measurement time, have been introduced to LC-NMR. A Hadamard-based real-time 2D TOCSY sequence was implemented in the on-flow LC-NMR experiment [122]. This example demonstrates that, in addition to hardware development, progress in the development of new NMR pulse sequences will likely speed up NMR measurements and improve sensitivity.

Improvements are anticipated not only in the ability to couple LC to ^{1}H and ^{1}H-detected 2D NMR but also to heteronuclear NMR. Coupling LC to ^{19}F NMR has been frequently used, especially in the analysis of biofluids, but so far there are few examples of using direct ^{13}C NMR and other nuclei in hyphenated methods. NMR-based hyphenation will continue to be extended to electrophoretic separation techniques [143], which are not the subject of the present review, and separation methods such as capLC and CE will continue to be miniaturized. In particular UPLC, which is on the way to becoming a standard separation method when coupled with MS [60,67,144], has the potential for hyphenation to NMR. Furthermore, LC-NMR-MS coupling is already an established method. Hypernation, which further extends LC coupling to multiple analytical methods, is also developing rapidly [2,145]. Combining LC-NMR with methods for estimating properties connected to biological activity [146] and with bioassays promises to make future developments interesting. As LC-NMR equipment is developed, applications in new disciplines will emerge.

ACKNOWLEDGMENTS

The authors wish to thank Dr. Michael Wenzler, Jena, for critical reading and helpful comments on this chapter, and Emily Wheeler, Jena, for editorial assistance. Cristina Daolio gratefully acknowledges financial support by the Alexander von Humboldt Foundation, Bonn.

REFERENCES

1. Albert, K.; *On-line LC–NMR and Related Techniques*, John Wiley & Sons Ltd., Chichester, U.K., 2004.
2. Wilson, I.D.; Brinkman, U.A.T.; *Trends in Analytical Chemistry*, 2007, 26, 847–854.
3. Wolfender, J.-L.; Rodriguez, S.; Hostettmann, K.; *Journal of Chromatography A*, 1998, 794, 299–316.

4. Lee, M.S.; Kerns, E.H.; *Mass Spectrometry Reviews*, 1999, 18, 187–279.
5. Strohschein, S.; Schlotterbeck, G.; Richter, J.; Pursch, M.; Tseng, L.-H.; Händel, H.; Albert, K.; *Journal of Chromatography A*, 1997, 765, 207–214.
6. Watanabe, N.; Niki, E.; *Proceedings of the Japan Academy Series B*, 1978, 54, 194–199.
7. Bayer, E.; Albert, K.; Nieder, M.; Grom, E.; Keller, T.; *Journal of Chromatography*, 1979, 186, 497–507.
8. Haw, J.F.; Glass, T.E.; Hausler, D.W.; Motell, E.; Dorn, H.C.; *Analytical Chemistry*, 1980, 52, 1135–1140.
9. Albert, K.; *Journal of Chromatography A*, 1999, 856, 199–211.
10. Albert, K.; *Journal of Chromatography A*, 1995, 703, 123–147.
11. Nicholson, J.K.; Foxall, P.J.D.; Spraul, M.; Farrant, R.D.; Lindon, J.C.; *Analytical Chemistry*, 1995, 67, 793–811.
12. Smallcombe, S.H.; Patt, S.L.; Keifer, P.A.; *Journal of Magnetic Resonance Series A*, 1995, 117, 295–303.
13. Kovacs, H.; Moskau, D.; Spraul, M.; Cryogenically cooled probes—A leap in NMR technology, *Progress in Nuclear Magnetic Resonance Spectroscopy*, 2005, 46, 131–155.
14. Olson, D.L.; Norcross, J.A.; O'Neil-Johnson, M.; Molitor, P.F.; Detlefsen, D.J.; Wilson, A.G.; Peck, T.L.; *Analytical Chemistry*, 2004, 76, 2966–2974.
15. Schroeder, F.C.; Gronquist, M.; *Angewandte Chemie International Edition*, 2006, 45, 7122–7131.
16. Griffiths, L.; Horton, R.; *Magnetic Resonance in Chemistry*, 1998, 36, 104–109.
17. Nyberg, N.T.; Baumann, H.; Kenne, L.; *Magnetic Resonance in Chemistry*, 2001, 39, 236–240.
18. Pullen, F.S.; Swanson, A.G.; Newman, M.J.; *Rapid Communications in Mass Spectrometry*, 1995, 9, 1003–1006.
19. Taylor, S.D.; Wright, B.; Clayton, E.; Wilson, I.D.; *Rapid Communications in Mass Spectrometry*, 1998, 12, 1732–1736.
20. Snyder, L.R.; Van der Wal, S.; *Analytical Chemistry*, 1981, 53, 877–884.
21. Snyder, L.R.; Kirkland, J.J.; Glajch, J.L.; *Practical HPLC Method Development*, John Wiley & Sons, New York, 1997.
22. Pavlović, D.M.; Babić, S.; Horvat, A.J.M.; Kaštelan-Macan, M.; *Trends in Analytical Chemistry*, 2007, 26, 1062–1075.
23. Mars, C.; Smit, H.C.; *Analytica Chimica Acta*, 1990, 228, 193–208.
24. de Koning, J.A.; Hogenboom, A.C.; Lacker, T.; Strohschein, S.; Albert, K.; Brinkman, U.A.Th.; *Journal of Chromatography A*, 1998, 813, 55–61.
25. Jayawickrama, D.A.; Wolters, A.M.; Sweedler, J.V.; *The Analyst*, 2003, 128, 421–426.
26. Griffiths, L.; *Analytical Chemistry*, 1995, 67, 4091–4095.
27. Albert, K.; Schlotterbeck, G.; Tseng, L.-H.; Braumann, U.; *Journal of Chromatography A*, 1996, 750, 303–309.
28. Pusecker, K.; Schwitz, J.; Gfrörer, P.; Tseng, L.-H.; Albert, K.; Bayer, E.; *Analytical Chemistry*, 1998, 70, 3280–3285.
29. Albert, K.; Dachtler, M.; Glaser, T.; Händel, H.; Lacker, T.; Schlotterbeck, G.; Strohschein, S.; Tseng, L.-H.; Braumann, U.; *Journal of High Resolution Chromatography*, 1999, 22, 135–143.
30. Rückert, M.; Wohlfarth, M.; Bringmann, G.; *Journal of Chromatography A*, 1999, 840, 131–135.
31. Godejohann, M.; *Journal of Chromatography A*, 2007, 1156, 87–93.
32. Albert, K.; *Journal of Chromatography A*, 1997, 785, 65–83.
33. Korhammer, S.A.; Bernreuther, A.; *Analytical Chemistry*, 1996, 354, 131–135.
34. Lindon, J.C.; Nicholson, J.K.; Wilson, I.D.; *Journal of Chromatography B*, 2000, 748, 233–258.

35. Gronquist, M.; Meinwald, J.; Eisner, T.; Schroeder, F.C.; *Journal of American Chemical Society*, 2005, 123, 10810–10811.

36. Hu, J.F.; Garo, E.; Yoo, H.D.; Cremin, P.A.; Zeng, L.; Goering, M.; O'Neil-Johnson, M.; Eldridge, G.R.; *Phytochemical Analysis*, 2005, 16, 127–133.

37. Lambert, M.; Wolfender, J.-L.; Stærk, D.; Christensen, S.B.; Hostettmann, K.; Jaroszewski, J.W.; *Analytical Chemistry*, 2007, 79, 727–735.

38. Jansma, A.; Chuan, T.; Albrecht, R.W.; Olson, D.L.; Peck, T.L.; Geierstanger, B.H.; *Analytical Chemistry*, 2005, 77, 6509–6515.

39. Spraul, M.; Freund, A.S.; Nast, R.E.; Withers, R.S.; Maas, W.E.; Corcoran, O.; *Analytical Chemistry*, 2003, 75, 1546–1551.

40. Exarchou, V.; Godejohann, M.; Van Beek, T.A.; Gerothanassis, I.P.; Vervoort, J.; *Analytical Chemistry*, 2003, 75, 6288–6294.

41. Wolfender, J.-L.; Rodriguez, S.; Hostettmann, K.; Hiller, W.; *Phytochemical Analysis*, 1997, 8, 97–104.

42. Wolfender, J.-L.; Verotta, L.; Belvisi, L.; Fuzzati, N.; Hostettmann, K.; *Phytochemical Analysis*, 2003, 14, 290–297.

43. Queiroz, E.F.; Wolfender, J.-L.; Atindehoua, K.K.; Traore, D.; Hostettmann, K.; *Journal of Chromatography A*, 2002, 947, 123–134.

44. Godejohann, M.; Preiss, A.; Mügge, C.; Wünsch, G.; *Analytical Chemistry*, 1997, 69, 3832–3837.

45. Sharman, G.J.; Jones, I.C.; *Magnetic Resonance in Chemistry*, 2003, 41, 448–454.

46. Zhou, Z.; Lan, W.; Zhang, W.; Zhang, X.; Xia, S.; Zhu, H.; Ye, C.; Liu, M.; *Journal of Chromatography A*, 2007, 1154, 464–468.

47. Griffiths, L.; *Magnetic Resonance in Chemistry*, 1997, 35, 257–261.

48. Schefer, A.B.; Braumann, U.; Tseng, L.-H.; Spraul, M.; Soares, M.G.; Fernandes, J.B.; Silva, M.F.G.F.; Vieira, P.C.; Ferreira, A.G.; *Journal of Chromatography A*, 2006, 1128, 152–163.

49. Elipe, M.V.S.; *Analytica Chimica Acta*, 2003, 497, 1–25.

50. Tran, B.Q.; Lundanes, E.; Greibrokk, T.; *Chromatographia*, 2006, 64, 1–5.

51. Miliauskas, G.; Van Beek, T.A.; Waard, P.; Venskutonis, R.P.; Sudhölter, E.J.R.; *Journal of Chromatography A*, 2006, 1112, 276–284.

52. Sturm, S.; Seger, C.; Godejohann, M.; Spraul, M.; Stuppner, H.; *Journal of Chromatography A*, 2007, 1163, 138–144.

53. Clarkson, C.; Sibum, M.; Mensen, R.; Jaroszewski, J.W.; *Journal of Chromatography A*, 2007, 1165, 1–9.

54. Sandvoss, M.; Bardsley, B.; Beck, T.L.; Lee-Smith, E.; North, S.E.; Moore, P.J.; Edwards, A.J.; Smith, R.J.; *Magnetic Resonance in Chemistry*, 2005, 43, 762–770.

55. Clarkson, C.; Stærk, D.; Hansen, S.H.; Smith, P.J.; Jaroszewski, J.W.; *Journal of Natural Products*, 2006, 69, 1280–1288.

56. Lee, S.-S.; Lai, Y.-C.; Chen, C.-K.; Tseng, L.-H.; Wang, C.-Y.; *Journal of Natural Products*, 2007, 70, 637–642.

57. Lambert, M.; Stærk, D.; Hansen, S.H.; Jaroszewski, J.W.; *Magnetic Resonance in Chemistry*, 2005, 43, 771–775.

58. Olson, D.L.; Peck, T.L.; Webb, A.G.; Magin, R.L.; Sweedler, J.V.; *Science*, 1995, 270, 1967–1970.

59. Wu, N.; Peck, T.L.; Webb, A.G.; Magin, R.L.; Sweedler, J.V.; *Journal of the American Chemical Society*, 1994, 66, 3849–3854.

60. Behnke, B.; Schlotterbeck, G.; Tallarek, U.; Strohschein, S.; Tseng, L.-H.; Keller, T.; Albert, K.; Bayer, E.; *Analytical Chemistry*, 1996, 68, 1110–1115.

61. Kautz, R.A.; Lacey, M.E.; Wolters, A.M.; Foret, F.; Webb, A.G.; Karger, B.L.; Sweedler, J.V.; *Journal of the American Chemical Society*, 2001, 123, 3159–3160.

62. Wolters, A.M.; Jayawickrama, D.A.; Larive, C.K.; Sweedler, J.V.; *Analytical Chemistry*, 2002, 74, 2306–2313.
63. Wolters, A.M.; Jayawickrama, D.A.; Larive, C.K.; Sweedler, J.V.; *Analytical Chemistry*, 2002, 74, 4191–4197.
64. Lacey, M.E.; Subramanian, R.; Olson, D.L.; Webb, A.G.; Sweedler, J.V.; *Chemical Reviews*, 1999, 99, 3133–3152.
65. Xiao, H.B.; Krucker, M.; Putzbach, K.; Albert, K.; *Journal of Chromatography A*, 2005, 1067, 135–143.
66. Sandvoss, M.; Application of LC-NMR: LC-NMR–MS hyphenation to natural products analysis. In: Albert, K. (ed.), *On-Line LC–NMR; Related Techniques*, John Wiley & Sons Ltd., Chichester, U.K., 2002.
67. Krucker, M.; Lienau, A.; Putzbach, K.; Grynbaum, M.D.; Schuler, P.; Albert, K.; *Analytical Chemistry*, 2004, 76, 2623–2628.
68. Korir, A.K.; Larive, C.K.; *Analytical and Bioanalytical Chemistry*, 2007, 388, 1707–1716.
69. Webb, A.; *Magnetic Resonance in Chemistry*, 2005, 43, 688–696.
70. Schneider, B.; Zhao, Y.; Blitzke, T.; Schmitt, B.; Nookandeh, A.; Sun, X.; Stöckigt, J.; *Phytochemical Analysis*, 1998, 9, 237–244.
71. Wolfender, J.-L.; Ndjoko, K.; Hostettmann, K.; *Current Organic Chemistry*, 1998, 2, 575–596.
72. Hölscher, D.; Schneider, B.; *Phytochemistry*, 1999, 50, 155–161.
73. Pusecker, K.; Albert, K.; Bayer, E.; *Journal of Chromatography A*, 1999, 836, 245–252.
74. Zhao, Y.; *Journal of Chromatography A*, 1999, 837, 83–91.
75. Schneider, B.; Natural products, liquid chromatography-nuclear magnetic resonance. In: I.D. Wilson (ed.), *Encyclopedia of Separation Science*, Academic Press, London, 1999.
76. Wolfender, J.-L.; Ndjoko, K.; Hostettmann, K.; *Journal of Chromatography A*, 2003, 1000, 437–455.
77. Nyberg, N.T.; Baumann, H.; Kenne, L.; *Analytical Chemistry*, 2003, 75, 268–274.
78. Bieri, S.; Varesio, E.; Veuthey, J.L.; Munõz, O.; Tseng, L.-H.; Braumann, U.; Spraul, M.; Christen, P.; *Phytochemical Analysis*, 2006, 17, 78–86.
79. Lam, S.-H.; Wang, C.-Y.; Chen, C.-K.; Lee, S.-S.; *Phytochemical Analysis*, 2007, 18, 251–255.
80. Lambert, M.; Stærk, D.; Hansen, S.H.; Sairafianpour, M.; Jaroszewski, J.W.; *Journal of Natural Products*, 2005, 68, 1500–1509.
81. Tatsis, E.C.; Boeren, S.; Exarchou, V.; Troganis, A.N.; Vervoort, J.; Gerothanassis, I.P.; *Phytochemistry*, 2007, 68, 383–393.
82. Exarchou, V.; Krucker, M.; van Beek, T.A.; Vervoort, J.; Gerothanassis, I.P.; Albert, K.; *Magnetic Resonance in Chemistry*, 2005, 43, 681–687.
83. Wang, C.-Y.; Lee, S.-S.; *Phytochemical Analysis*, 2005, 16, 120–126.
84. Sørensen, D.; Raditsis, A.; Trimble, L.A.; Blackwell, B.A.; Sumarah, M.W.; Miller, J.D.; *Journal of Natural Products*, 2007, 70, 121–123.
85. Hansen, S.H.; Jensen, A.G.; Cornett, C.; Bjørnsdottir, I.; Taylor, S.; Wright, B.; Wilson, I.D.; *Analytical Chemistry*, 1999, 71, 5235–5241.
86. Jensen, A.G.; Cornett, C.; Gudiksen, L.; Hansen, S.H.; *Phytochemical Analysis*, 2000, 11, 387–394.
87. Clarkson, C.; Stærk, D.; Hansen, S.H.; Jaroszewski, J.W.; *Analytical Chemistry*, 2005, 77, 3547–3553.
88. Jayawickrama, D.A.; Sweedler, J.V.; *Journal of Chromatography A*, 2003, 1000, 819–840.
89. Kuzovkina, I.; Al'terman, I.; Schneider, B.; *Phytochemistry*, 2004, 65, 1095–1100.
90. Schneider, B.; Paetz, C.; Hölscher, D.; Opitz, S.; *Magnetic Resonance in Chemistry*, 2005, 43, 724–728.

91. Spring, O.; Chemotaxonomy based on metabolites from glandular trichomes. In: J.A. Callow (ed.), *Advances in Botanical Research*, Academic Press, London, 2000.
92. Schneider, B.; Gershenzon, J.; Graser, G.; Hölscher, D.; Schmitt, B.; *Phytochemistry Reviews*, 2003, 2, 31–43.
93. Schneider, B.; *Progress in Nuclear Magnetic Resonance Spectroscopy*, 2007, 51, 155–198.
94. Schmitt, B.; Schneider, B.; *Phytochemistry*, 1999, 52, 45–53.
95. Schmitt, B.; Schneider, B.; *Phytochemical Analysis*, 2001, 12, 43–47.
96. Bailey, N.J.C.; Stanley, P.D.; Hadfield, S.T.; Lindon, J.C.; Nicholson, J.K.; *Rapid Communication in Mass Spectrometry*, 2000, 14, 679–684.
97. Lenz, E.M.; Lindon, J.C.; Nicholson, J.K.; Weeks, J.M.; Osborn, D.; *Xenobiotica*, 2002, 32, 535–546.
98. Sleiman, M.; Ferronato, C.; Fenet, B.; Baudot, R.; Jaber, F.; Chovelon, J.M.; *Analytical Chemistry*, 2006, 78, 2957–2966.
99. Benfenati, E.; Pierucci, P.; Fanelli, R.; Preiss, M.; Godejohann, M.; Astratov, M.; Levsen, K.; Barceló, D.; *Journal of Chromatography A*, 1999, 831, 243–256.
100. Preiss, A.; Sänger, U.; Karfich, N.; Levsen, K.; Mügge, C.; *Analytical Chemistry*, 2000, 72, 992–998.
101. Simpson, A.J.; Tseng, L.-H.; Simpson, M.J.; Spraul, M.; Braumann, U.; Kingery, W.L.; Kelleher, B.P.; Hayes, M.H.B.; *The Analyst*, 2004, 129, 1216–1222.
102. Weißhoff, H.; Preiss, A.; Nehls, I.; Win, T.; Mügge, C.; *Analytical and Bioanalytical Chemistry*, 2002, 373, 810–819.
103. Cardoza, L.A.; Almeida, V.K.; Carr, A.; Larive, C.K.; Graham, D.W.; *Trends in Analytical Chemistry*, 2003, 22, 766–775.
104. Corcoran, O.; Spraul, M.; *Drug Discovery Today*, 2003, 8, 624–631.
105. Lindon, J.C.; Nicholson, J.K.; Sidelmann, U.G.; Wilson, I.D.; *Drug Metabolism Reviews*, 1997, 29, 705–746.
106. Alexander, A.J.; Xu, F.; Bernard, C.; *Magnetic Resonance in Chemistry*, 2006, 44, 1–6.
107. Novak, P.; Cindrić, M.; Tepes, P.; Dragojević, S.; Ilijas, M.; Mihaljević, K.; *Journal of Separation Science*, 2005, 28, 1442–1447.
108. Segmuller, B.E.; Armstrong, B.L.; Dunphy, R.; Oyler, A.R.; *Journal of Pharmaceutical and Biomedical Analysis*, 2000, 23, 927–937.
109. Novak, P.; Tepeš, P.; Fistrić, I.; Bratoš, I.; Gabelica, V.; *Journal of Pharmaceutical and Biomedical Analysis*, 2006, 40, 1268–1272.
110. Pan, C.; Liu, F.; Ji, Q.; Wang, W.; Drinkwater, D.; Vivilecchia, R.; *Journal of Pharmaceutical and Biomedical Analysis*, 2006, 40, 581–590.
111. Nookandeh, A.; Frank, N.; Steiner, F.; Ellinger, R.; Schneider, B.; Gerhäuser, C.; Becker, H.; *Phytochemistry*, 2004, 65, 561–570.
112. Eldridge, S.L.; Almeida, V.K.; Korir, A.K.; Larive, C.K.; *Analytical Chemistry*, 2007, 79, 8446–8453.
113. Lenz, E.M.; D'Souza, R.A.; Jordan, A.C.; King, C.D.; Smith, S.M.; Phillips, P.J.; McCormick, A.D.; Roberts, D.W.; *Journal of Pharmaceutical and Biomedical Analysis*, 2007, 43, 1065–1077.
114. Shockcor, J.P.; Unger, S.H.; Savina, P.; Nicholson, J.K.; Lindon, J.C.; *Journal of Chromatography B*, 2000, 748, 269–279.
115. Tugnait, M.; Lenz, E.M.; Hofmann, M.; Spraul, M.; Wilson, I.D.; Lindon, J.C.; Nicholson, J.K.; *Journal of Pharmaceutical and Biomedical Analysis*, 2003, 30, 1561–1574.
116. Godejohann, M.; Tseng, L.-H.; Braumann, U.; Spraul, M.; *Journal of Chromatography A*, 2004, 1058, 191.
117. Kammerer, B.; Scheible, H.; Zurek, G.; Godejohann, M.; Zeller, K.P.; Gleiter, C.H.; Albrecht, W.; Laufer, S.; *Xenobiotica*, 2007, 37, 280.

118. Hentschel, P.; Krucker, M.; Grynbaum, M.D.; Putzbach, K.; Bischoff, R.; Albert, K.; *Magnetic Resonance in Chemistry*, 2005, 43, 747–754.
119. Ohtani, I.; Kusumi, T.; Kashman, Y.; Kakisawa, H.; *Journal of American Chemical Society*, 1991, 113, 4092–4096.
120. Seger, C.; Godejohann, M.; Spraul, M.; Stuppner, H.; Hadacek, F.; *Journal of Chromatography A*, 2006, 1136, 82–88.
121. Schlauer, J.; Rückert, M.; Wiesen, B.; Herderich, M.; Assi, L.A.; Haller, R.D.; Bär, S.; Fröhlich, K.U.; Bringmann, G.; *Archives of Biochemistry and Biophysics*, 1998, 350, 87–94.
122. Zhou, C.Z.; Hill, D.R.; *Magnetic Resonance in Chemistry*, 2007, 45, 128–132.
123. Lindon, J.C.; Farrant, R.D.; Sanderson, P.N.; Doyle, P.M.; Gough, S.L.; Spraul, M.; Hofmann, M.; Nicholson, J.K.; *Magnetic Resonance in Chemistry*, 1995, 33, 857–863.
124. Chin, J.; Fell, J.B.; Jarosinski, M.; Shapiro, M.J.; Wareing, J.R.; *Journal of Organic Chemistry*, 1998, 63, 386–390.
125. Pasch, H.; Trathnigg, B.; *HPLC on Polymers*, Springer, New York, 1998.
126. Hatada, K.; Ute, K.; Okamoto, Y.; Imanari, M.; Fujii, N.; *Polymer Bulletin*, 1988, 20, 317–321.
127. Hatada, K.; Ute, K.; Kashiyama, M.; Imanari, M.; *Polymer Journal*, 1990, 22, 218–222.
128. Ute, K.; Niimi, R.; Hongo, S.; Hatada, K.; *Polymer Journal*, 1998, 30, 439–443.
129. Hatada, K.; Ute, K.; Kitayama, T.; Nishimura, T.; Kashiyama, M.; Fujimoto, N.; *Polymer Bulletin*, 1990, 23, 549–554.
130. Hatada, K.; Ute, K.; Kitayama, T.; Yamamoto, M.; Nishimura, T.; Kashiyama, M.; *Polymer Bulletin*, 1989, 21, 489–495.
131. Ute, K.; Niimi, R.; Hatada, K.; Kolbert, A.C.; *International Journal of Polymer Analysis and Characterization*, 1999, 5, 47–49.
132. Hiller, W.; Pasch, H.; Macko, T.; Hofmann, M.; Ganz, J.; Spraul, M.; Braumann, U.; Streck, R.; Mason, J.; Van Damme, F.; *Journal of Magnetic Resonance*, 2006, 183, 290–302.
133. Kitayama, T.; Janco, M.; Ute, K.; Niimi, R.; Hatada, K.; Berek, D.; *Analytical Chemistry*, 2000, 72, 1518–1522.
134. Pasch, H.; Hiller, W.; *Macromolecules*, 1996, 29, 6556–6559.
135. Pasch, H.; Hiller, W.; Haner, R.; *Polymer*, 1998, 39, 1515–1523.
136. Krämer, I.; Pasch, H.; Händel, H.; Albert, K.; *Macromolecular Chemistry and Physics*, 1999, 200, 1734–1744.
137. Krämer, I.; Hiller, W.; Pasch, H.; *Macromolecular Chemistry and Physics*, 2000, 201, 1662–1666.
138. Schlotterbeck, G.; Pasch, H.; Albert, K.; *Polymer Bulletin*, 1997, 38, 673–679.
139. Hiller, W.; Brüll, A.; Argyropoulos, D.; Hoffmann, E.; Pasch, H.; *Magnetic Resonance in Chemistry*, 2005, 43, 729–735.
140. Hiller, W.; Sinha, P.; Pasch, H.; *Macromolecular Chemistry and Physics*, 2007, 208, 1965–1978.
141. Kautz, R.A.; Goetzinger, W.K.; Karger, B.L.; *Journal of Combinatorial Chemistry*, 2005, 7, 14–20.
142. Kupce, E.; Nishida, T.; Freeman, R.; *Progress in Nuclear Magnetic Resonance Spectroscopy*, 2003, 42, 95–122.
143. Wolters, A.M.; Jayawickrama, D.A.; Webb, A.G.; Sweedler, J.V.; *Analytical Chemistry*, 2002, 74, 5550–5555.
144. Lacey, M.E.; Tan, Z.J.; Webb, A.G.; Sweedler, J.V.; *Journal of Chromatography A*, 2001, 922, 139–149.
145. Wilson, I.D.; Brinkman, U.A.T.; *Journal of Chromatography A*, 2003, 1000, 325–356.
146. Pukalskas, A.; van Beek, T.A.; de Waard, P.; *Journal of Chromatography A*, 2005, 836, 245–252.

4 Application of Infrared and Raman Spectroscopy for Detection in Liquid Chromatographic Separations

Peter J. Mahon

CONTENTS

4.1   Introduction 100
4.2   Spectroscopic Theory 101
      4.2.1   Raman Effect 102
      4.2.2   Resonance Enhancement 104
4.3   Instrumental Factors 108
4.4   Interfacing Liquid Chromatography with Infrared Detection 108
      4.4.1   Flow Cells 109
      4.4.2   Flow Cell Detection Based on Solvent Exchange 113
      4.4.3   Solvent Elimination 114
      4.4.4   Evaporation Interfaces 116
      4.4.5   Thermospray Interfaces 118
      4.4.6   Pneumatic Interfaces 120
      4.4.7   Concentric Flow Interfaces 121
      4.4.8   Ultrasonic Interfaces 124
      4.4.9   Electrospray Interfaces 126
      4.4.10 Particle Beam Interfaces 126
4.5   Interfacing Liquid Chromatography with Raman Detection 129
      4.5.1   Flow Cells 129
      4.5.2   Waveguides 131
      4.5.3   Enhanced Methods 133
      4.5.4   Solvent Elimination Methods 135
      4.5.5   Evaporative Methods 136
      4.5.6   Enhanced Methods 136
      4.5.7   Concentric Flow Interface: Enhanced Methods 137
4.6   Conclusions 138
References 139

4.1   INTRODUCTION

The coupling of molecular spectroscopy with chemical separation procedures has often been demonstrated to be a very powerful combination because it includes the ability to isolate a particular molecule and then characterize it based upon its unique spectroscopic signature. Applications range from using simple photometric detection tuned for a specific attribute of the analyte to very detailed *in situ* spectroscopic measurements of chromatographically separated mixtures when other molecular isolation methods fail. Just as separation science is multifaceted where an expert can identify and apply the most suitable chromatographic technique for each unique situation, the analytical spectroscopist can also draw upon an extensive array of techniques that can selectively exploit a particular interaction between a molecule and the appropriate electromagnetic (EM) radiation. Therefore, the coalescence of these quite different areas of expertise enables a wide range of the potential applications to be investigated and readily resolved.

Optical spectroscopy encompasses the numerous modes of interaction between a molecule and photons with the specific energies that stimulate vibrational and electronic changes within the molecule. The energy of a photon, E, can be related to its frequency, ν, wavelength, λ, and wavenumber, $\bar{\nu}$, based on the following relations

$$E = h\nu = \frac{hc}{\lambda} = hc\bar{\nu} \tag{4.1}$$

where
 h is Planck's constant equal to 6.626×10^{-34} J s
 c is the speed of light equal to 3.00×10^{8} m s^{-1}

Typically, electronic transitions can be correlated with photons from the ultraviolet/visible (UV/Vis) region of the EM spectrum. Vibrational motions within a molecule, which includes bond rotation, stretching, and bending require less energy and result from interaction with photons from the infrared (IR) region.

Historically, the ease of visually detecting color and intensity variation of samples by comparison to standard materials was the basis for the first spectroscopic methods of analysis. As a result of this initial advantage and additional developments in instrumentation, UV/Vis detectors have found wide application when coupled to chromatographic systems. One advantage of probing vibrational changes compared to electronic transitions is that greater spectral information is available due to the larger number of discrete motions within a molecule and this enables substantial qualitative discrimination that can lead to the specific characterization of an individual molecule. Rigorous identification also requires that the observed characteristics can be attributed to one molecule and therefore the separation of molecule mixtures is of critical importance.

It will be seen that there are two ways for identical vibrational changes to be induced and the spectroscopy associated with the origin for these interactions is separated into two categories known as IR and Raman spectroscopy. Most chemists will be familiar with IR spectroscopy because it is routinely employed as the starting point for any effort to identify a molecule based on well-established tabulations that correlate specific absorption frequencies with energies that are characteristic for particular functional groups that constitute a molecule. It is therefore obvious that coupling of an IR detector with a chromatographic system is a powerful combination but there are many technical issues that need to be optimized in order to ensure that the analyte that emerges from the chromatography column is compatible with the detector.

The confluence of a number of factors has lead to a realization that there are many advantages in exploiting the Raman effect that can provide a great deal of both qualitative and quantitative spectroscopic information. The applicability of Raman spectroscopy has benefited from recent advances in instrumentation and this has lead to a considerable increase in the body of knowledge associated with the characterization and theoretical description of molecular behavior that gives rise to the Raman effect [1].

Recently, reviews on the use of IR [2] and Raman [3] spectroscopies for detection in flow techniques (i.e., flow injection analysis, liquid chromatography, and capillary electrophoresis) have been published that evaluate the current state in the development of these hyphenated techniques. In addition, a monograph [1] focusing on general aspects of the application of Raman spectroscopy for chemical analysis that contains detailed descriptions of the many factors that need to be considered during analyses where quantitative and qualitative characterization are the goal has been published.

In this chapter, a brief overview of the aspects essential to develop a working understanding of the application of IR and Raman spectroscopy for detection and characterization in chromatographic system will be presented. Approaches that have been adopted to take advantage of the benefits of applying these spectroscopies will be described, but this will be balanced against some of the substantial problems that do exist and are an inherent drawback of analytical applications for the detection systems that have been developed.

4.2 SPECTROSCOPIC THEORY

The most general description of spectroscopy involves the directed generation of photons for exposure to matter that will stimulate energetic processes that can occur within molecules and/or within individual atoms. The specific interactions that occur between the photons and the matter are dependent upon the energy of the photons as well as the characteristics and the arrangement of atoms within the exposed matter. The energy correlated with changes in bond lengths and the angles between atoms is typically associated with photons from the IR region of the EM spectrum. Figure 4.1 indicates various energy transitions that can occur within a molecule and the most familiar process is the direct absorption of a photon with a particular energy that raises the overall energy of the molecule by one vibrational level.

This process is the basis of IR spectroscopy, which is routinely used qualitatively for characterization purposes because the energy absorbed can be associated with specific arrangements and interactions of atoms within a molecule, such as bond rotation, translation, and vibration. These experiments are typically performed in a

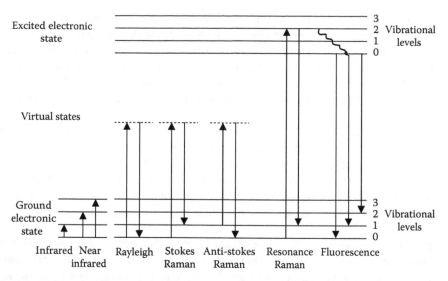

FIGURE 4.1 Energy level diagram demonstrating the various energy transitions due to the interaction of photons with a molecule.

transmission mode where the incident light, with the all appropriate energies, passes through a space containing the molecules and the components of energy equal to the rotational, translational, and vibrational processes within the molecule are removed from the light that is then detected. IR spectra are therefore presented as wavelength-dependent transmission plots.

4.2.1 Raman Effect

It has also been observed that photons with energy greater than is required to match the discrete amount of energy to cause a similar vibrational transition can also interact with a molecule to induce the vibrational energy transition and this process is known as the Raman effect and was discovered in 1928 by Krishnan and Raman [4]. This interaction involves an energy transition through a "virtual state" generated during a momentary distortion of the molecule's electron cloud induced by the oscillating electric field of the incident light. The process results from an inelastic interaction whereby the incident photon is absorbed and then a photon of lower energy is emitted with the difference being equal to a change in the vibrational energy equivalent to that observed during the direct absorption process in conventional IR spectroscopy. Photons emitted with a lower energy than the incident photons are described as Stokes shifted and these are shown in Figure 4.1. Although the energy changes are identical, the mechanism and probability of the inelastic Raman process is quite distinct from the direct absorption of a photon. It is considered that Raman and IR spectroscopy are complimentary methods where the identical vibrational states can be accessed using two different processes, each with their individual advantages and disadvantages. Figure 4.2 shows Raman and IR spectra for the same compound, many vibrational modes are active for both types of spectroscopy and it can be seen that there are many corresponding peaks but the magnitudes are quite different.

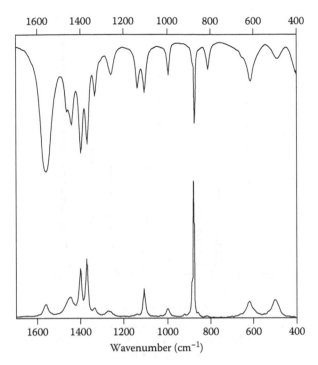

FIGURE 4.2   Infrared and Raman spectra for nitroethane.

The overriding disadvantage of Raman spectroscopy is that only 1 of 100 million photons are influenced by the Raman effect with the majority of the inelastically scattered photons being reemitted without an energy change as described by the Rayleigh transitions in Figure 4.1. From an analytical perspective, a further complication with the Raman effect is that the emitted photons with altered wavelengths are randomly scattered and therefore the collection efficiency of the detection for these photons can be an important factor. A proportion of the molecules may absorb the incident photons while in a higher vibrational energy state and then emit a photon with an increased amount of energy during a return to the ground state. These energy shifts as demonstrated in Figure 4.1 are also described by the Raman effect and are known as the anti-Stokes shifted photons. In general, the population of the higher vibrational energy states is less than the ground state and therefore the anti-Stokes lines are less intense than the Stokes lines. Figure 4.3 demonstrates the symmetry of the Stokes and anti-Stokes line centered on the excitation wavelength with a noticeable decrease in peak intensity for the anti-Stokes lines.

Figure 4.3 also shows how the Raman shift convention is used so that the complementary nature of Raman and IR spectroscopies is demonstrated where the equivalence of an IR absorption wavenumber and the Raman shift are both correlated with the same change in the vibrational energy state for a molecule. The excitation photons for Raman spectroscopy are generally in the UV-Visible-NIR region of the spectrum where the convention is to use nanometers to describe the radiation whereas the convention in IR spectroscopy is to refer to the frequency as wavenumbers with the

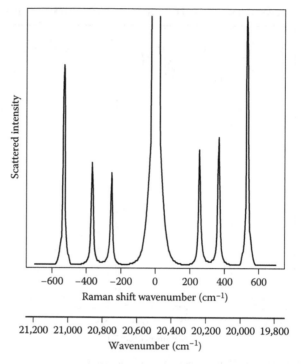

FIGURE 4.3 Spectrum of CCl_4 showing both anti-Stokes and Stokes lines relative to the excitation wavenumber of the 488 nm laser source.

unit cm^{-1} and this is also the unit used for describing the Raman shift. The complementarity extends to the origin of each effect with IR absorption being more intense for bonds with a high dipole moment such as C=O, O–H, C–H, etc., and the Raman effect being more pronounced for symmetrical bonds such as C–C, C–S, S–S, etc., because the mode of action is dependent on polarization changes associated with the bond. This trend is obvious in Table 4.1 that contains a listing of the relative intensities and wavenumbers corresponding to individual functional groups for both Raman and IR. A consequence of the differing sensitivities is that the common restrictions associated with using materials such as glass for sample preparation and contamination due to water when applying IR spectroscopy are not as critical during Raman spectroscopic analysis.

4.2.2 RESONANCE ENHANCEMENT

If the excitation photon has sufficient energy to enable an electronic transition to occur, then there are two possible paths for returning to the ground state and they are either by a Raman route or by a fluorescence process. The Raman route is actually a special case that has major consequences for analytical purposes and it is known as the resonance Raman (RR) effect. If the excitation energy is sufficient, then an electronic transition is more highly favored than the formation of a "virtual state" under nonresonant conditions and this massively increases the number of

TABLE 4.1
Comparison of Raman and IR Intensities for Common Functional Groups

Functional Group Vibration	Region (cm⁻¹)	Raman	IR
Lattice vibrations in crystals	10–200	Strong	Strong
δ(C–C) aliphatic chains	250–400	Strong	Weak
υ(Se–Se)	290–330	Strong	Weak
υ(S–S)	430–550	Strong	Weak
υ(Si–O–Si)	450–550	Strong	Weak
υ(X$_{metal}$–O)	150–450	Strong	Medium to weak
υ(C–I)	480–660	Strong	Strong
υ(C–Br)	500–700	Strong	Strong
υ(C–Cl)	550–800	Strong	Strong
υ(C–S) aliphatic	630–790	Strong	Medium
υ(C–S) aromatic	1080–1100	Strong	Medium
υ(O–O)	845–900	Strong	Weak
υ(C–O–C)	800–970	Medium	Weak
υ(C–O–C)$_{asym}$	1060–1150	Weak	Strong
υ(CC) alicyclic, aliphatic chain vibrations	600–1300	Medium	Medium
υ(C=S)	1000–1250	Strong	Weak
υ(CC) aromatic ring chain vibrations	*1580, 1600	Strong	Medium
	*1450, 1500	Medium	Medium
	*1000	Strong/medium	Weak
δ(CH$_3$)	1380	Medium	Strong
δ(CH$_2$) δ(CH$_3$)$_{asym}$	1400–1470	Medium	Medium
υ(C–(NO$_2$))	1340–1380	Strong	Medium
υ(C–(NO$_2$))$_{asym}$	1530–1590	Medium	Strong
υ(N=N) aromatic	1410–1440	Medium	—
υ(N=N) aliphatic	1550–1580	Medium	—
δ(H$_2$O)	~1640	Weak broad	Strong
υ(C=N)	1610–1680	Strong	Medium
υ(C=C)	1500–1900	Strong	Weak
υ(C=O)	1680–1820	Medium	Strong
υ(C≡C)	2100–2250	Strong	Weak
υ(C≡N)	2220–2255	Medium	Strong
υ(S–H)	2550–2600	Strong	Weak
υ(C–H)	2800–3000	Strong	Strong
υ(=(C–H))	3000–3100	Strong	Medium
υ(≡(C–H))	3300	Weak	Strong
υ(N–H)	3300–3500	Medium	Medium
υ(O–H)	3100–3650	Weak	Strong

Source: Reproduced with permission from HORIBA Jobin Yvon Ltd. and available as a web document from http://www.horiba.com/fileadmin/uploads/Scientific/Documents/Raman/bands.pdf

Raman-shifted photons by factors around 10^5. An important aspect of this, which is particularly relevant to large biomolecular systems, is that only the localized structure associated with a particular electronic transition will experience the resonance effect and therefore only these specific structures within the larger molecule will be detected due to the resonance enhancement. In this regard, the advent of resonance increases selectivity over nonresonant Raman signals because the relative intensities of signature peaks will be different and in most cases by many orders of magnitude.

Unfortunately, from the perspective of observing Raman-shifted photons, fluorescent processes can be a major interference that swamps the Raman signal. Even a weakly fluorescent process will prevent a strong Raman signal from being detected. The source of the fluorescence may be due to a trace impurity and a strategy to improve the ability to detect the Raman response is to use an excitation source that is less energetic (i.e., higher wavelength) to avoid the electronic transitions that give rise to the fluorescence. It is common for commercial instruments to be available with a number of excitation laser wavelengths. The intensity of the Raman response is known to depend on the frequency of the excitation source raised to the fourth power (i.e., ν^4), therefore there will be a decrease in the intensity of the Raman signal when using a higher wavelength source. Although the Raman signal could be decreased by a factor of 10, the magnitude of the Raman signal relative to the greatly diminished fluorescence could be sufficient for useful spectra to be obtained.

Alternatively, it has been observed that the use of lasers with wavelengths in the UV region can also minimize fluorescence because many fluorophores do not significantly absorb these higher energy photons [5]. Figure 4.4 shows the laser excitation wavelength dependence of the fluorescence that demonstrates

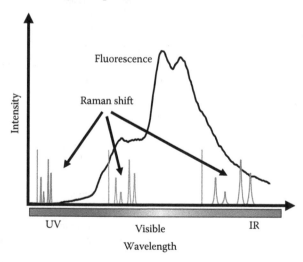

FIGURE 4.4 Comparison of the Raman signal intensity for different laser excitation wavelengths compared to the fluorescent background. (Copyright Renishaw plc. Reproduced with permission.)

how selection of the laser source can significantly affect the ratio of the Raman signal relative to the fluorescence background.

The nature of the random scattering process associated with the Raman effect suits the study of surfaces and interfaces. It is well established that under certain conditions the Raman intensity can also be greatly enhanced due to the interaction of the excitation source and particular properties associated with the surface [6]. The electric field of the excitation source can be amplified by an interaction of the light with the electronic structure of a surface and the increase in intensity of the electric field leads to an enhancement in the observed Raman signal by as much as a factor of 10^6. Surface enhanced Raman spectroscopy (SERS) exploits this effect and it is often seen with metallic surfaces such as copper, silver, and gold with roughness or curvature also being an important contributing factor. It is considered that the enhancement can have two contributions with one known as *chemical enhancement* and the other component known as *field enhancement* [1].

A factor of $\sim 10^2$ is attributed to a chemical enhancement process where there is an electronic interaction between the adsorbate and the metal that decreases the energy required to form an excited electronic state for the adsorbate due to a charge transfer process, which is a form of vibronic coupling. This enables a resonance process to occur more readily because the electronic excited state energy is lowered and therefore is closer to the energy of the incident laser.

The field enhancement results from an amplification of the electric field of laser photons due to an interaction with the enhanced optical fields of the surface being illuminated at plasmon resonance frequencies. Irregularities in the surface structure of the metals (i.e., roughness) have been demonstrated to be an important factor that affects the magnitude of the surface enhanced effect. The increase in local electric field intensity at the roughened surface is equivalent to an increase in power density of the laser and therefore leads to an increase in the detected Raman intensity. Adsorption onto colloidal metallic particles has often been used for the characterization of low concentrations of molecules down to the level that single molecules can be detected [7,8].

When the molecules that adsorb also have intense electronic absorption bands in the same wavelength region as the metal surface plasmon resonance, the combination of surface enhancement and resonance (i.e., SERRS) can result in an overall enhancement as large as 10^{12} or the product of the two individual effects. A further advantage obtained with the surface-enhanced effect is that fluorescence quenching occurs due to the possibility that the excited state of the molecule can return to the ground state via a charge transfer mechanism at the metal surface rather than emitting a photon. This increases the relative amount of Raman scattering compared with the fluorescence and therefore improves the detectability of the Raman signal.

Surface enhanced infrared absorption spectroscopy (SEIRA) has also been demonstrated but the relative enhancement is insignificant compared to the effect observed for much weaker Raman scattering [9]. Therefore, the analytical advantage of signal enhancement is greatly outweighed by the difficulties associated with preparing the sample in a manner that promotes the enhancement process.

4.3 INSTRUMENTAL FACTORS

The emergence of Fourier transform infrared (FT-IR) instruments has greatly improved the speed that an IR spectrometer can acquire a spectrum. The ability to accumulate numerous spectra has substantially improved the sensitivity, which reduces the sample volume that is needed for analysis and this has enabled the direct coupling of FT-IR detectors to chromatographic systems.

The recent renaissance of Raman spectroscopy is due to a number of technological developments. The application of lasers as the excitation source overcomes the low probability of creating a Raman-shifted photon due to the massive increase in photon intensity compared to conventional light sources. The development of sensitive silicon-based detectors that operate in a similar spectral region as the commonly available lasers has greatly improved the signal-to-noise (S/N) ratio. The use of multichannel imaging arrays that enable the simultaneous detection of a range of wavelengths after the scattered light is dispersed allows for the rapid acquisition of spectra compared with scanning monochromators. The combination of high photon flux and increased detector sensitivity has enabled the development of commercial instruments for Raman spectroscopy that are available for reliable, and even routine, use. When these instruments are combined with confocal microscopes, it is possible to have high spatial resolution (i.e., ~1 μm), which enables very small quantities of a sample to be analyzed and Raman microspectroscopy has become a powerful analytical tool [1].

The application of hyphenated chromatographic techniques using IR and Raman spectroscopies has involved both online and at-line methods of detection but for different reasons. An overriding problem with the use of IR detectors for eluents from a chromatographic system is the detection of a dilute analyte in a matrix (i.e., solvent, buffers, ion-pair reagents, etc.) that also has strong absorptions in the same spectral region and solvent elimination strategies have often been employed. However, when employing Raman spectroscopy, the elimination of the solvent enables specific interaction between the analyte and an engineered surface that can result in resonance enhancement and therefore greater analytical sensitivity. Although there is great deal of similar or complementary information available from the use of IR and Raman spectroscopies, the technical issues associated with the application of each technique require that they are discussed separately so that specific attributes can be highlighted.

4.4 INTERFACING LIQUID CHROMATOGRAPHY WITH INFRARED DETECTION

When considering the coupling of IR spectroscopy with liquid chromatography (LC) separations to take advantage of the potential benefits of obtaining simultaneous qualitative and quantitative information, it is necessary to consider the physical and spectral properties of the analytes and the mobile phase that emerge from the column. Just as there is not one type of chromatography, there is not a truly general method for interfacing the two techniques and specific factors such as thermal stability, volatility, solubility, and spectral characteristics of all components that constitute the eluent determine the suitability of the type of interface employed.

Two strategies exist for extracting useful spectra from separated analytes, the most straightforward method is to use a flow cell as used for most convention online chromatography detection methods but constructed from materials suited to the IR region of the spectrum. The alternative method is to collect the analytes at-line as they are deposited on an appropriate substrate and transferred in a modified form into the detection beam of a spectrometer either directly after deposition or delyed until the chromatogram is fully collected.

4.4.1 Flow Cells

The general advantages of using flow cells are based on their ease of operation, reliability, real-time application, and the nondestructive passage of the sample through the detector so that alternative detection techniques [10–12] can be utilized also apply when using IR detection. The earliest applications using convention variable wavelength IR spectrophotometers demonstrated the expected high spectral specificity that could be obtained when applied to the detection of various coal-derived products separated by gel permeation (GP) chromatography [13]. The use of a stopped-flow procedure enabled reasonable spectral identification of chromatographically separated lipids to be obtained after direct subtraction of the recorded spectrum due to the mobile phase, which was purified methylene chloride [14]. The use of a pure solvent as the mobile phase greatly reduced potential subtraction problems when using a mixed solvent where maintaining sufficient control of the mobile phase composition is required to minimize subtraction errors and it was observed that the technique was very sensitive to impurities in the mobile phase.

The advent of FT-IR instruments enabled better quality and faster spectra to be measured so that real-time or "on-the-fly" spectral detection in chromatographic systems could be achieved. However, the extraction of the solute spectra from the total eluent spectra continued to be problematic and only separations using size exclusion (SE) chromatography were initially investigated [13,15–17]. In these cases, the importance of the spectral characteristics of the mobile phase takes precedence over the chromatographic properties because the separation mechanism is based on mobility rather than partitioning or adsorption.

It was demonstrated that the spectral overlap between the solutes and common solvents used for normal phase (NP) chromatography could be minimized by the use of halogenated alkanes such as Freon 113 [18] and $CHCl_3$ [19]. Optimization of the cell pathlength is a balance between an extended spectral window and solute sensitivity [18]. The use of deuterated solvents where the absorption bands are shifted away from the equivalent non-deuterated processes that occur in the solutes was demonstrated using $CDCl_3$ as the solvent for the NP separation of a coal-derived material of intermediate polarity [20]. A comprehensive demonstration of this concept employed a number of deuterated solvents as mobile phases for microcolumn separations using revered phase (RP), NP, and SE modes [21]. It was shown that the most common spectral responses for organic molecules due to C–H, O–H, C–O, and C=O could be used to identify specific solutes because of shifts due to deuteration of the equivalent O–H and C–H bonds in the solvents and also due to lower background absorbances for the C–O and C=O regions when deuterated solvents are used. The expense of using such solvents was reduced through the use of a microbore column which had the additional advantages of being able to handle higher eluent concentrations and greater efficiency compared to conventional analytical columns [22]. Subsequently, the use of zero-dead-volume flow cells constructed by drilling a channel with the same internal diameter as the delivery tubing through IR transparent blocks of KBr and CaF$_2$ was demonstrated to give superior detection limits compared to the previously employed parallel plate "ultramicro" design [19].

There are compatibility difficulties associated with the materials generally used to construct the cell windows due to the demands of wide spectral transparency and stability in the presence of aqueous-based mobile phases. An early example by Jinno and coworkers used a pinched PTFE tube as a flow cell [23] and when combined with perfluoronated or deuterated solvents, there was adequate spectral transparency in the region 2600–4000cm^{-1} to detect the NP separation of hydrocarbons in a gasoline sample [24]. However, inorganic materials are more widely favored and Table 4.2 is a compilation of the properties of the most commonly used window materials that have been used in flow analysis-IR techniques [25]. The separation of sugars contained in samples of nonalcoholic beverages by ion exchange (IE) chromatography using an aqueous mobile phase has been achieved using CaF$_2$ windows in a commercially available flow cell [26]. A cell pathlength of 25 μm was necessary to limit the background absorption due to water at 1100cm^{-1}. In a subsequent analysis of

TABLE 4.2
Compilation of Common Optical Materials Used as Windows
in IR Flow Cells

Material	Spectral Transparency (cm^{-1})	Solubility in Water (g/100 mL)	Precautions Attacked by
KBr	40,000–400	Very soluble—53.5	H$_2$O, lower alcohols
NaCl	40,000–625	Very soluble—35.7	H$_2$O, lower alcohols
CsI	40,000–200	Very soluble—44.4	H$_2$O, lower alcohols
KRS-5	20,000–250	Slightly soluble—0.05	Complexing agents
AgCl	25,000–360	0.00015	Complexing agents
Quartz	25,000–2,500	Insoluble	Hot H$_2$SO$_4$, aqua regia
Polyethylene	625–33	Insoluble	Chlorinated solvents
AMTIR	11,000–725	Insoluble	Alkalis
BaF$_2$	50,000–740	Insoluble	NH$_4^+$, acids, salts
CaF$_2$	50,000–1,110	Insoluble	NH$_4^+$, acids, salts
CdTe	20,000–360	Insoluble	Acids, HNO$_3$
Ge	5500–600	Insoluble	Hot H$_2$SO$_4$, aqua regia
Si	8300–660	Insoluble	HF, HNO$_3$
	360–70		
ZnS	17,000–720	Insoluble	Acids
ZnSe	20,000–454	Insoluble	Acids, strong alkalis, amines

Source: Reprinted from *Talanta*, 64, Gallignani, M. and Brunetto, M.D., 1127, Copyright 2004, with permission from Elsevier.

carbohydrates, alcohols, and organic acids using IE chromatography where the mobile phase was 0.005 M H_2SO_4 that ensured that all the organic acids dissociated, it was found that it was necessary to dip-coat the CaF_2 windows with a thin layer of low-density polyethylene to counteract the slight solubility of CaF_2 in acidic conditions [27]. In a RP analysis of peroxide-based explosives also using the 25 μm CaF_2 flow cell, it was demonstrated that selective direct detection using FT-IR was possible compared to the alternative indirect detection methods based on the detection of hydrogen peroxide resulting from the photolysis of the explosives [28].

It is apparent from Table 4.2 that ZnSe is a material with good spectral and stability characteristics; it has been used in a commercial flow cell that enabled the detection of cholesterol in animal greases using RP chromatography [29]. However, a stopped-flow method was necessary to obtain good-quality spectra due to the poor sensitivity associated with using flow cells for FT-IR detection.

The use of ZnSe as a window material is greatly overshadowed by the numerous applications where it has been used in flow cells where attenuated total reflectance (ATR) increases the effective pathlength of the incident light due to multiple internal reflections. The measurement principle involves light passing through a highly refractive material with an angle of incidence greater than the critical angle and on each reflection the light (or *evanescent wave*) penetrates the external medium by half a wavelength of the propagating light. When used in a flow cell, the external medium is the eluent and the pathlength through the sample is dependent on the number of reflections and the wavelength of light. It is usual for spectra obtained with an ATR accessory to be corrected for the wavelength-dependent pathlength. In flow cell applications with FT-IR detection, cylindrical internal reflectance cells (CIRCLE) are commercially available where a cylindrical-shaped ZnSe crystal with cone-shaped ends is mounted in a stainless steel cell body. Figure 4.5 contains a diagram [30] of a recent incarnation of this method [31] using an ultramicro-CIRCLE flow cell where the sample volume is 1.75 μL. The limited penetration of the light into the sample volume could be considered as inefficiency but as demonstrated when using windowed cells, the pathlength of flow cells is always a compromise based on the maximum absorbance of the mobile phase and the concentration of the solutes regardless of the cell configuration.

Sabo and coworkers were first to demonstrate the use of the ATR effect in a flow cell with FT-IR detection using both NP and RP-HPLC for the analysis of a mixture of acetophenone, ethyl benzoate, and nitrobenzene [32]. They used a micro-CIRCLE cell with a nominal volume of 24 μL and an effective pathlength of 4–22 μm over the spectral range. Louden and coworkers used a macro-CIRCLE cell with a volume of 400 μL as part of a multiple hyphenated detection system that also included UV diode array, ^1H nuclear magnetic resonance, and mass spectroscopy (MS) detection for the analysis of nonsteroidal antiinflammatory drugs [11] and a plant extract [12]. This system was then modified by the use of a micro-CIRCLE flow cell [10] to increase sensitivity and decrease band broadening.

As a continuation of a previous polymer-coated CaF_2 windowed flow cell analysis of carbohydrates, alcohols, and organic acids in wine [27], a method using a diamond-based ATR cell was employed [33]. The inertness of the diamond overcame the problem of frequent replacement of the polymer coating that degraded

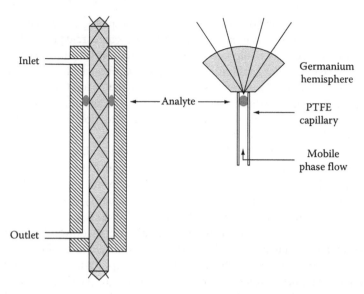

FIGURE 4.5 Comparison of two configurations using ATR crystals with the left diagram showing an ultramicro-CIRCLE cell and the right diagram showing the germanium micro-ATR interface. (Reprinted with permission from Patterson, B.M., Danielson, N.D., and Sommer, A.J., *Anal. Chem.*, 75, 1418, 2003. Copyright 2003 American Chemical Society.)

during routine application of the technology. It was noted that although chromatographic efficiency is greatly aided by using gradient elution, the difficulty of using background subtraction with a changing mobile phase composition requires isocratic conditions are used which results in a compromised separation with co-eluting solutes. Full-spectrum recording enabled a multivariate curve resolution technique to be successfully used for the quantitative analysis of overlapping solutes.

The application of ATR microspectroscopy where the incident beam is focused into an internal reflective element (IRE) so that the light is collected after a single reflection for FT-IR detection has been demonstrated and compared to multibounce CIRCLE flow cell analyses [30]. Figure 4.5 shows the geometry of the IRE relative to the outlet capillary of a HPLC system. The high spatial resolution of the IRE compared to the size of the solute volume in the eluent greatly improves the ratio of the sampling volume to the cell volume compared to other CIRCLE geometries and resulted in a mass sensitivity of 180 fg. However, the high spatial resolution compared to the diameter of the capillary allowed some solute to pass around the evanescent cone without detection and it was concluded that this sampling geometry would be more useful if the capillary diameter was 5.5 μm, which is more compatible with capillary electrophoresis.

Prompted by the improvements in FT-IR spectrometers, a recent investigation has revisited the use of micro-flow cells in an application based on a conventional NP separation of the insecticide β-cypermethrin [34]. Four solvent systems based on acetonitrile, dichloromethane, *n*-hexane, 2-propanol, and tetrahydrofuran were chosen based primarily on their chromatographic properties rather than their spectral characteristics. A flow rate of 2 mL min^{-1} was used with a silica packed column and a 10.8 μL flow cell with NaCl windows. It was found that significant spectral overlap was present but there were spectral windows where the solutes could be differentiated from the solvents and these restricted regions were used to produce chemigrams. The chemigrams had significantly better S/N ratios compared to the usual Gram–Schmidt chromatograms that are often presented because the Gram–Schmidt reconstruction is based on the full spectrum which includes the dominating solvent contribution. Based on the chemigrams, good linearity between the peak heights and concentrations in the range 0.3–4.0 mg mL^{-1} was found for all mobile phases, an identification limit of 1.0 mg mL^{-1} allowed spectral discrimination of the two diastereomers of β-cypermethrin and a detection limit of 0.3 mg mL^{-1} was generally obtained but was estimated to be 0.1 mg mL^{-1} when acetonitrile/n-hexane is the mobile phase.

4.4.2 Flow Cell Detection Based on Solvent Exchange

The acknowledged difficulties of having a substantial proportion of water in the mobile phase are due to its high infrared absorptivity and incompatibility with common window materials. These limitations have been overcome using strategies where the solute is transferred from an aqueous-based chromatographically compatible solvent to a more spectroscopically compatible solvent that does not contain water.

An initial study was undertaken where the eluent from the reversed-phase separation of a ketone model mixture was detected using a parallel plated window cell after the solutes had been extracted into chloroform [35]. A more extensive study based on the separation of a seven-component mixture that contained four ketones was then reported [36]. Figure 4.6 contains diagrams of the key elements that enable the online extraction of the solutes. The aqueous eluent and chloroform are delivered into two arms of a tee junction, which results in a segmented stream containing two immiscible liquids. The solutes partition between the two liquids so that a portion of the solute is transported into the chloroform segments. The segmented stream enters a membrane separator constructed from layers of hydrophobic PTFE so that the solute containing chloroform segments passes through the membrane and this stream is directed to the flow cell for detection. Efficient extraction is dependent on careful control of the differential backpressure of the exiting streams. It was concluded that the technique provided enhanced spectral performance due to the elimination of water and the use of ZnSe windows that enabled the solutes to be easily identified. Band broadening due to the membrane separator was minimal (~15%) when compared with a refractive index (RI) detector.

DeNunzio demonstrated a technique where the RP-separated solutes were adsorbed into solid-phase cartridges, which were then flushed dry using nitrogen gas with final elution of the solutes into CCl$_4$ enabling transport to a flow cell located in the beam path of an FT-IR [37]. The arrangement is shown in Figure 4.7 where it can be seen that the extraction efficiency was enhanced by diluting the eluent stream with water before the solutes were adsorbed on the hydrophobic solid-phase media.

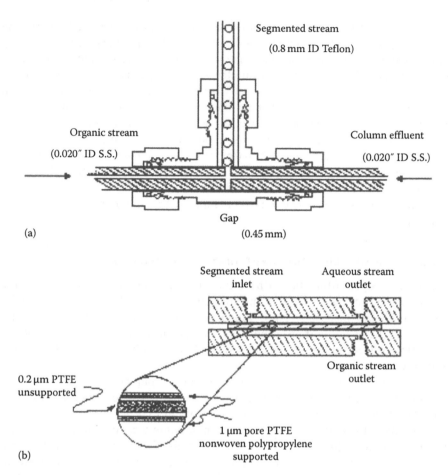

FIGURE 4.6 (a) Diagram of the segmented stream generator manufactured from a Swaglock SS-100-3BT union. (b) Cross-sectional view of the membrane phase separator with an expanded view of the multilayered membrane. The membrane consists of layers of Gore-Tex films with two layers of 0.2 μm pore unsupported PTFE surrounded by single layers of 1 μm pore PTFE with an external nonwoven polypropylene fabric. (Reprinted with permission from Hellgeth, J.W. and Taylor, L.T., *Anal. Chem.*, 59, 295, 1987. Copyright 1987 American Chemical Society.)

4.4.3 SOLVENT ELIMINATION

Elimination of the solvent from the eluent requires that the mobile-phase solvents are more volatile than the solutes and that the solutes can be collected in a way that is suitable for spectral analysis, which can also be referenced to the timeline of the chromatographic separation. It would be expected that volatile solutes are more readily analyzed using gas chromatography. It follows that deposition methods are the primary way that a dissolved solute can be removed from the spectral interferences due to components of the mobile phase. From a spectroscopic perspective, the advantage of analyzing the deposit is that the spectral analysis timescale is no longer

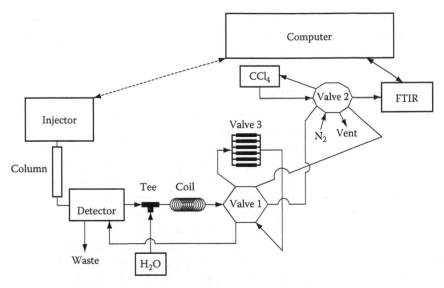

FIGURE 4.7 Diagram of the automated solid-phase extraction system. (Reprinted from *J. Chromatogr.*, 626, DiNunzio, J.E., 97, Copyright 1992, with permission from Elsevier.)

tied to the chromatographic timescale and this enables better quality spectra to be obtained because longer acquisition times can be employed.

Whereas transmission is the mode of IR detection that is most widely used when using a flow cell as the detector, alternative modes of detection are often employed when solvent elimination interfaces are used. A detection configuration based on diffuse reflectance infrared Fourier transform (DRIFT) enables the spectral properties of a solid to be detected. The light that impinges on the sample is partially reflected by the surface and partially passes into the sample where the unabsorbed light may be reflected back out of the solid due to internal reflection. The detected radiation will contain both the surface reflected and reemitted light. In general, a thin layer of analyte is deposited on the outside of powdered substrates which are used to maximize the quantity of unabsorbed light that is reflected back toward the collecting optics.

Another mode of detection involves light passing through a solid and reflected by a highly polished surface that passes the light back through the solid so that the unabsorbed light can be detected after two passes through the sample. This configuration is known as reflection-absorption (R-A) [38].

Many alternative solute deposition modes have been employed and have been gradually increasing in effectiveness and complexity as the field evolves. Some of the ideas that have been applied for interfacing FT-IR to liquid chromatographic systems have an origin in the interfaces used to combine the more mature linking of MS with LC but the different sample requirements has always resulted in some modifications being necessary. MS requires only a small percentage of the total eluent and the interface is optimized to ensure that the solutes are selectively sampled over the solvent molecules in the gas phase but FT-IR requires that the solutes are fully deposited on a substrate in the smallest volume possible. The efficiency of modern interfaces is based on the ability to rapidly remove the volatile solvents from the eluent by increasing the surface area to volume ratio of the emerging droplets by the generation of sprays and the different methods that have been developed will be discussed separately.

4.4.4 EVAPORATION INTERFACES

The early solvent elimination interfaces relied on simple evaporation of the highly volatile mobile phase solvents that are commonly used in NP-LC. The original work of Kuehl and Griffiths [39,40] relied on the eluent being heated in a concentrator tube with a stream of nitrogen flowing through it to remove the majority of the solvent and then having drops of eluent falling into small cups containing powdered KCl suspended on a wire mesh. The remaining solvent was then evaporated using another stream of nitrogen and the carousel of cups containing solutes rotated into the sample compartment for DRIFT detection. The high sensitivity of the DRIFT measurement enabled good spectra to be obtained in sub-μg quantities but atmospheric adsorption of water onto the KCl was found to be a difficult interferent to control. Although the authors thought that the method would be limited to NP and SE-LC, a subsequent refinement of the method based on the online extraction of the solutes into chloroform enabled RP-HPLC to be successfully performed [41]. Expansion of the carousel to 160 cups allowed the full chromatogram to be collected rather than just when the solute peaks eluted as detected using a UV flow cell detector for collection control. The rotating carousel created a "real-time" link between the eluting solutes and the spectral analysis albeit with a slight delay, it was also possible to improve sensitivity by increasing the spectrum acquisition time but only after the complete chromatogram had been deposited. Adaptation of the method for use with microbore columns eliminated the concentrator stage and enabled direct deposition into the cups because of the decreased eluent volumes [42].

The chemical elimination of water using the acid catalyzed reaction of 2,2-dimethoxy propane with water to form more volatile acetone and methanol has been demonstrated for use with microbore columns [43]. Both NP and RP separations with subsequent solvent evaporation and DRIFT analysis was achieved with up to 80% water being present in the mobile phase.

A further example of the simple evaporative interface was demonstrated with solutes being separated in packed microcolumns and then deposited onto IR transparent surfaces with the solvents evaporated by a continuous stream of nitrogen gas. FT-IR transmission spectra were then obtained for the mounted substrates. Solutes separated using NP and SE conditions were deposited on smooth KBr crystals [44–46] and a strip of stainless steel mesh was used for an aqueous-based RP separation [47]. This method is described as the "buffer memory" technique because the complete chromatogram is stored on the deposition substrate before any spectral analysis is undertaken. Although the total chromatograph is collected, the spectral analysis is no longer "on-the-fly" and the technique is considered to be at-line rather than online. The small capacity of the linear deposition substrates was overcome using a rotating 50 mm diameter KBr disk mounted on a stepper motor [48]. Identical drive units were constructed for use during chromatographic deposition and inside the sample compartment of the spectrometer for recording FT-IR spectra.

Dekmezian and Morioka developed an interface that enabled higher boiling solvents to be evaporated from polymer solutes after they were separated using GP chromatography [49]. The deposition substrate was placed in a vacuum oven and the temperature and pressure were optimized to ensure that the 1,2,4-trichlorobenzene (TCB) (BP 213°C), which was used as the solvent, flash evaporated from the eluent droplet as it passed to the KBr collecting dish.

The addition of a microdispenser to the outlet of a flow splitter connected to a microbore column has demonstrated that reducing the size of the droplet volume to 60 pL results in natural evaporation of the aqueous 1% acetic acid:15% methanol mobile phase occurring quick enough that deposits of 50 μm diameter could be collected [50]. The geometry of the flow path and the controlled piezoelectric element are shown in Figure 4.8.

A deposit consisted of approximated 30 repeated drops and the size of these deposits is compatible with the diameter of the focused beam of an IR microscope. Considering that only 0.01% of the eluent enters the microdispenser with 0.001% of the actual eluent being deposited on the CaF_2 substrate, good reproducibility for the FT-IR measurements was found. The interface was effective under both isocratic and gradient conditions; however, it was observed that proper degassing of the solvents was necessary to avoid the formation of air bubbles during the gradient mixing of water with organic solvents that interfere with the metering and ejection of the droplets.

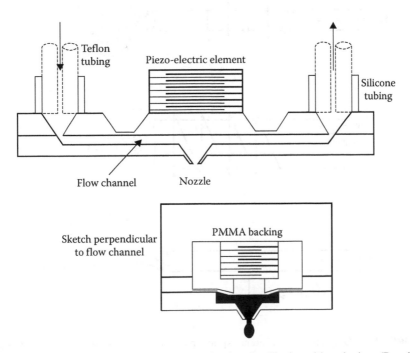

FIGURE 4.8 Diagram of the piezo-actuated microdroplet deposition device. (Reprinted from *J. Chromatogr. A*, 1080, Surowiec, I., Baena, J.R., Frank, J., Laurell, T., Nilsson, J., Trojanowicz, M., and Lendl, B., 132, Copyright 2005, with permission from Elsevier.)

4.4.5 THERMOSPRAY INTERFACES

The desire to rapidly vaporize the volume of eluent produced from analytical-sized columns and being able to remove water, which has a high heat capacity, has led to the development of more sophisticated elimination interfaces. Thermospray is one method that is used to increase the efficiency of solvent elimination and involves passing the eluent though a heated narrow bore tube so that the majority of the solvent vaporizes. The expansion within the tube results in an acceleration of the remaining liquid out of the tube as a fine mist. The spray characteristics of a thermospray along with other nebulizer designs [51] are shown in Figure 4.9 and it can be seen that a divergent spray emerges that is much larger than the diameter of the internal diameter of the tube from where it originated for the thermospray design.

Jansen developed a moving belt interface where the eluent passed through a heated stainless steel capillary to generate a thermospray that deposited solutes of polymers and additives on the surface of a stainless steel belt as demonstrated in Figure 4.10 [52]. Further evaporation was possible using an infrared heater and the deposited solutes then entered the sample compartment of an FT-IR spectrometer where diffuse reflectance spectra were obtained from a sampling spot with a

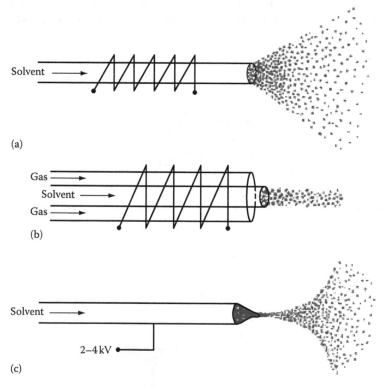

FIGURE 4.9 Diagram of different spray patterns for various nebulizer designs. (a) Thermospray, (b) concentric flow, and (c) electrospray. (From Raynor, M.W., Bartle, K.D., and Cook, B.W.: *J. High Resolut. Chromatogr.* 1992. 15. 361. Copyright Wiley-VCH Verlag GmbH & Co. KGaA. Reproduced with permission.)

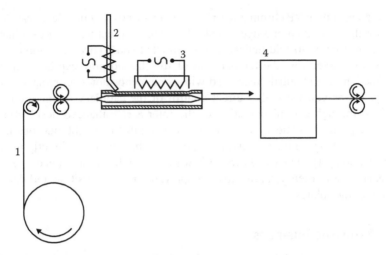

FIGURE 4.10 Diagram of the moving belt LC-FT-IR interface that comprises (1) moving stainless steel tape; (2) heated outlet of liquid chromatograph; (3) infrared lamp; (4) diffuse reflectance cell mounted in the sample compartment of the FT-IR spectrometer. (With kind permission from Springer Science+Business Media: *Fresen. J. Anal. Chem.*, 337, 1990, 398, Jansen, J.A.J.)

2 mm diameter. GP chromatography enabled the separation of a mixture containing a polystyrene calibration standard, unsaturated polyester, and polymethylmethacrylate with clearly identifiable spectra to be obtained from injected quantities of 20–80 μg of sample in dichloromethane. Further polymer examples were demonstrated and it was noted that the temperature of the thermospray capillary should be optimized because if it is too high, then the more volatile low-molecular-weight components also vaporize, but if the temperature is too low then the solvent condenses on the belt and separation resolution decreases due to excessive spreading of the deposit. A final example using a RP separation of some polymer stabilizer additives demonstrated that aqueous mobile phases could be used and good spectra were obtained for sample quantities of 100 μg.

The thermospray moving belt interface has also been employed by Robertson and coworkers for the detection of RP separated amino acids [53]; saccharides, carboxylic acids and identify the antioxidant Irganox 565 [54]; and a complex industrial sample containing polyadipate plasticizers [55]. Mottaleb and coworkers used a slightly modified version of the Robertson design, where an external flow of heated nitrogen gas envelopes the spray as it passes out of the small orifice, for the detection of polystyrene solutes separated using GP chromatography [56]. A further application involving the RP separation of phenolic antioxidants was demonstrated [57]. The effect of mobile-phase modifiers used for the RP separation of alkylbenzene sulfonates was investigated and it was found that although the addition of 10 g L^{-1} NaClO$_4$ enabled better chromatography, it interfered with the operation of the thermospray deposition process. The use of more volatile acetic acid at a concentration of 50 mg L^{-1} enabled adequate separation and allowed the deposition interface to operate unimpeded so that FT-IR spectra could be obtained [58,59]. The viability of using gradient elution RP chromatography with a thermospray interface was demonstrated for the separation of a dye sample [60]. The gradient was a linear change in composition from 90:10 acidified water:acetonitrile to 80:20 over a period of 10 min and it was observed that the temperature of the thermospray capillary increased as the fraction of acetonitrile increased with some deposition occurring within the capillary. It was concluded that direct control of the capillary temperature could be avoided through careful operation of the interface. Enhanced desolvation and improved deposition efficiency was obtained using a further slight modification that increased the nitrogen flow rate and incorporated a heating plate directly under the deposition point [61]. Increased nitrogen flow improved the desolvation of the solutes which enabled lower temperatures to be employed for the deposition and decreased spreading of the solutes.

4.4.6 PNEUMATIC INTERFACES

The basic operating principle of a pneumatic nebulizer involves the mixing of a liquid with a pressurized gas so that a jet of spray results when they are released through a small opening. Gagel and Biemann pioneered the development of a pneumatic nebulizer as a solvent elimination interface for use with NP chromatography and FT-IR detection with a R-A sampling configuration [38]. The original design (Figure 4.11) had the LC eluent mixed with pressurized nitrogen in a high pressure tee piece so that the eluent emerged as a fine spray on to the reflective surface of a mirror disk. A gentle stream of nitrogen across the surface of the mirror was also employed to rapidly remove any remaining solvent. After the chromatogram was deposited, the mirror was placed in the sample compartment of an FT-IR equipped with an R-A accessory. A silica-packed microbore column was used to separate a mixture of polycyclic aromatic hydrocarbons using a mobile phase containing hexane and dichloromethane. Wavelength-selective chromatograms and complete spectra of individual components were obtained which compared favorably with reference spectra. An improved

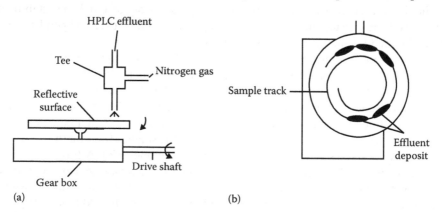

FIGURE 4.11 Diagram of the continuous collection device showing (a) the deposition of the HPLC eluent and (b) the view of the deposited components. (Reprinted with permission from Gagel, J.J. and Biemann, K., *Anal. Chem.*, 58, 2184, 1986. Copyright 1986 American Chemical Society.)

design where a heated stream of nitrogen gas forms a sheath around the spray from the nebulizer was demonstrated to be effective enough that applications based on RP isocratic and gradient elution separations could be developed [62]. An example based on the separation of positional isomers of five naphthalenediols (~500 ng injected) where the mobile phase was 55% water with methanol demonstrated how this low volatility solvent could be easily removed. A further separation example requiring the use of a gradient where the water composition was reduced from 55% to 5% over a 30 min period was presented but it was necessary to simultaneously control the gas temperature so that a constant vaporization rate could be maintained. Although as little as 16 ng injected could be detected, it was found that good-quality qualitative data required 31 ng of sample to be injected. It was discovered that spectral abnormalities due to the use of a reflective aluminum disk was overcome by changing the deposition substrate to an aluminum back coated germanium disk [63].

A commercialized version of this interface is available from Lab Connections (Carrboro, NC, USA) [64]. Various instrument configurations based on a collection module, an optics module, and appropriate collection media can be assembled depending on specific demands. The LC-Transform instruments have been used for applications such as the separation of polymers [65–68], polymer additives [69], complex mixtures using RP-HPLC [70,71], and supercritical fluid chromatography [72]. An in-house modification of an LC-Transform instrument was made so that the low flow (1–10 μL min⁻¹) from a packed capillary column could be nebulized more efficiently by reducing the amount of dead volume that was a problem with the original commercial design [73]. Temperature programming was used for the characterization of the commercially manufactured antioxidant Irgafos P-EPQ that contains a mixture of isomers with the same molecular mass, which makes FT-IR the preferred identification technique. A mass sensitivity of 40 ng was determined using the R-A sampling configuration and it was speculated that the sensitivity could be improved by a factor of 5 by changing the FT-IR detector from a deuterated triglycerine sulfate sensor to a mercury cadmium telluride. The variation in temperature of the eluent did not alter the characteristics of the deposition interface and it was proposed that a temperature gradient is more easily handled than a solvent gradient when using a pneumatic nebulizer.

An early example of a pneumatic nebulizer was also investigated by Kalasinsky and coworkers during a comparison of various solvent elimination configurations that also included the use of an eluent concentrator or a flow splitter and the chemical elimination of water technique that has already been described [74]. They were able to demonstrate that the eluent from both NP and RP analytical columns could be deposited with a similar sensitivity to other methods available at the time.

4.4.7 Concentric Flow Interfaces

Recognition that the deposition characteristics of the thermospray and simple pneumatic nebulizer generated a spray pattern that dispersed the solvent stream over a wide area lead to the application of the concentric flow nebulizer for FT-IR detection of RP-separated solutes [75]. The concentric flow nebulizer is shown as diagram B in Figure 4.9 and consists of two tubes where the inner tube contains a liquid stream and a gas flows between the walls of the two tubes. The flow rate of the outer gas sheath is much greater than the liquid flow rate and the momentum imparted to the liquid in the region where the two streams meet causes the liquid to break into smaller droplets. Heating the gas increases the vaporization rate of the liquid and prevents dispersion of the liquid spray. When the liquid is the eluent from a chromatography column, the solvent vaporizes and the less volatile solutes can be deposited on a suitable substrate. Therefore, the concentric flow nebulizer could be considered to be a combination of thermospray and pneumatic nebulizers because the inner capillary can be heated to temperatures above the boiling point of the liquid and the pressurized gas flow in the outer capillary produces a pressure gradient that disturbs the flow of liquid.

In a series of applications by Griffiths and coworkers [75–79], they described a rudimentary design where a stream of heated nitrogen passed between two fused silica tubes with identifiable spectra for deposits of 5 ng of aromatic quinones being detected [76]. A more comprehensive design (Figure 4.12) where helium is used as the gas due to its higher thermal conductivity and the interface was mounted in a vacuum chamber to increase vaporization efficiency was developed for RP separations using a microbore column [75,77]. An identification limit for an injection of 840 pg methyl violet 2B was determined using a flow injection deposition procedure where spots 78 × 81 μm were detected using transmission FT-IR microscopy.

A further refinement was to place the deposition interface directly within the sample compartment of an FT-IR microscope so that spectra could be directly obtained as the rotating ZnSe substrate passed from the deposition point to then be located in the 100 μm beam path of the spectrometer [78]. The effectiveness of using the online interface with buffered solutions was demonstrated for RP separations with a microbore column [79]. When a nonvolatile buffer such as phosphate was used, it was necessary to use background subtraction to remove spectral interference due to the buffer. This was not the case when low concentrations of ammonium acetate was used as the buffer. Identification limits in the low ng range were obtained.

In a series of investigations by Somsen and coworkers [80–87], a deposition interface containing a concentric flow nebulizer was developed as a modification of a device that had previously been used to interface LC with thin layer chromatography [88]. The spray jet interface was initially applied to an RP separation of a test mixture of polyaromatic hydrocarbons using a narrow-bore column with a flow rate of 20 μL min⁻¹ [80]. The eluent from the column was fed through a 50 μm internal diameter (ID) fused silica capillary into a 100 μm ID stainless steel syringe needle. The needle protruded from a concentric tube that had a flow of heated nitrogen that surrounded the needle and served both to heat the liquid eluent and also enhanced solvent vaporization of the spray that exited from the needle. The deposits were collected on a ZnSe substrate mounted on a computer-controlled microscope stage and the substrate was then transferred to an FT-IR microscope where the IR spectra of 13 ng pyrene could be identified in a 10 μL injection. The separation of a test mixture of quinones proved that equivalent results could also be achieved that enabled comparison to alternative methods that had been previously used for these compounds. This utility of this arrangement was demonstrated during the separation of impurities in a steroid drug where spectroscopic analysis enabled the contaminants to be

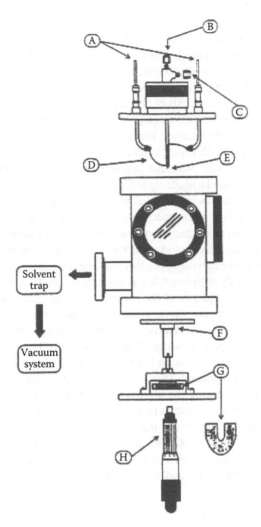

FIGURE 4.12 Exploded view of the concentric flow nebulizer interface that consists of (A) electrical feed-throughs; (B) inlet for chromatography eluent; (C) inlet for helium gas; (D) chromel heating wire; (E) concentric fused silica tubes; (F) rotatable stage for solute deposition; (G) horseshoe magnet and bar magnet used for stage rotation; and (H) micrometer for vertical stage positioning. (Reprinted with permission from Lange, A.J., Griffiths, P.R., and Fraser, D.J.J., *Anal. Chem.*, 63, 782, 1991. Copyright 1991 American Chemical Society.)

identified [81] and for the analysis of a chlorinated pyrene sample where individual isomers and congeners were characterized [82]. Application to the RP separation of polymer additives and the SE separation of polystyrene oligomers demonstrated the versatility of their deposition interface [84]. The discovery that higher eluent flow rates were possible for the SE separations (i.e., 200 μL min^{-1} vs. 20 μL min^{-1} for RP) resulted in the inclusion of a liquid–liquid extraction (LLE) module between the column and the spray jet interface for the RP separation of two test mixtures, one containing quinones and the other herbicides [85]. The performance of the LLE separator, with a groove volume of 8 μL, had been previously demonstrated to have an extraction efficiency of 1.0 when using 1,2-dichloroethane as the extracting solvent at a flow rate of 0.2 mL min^{-1} [89]. It was found to be just as effective in this situation with dichloromethane used as the extraction solvent. Online extraction also enabled the use of nonvolatile phosphate buffers that have interfered with other deposition interfaces [57,79]. The trace analysis of herbicides in river water with the addition of a pre-column solid-phase extraction stage greatly improved the concentration sensitivity so that 1 μg L^{-1} could be detected from a 50–100 mL sample of river water [86]. The problems associated with using solvent gradient elution for RP separations with solvent elimination interfaces was also investigated using the same spray jet interface [87]. The approach used a make-up solvent with a flow rate of 20 μL min^{-1} added at a tee piece to an eluent flow rate of 2 μL min^{-1} that carried solutes separated using a capillary column. The programmed eluent composition was varied from 68:32 water:methanol to 14:86 water:methanol with the make-up solvent being methanol and because the varying water component was only a small percentage of the total liquid passing through the nebulizer, the thermal and flow characteristics of the interface could be more readily optimized. It was also again observed that the morphology of the deposit impacts the quality of the spectra obtained with smooth and uniform being better than open and granular deposits as had been previously reported [80,81].

Geiger and coworkers have adapted the Somsen design just described for the detection of terpenoids separated using RP conditions [90]. They were able to employ a solvent gradient for their separation but found that optimization of the deposition process was a difficult balance between overdrying the solutes that then clog the inner capillary and overheating the already deposited spots that then vaporized to be deposited in cooler regions that were not in the normal deposition line. However, they did successfully identify two of the seven detected reaction products between ozone and α-pinene using FT-IR with the expectation that further analysis using alternative detectors such as NMR and MS would provide complementary information that would enable the total identification of all components.

4.4.8 ULTRASONIC INTERFACES

Thermal degradation is possible with thermospray, concentric flow, and pneumatic nebulizers if excessive heating is used to form an aerosol fine enough to be deposited without excessive spreading on the substrate. A solvent elimination approach that avoids the thermal degradation of solutes relies on the vibration of a transducer operating at ultrasonic frequencies to disperse any liquid that it contacts into a fine mist.

The introduction of the ultrasonic nebulizer to an LC-FT-IR interface was based on the need to replace KCl as the deposition substrate so that IR spectroscopy could be used with RP separations [91]. Diamond powder was chosen to replace KCl but it was easily disturbed by eluent droplets which resulted in matrix compensation errors when using DRIFT detection. This problem was overcome through the use of the ultrasonic nebulizer to break up the eluent and the application of a low vacuum to facilitate solvent vaporization. A moving trough combined with a modified optics bench enabled online detection of the separated test mixture of two nitroaromatic compounds eluting from a microbore column.

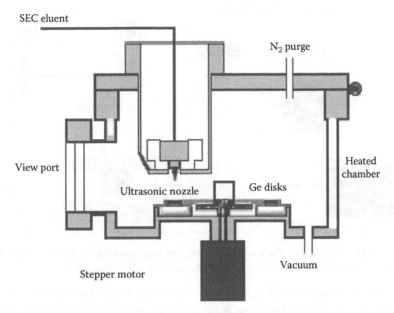

FIGURE 4.13 Diagram of the ultrasonic nebulizer used for the collection of polymer solutes after SE chromatography. (Reprinted from *J. Chromatogr. A*, 910, Torabi, K., Karami, A., Balke, S.T., and Schunk, T.C., 19, Copyright 2001, with permission from Elsevier; Reprinted from *J. Chromatogr. A*, 911, Karami, A., Balke, S.T., and Schunk, T.C., 27, Copyright 2001, with permission from Elsevier.)

An ultrasonic nebulizer [92] replaced the simple droplet tube used in an earlier solvent elimination interface developed by Dekmezian and coworkers [49]. The modification produced smaller droplets of eluent which improved the efficiency of the solvent vaporization that was controlled by altering the temperature and pressure of the collection chamber. A noted advantage over pneumatic devices is that "bounce back" due to the high-velocity jet generated by the pressurized liquid is prevented because the ultrasonic nozzle produces a low-velocity spray. They investigated the molecular weight distribution of polyester block copolymers using GP chromatography with TCB as the solvent. Micrometer thick spots of polymer deposits with diameters of 4–6 mm were obtained using this interface.

Schunk and coworkers used a similar collection device (Figure 4.13) for the SE separation of polystyrene and polymethylmethacrylate blends using tetrahydrofuran as the solvent [93,94]. After collection of the polymer deposits on the surface of polished germanium disks, the collection wheel was exposed to dichloromethane vapor which improved the spectral baseline because annealing of the sample decreased IR scatter.

The RP separation of water-soluble vitamins in multivitamin tablets with online FT-IR detection was achieved using the interface shown in Figure 4.14 [95]. Part of the eluent from the chromatographic column passes over an ultrasonic nebulizer and is carried by helium gas down a heated drift tube to be deposited on a ZnSe disk located within a vacuum chamber where the transmission FT-IR spectra are continuously recorded. The separation was achieved using gradient elution with a mobile

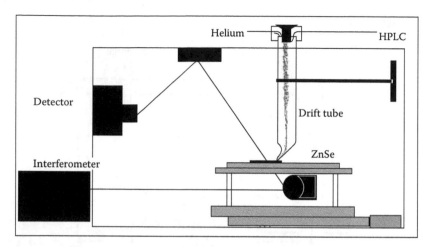

FIGURE 4.14 Diagram of the LC-FT-IR interface incorporating an ultrasonic nebulizer. (From Li, Y. and Brown, P.R., *J. Liq. Chrom. Relat. Tech.*, 26, 1769, 2003.)

phase contained ammonium acetate buffer and methanol. The vitamins could be identified based on either their functional group characteristics or from the use of a spectral library.

4.4.9 ELECTROSPRAY INTERFACES

A common interface used with MS because it results in the formation of desolvated charged molecules has also found use with IR spectroscopy [51]. Nebulization of a liquid jet occurs when the solution containing charged molecules experiences a high electric field so that the combined effects of solvent evaporation and coulombic repulsion forces disintegrate the liquid into a fine mist of charged molecules and a schematic of this process is given in Figure 4.15 [51]. The only description where the electrospray interface has been coupled with LC was for the RP separation of caffeine and barbital using packed microcolumns with an aqueous methanol mobile phase at a flow rate of $4\,\mu L\ min^{-1}$. An IR microscope with an aperture of $100\,\mu m$ was used to examine deposited spots that were $500\,\mu m$ in diameter, ideally these two diameters should match but deposit nonuniformity allowed for some clean substrate areas to dilute the signal. It was found that excessive spreading of the spray, which is exacerbated by the coulombic repulsion between particles, could be minimized by earthing the semiconducting ZnSe collection substrates. This also improved the stability of the electrospray because a direct conduction path between the nozzle and an earthing point is established rather than relying on randomly fluctuating voltage leakage processes.

4.4.10 PARTICLE BEAM INTERFACES

An interface that also originated for interfacing LC to MS that does not generate charged molecules and operates at room temperature is based on using a gas flow to disrupt a liquid jet that forms as the liquid is forced through an orifice with a reduced diameter.

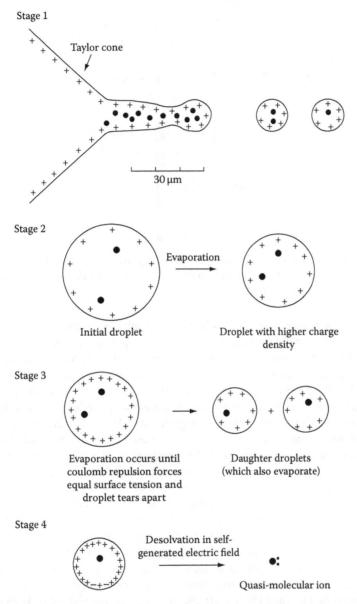

FIGURE 4.15 Schematic illustration of the four stages of electrospray nebulization: (1) Droplet formation from a liquid cone; (2) evaporation of the solvent producing droplets with higher charge density; (3) desolvation and formation of daughter droplets; and (4) further desolvation and molecular ion formation. (From Raynor, M.W., Bartle, K.D., and Cook, B.W.: *J. High Resolut. Chromatogr.* 1992. 15. 361. Copyright Wiley-VCH Verlag GmbH & Co. KGaA. Reproduced with permission.)

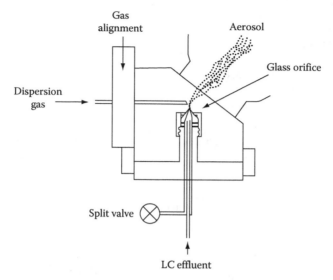

FIGURE 4.16   A schematic of the monodisperse aerosol generator (MAG). (Reprinted with permission from Willoughby, R.C. and Browner, R.F., *Anal. Chem.*, 56, 2626, 1984. Copyright 1984 American Chemical Society.)

The flows are arranged as in Figure 4.16 and the result is an essentially monodispersed aerosol where the size of the particles is primarily dependent on the size of the orifice and the gas flow disperses the particles to prevent coagulation [96]. The increase in surface area during aerosol formation increases the evaporation rate of the solvent which is stripped in the desolvation chamber and the particles are then transported through a two-stage aerosol beam separator where vacuum pumps reduce the pressure so that the particles can directly enter the sampling chamber. In a MS interface, the sampling chamber would contain an ion source but when used in conjunction with FT-IR, the particles deposit on a suitable substrate for transferal to an FT-IR spectrometer.

The inventors of this interface called it the monodispersed aerosol generation interface or MAGIC for use with MS [96] with de Haseth and coworkers developing a system suited for use with FT-IR [97–104]. An initial demonstration of the interface indicated that only about 10% of a 100 ng injection of methyl red was deposited and then detected using an FT-IR microscope [97]. It was subsequently demonstrated that a mobile phase containing 100% water could be removed for flow rates up to 0.3 mL min⁻¹ [98]. Detection of solutes in the presence of buffer components was investigated and it was demonstrated for the first time that identifiable spectra could be obtained [100]. Caffeine was used as the test solute because of its good solubility in RP solvents and very characteristic spectrum. It was observed that when ammonium acetate was used as the buffer, there were no spectral features due to the buffer for an injection of 26 μg of caffeine but interference was obvious when the quantity injected was decreased by a factor of 10. There was no evidence of ammonium acetate deposition in the absence of caffeine, which prevented proper background correction of the buffer spectrum. The use of nonvolatile buffers potassium hydrogen phthalate and potassium dihydrogen phosphate resulted in significant spectral contributions from the co-deposited buffer for injections of 130 μg of caffeine and background subtraction was effective for this amount of solute. However, background mismatch greatly distorted the FT-IR spectrum upon a 10-fold dilution of the caffeine because of the dominance of the buffer contribution to the spectrum.

The soft desolvation process that occurs during the deposition of solutes using the MAGIC technique has found great utility for the FT-IR study of proteins [101–104] with the solid state spectra strongly resembling solution structures [101]. FT-IR can be used to probe the secondary structure of proteins due to the sensitivity of the position of the amide I and II absorbance bands to conformational changes. A study of the influence of chromatographic conditions on protein conformation showed that column effects, nonaqueous solvents, and pressure produced structural changes that could be reversed with subsequent redissolution and evaporation of the deposited protein [102]. Confirmation that the degree of secondary structural integrity influences chromatographic separations was obtained by examining the effect of chemically induced conformational changes for a native globular protein on the retention time of the separated components [103]. Identification of each modified structure demonstrated that denatured proteins have longer retention times because of greater interactions between the stationary phase and the unfolded protein. In conjunction with mass spectral analysis, the complete structure of β-lactoglobulin was determined by identification of functional groups from peptide fragments formed after digestion of the protein [104].

The MAGIC-FTIR interface has been used with common NP and RP solvents, which also includes buffered solutions, and could be considered as a soft solvent elimination process as demonstrated by the deposition of proteins. However, there has been limited wider development of the particle beam interface and a complete chromatogram has never been recorded at-line using a moving substrate.

4.5 INTERFACING LIQUID CHROMATOGRAPHY WITH RAMAN DETECTION

In comparison to the development of hyphenated IR spectroscopy and LC techniques, there are relatively fewer attempts at utilizing Raman spectroscopy for the detection of solutes eluting from LC columns. The principal disadvantages are the poor detection limits associated with normal Raman detection and the reproducibility of surface enhanced Raman strategies because of the difficulties with creating a stable surface where the increased sensitivity is not strongly influenced uncontrollable environmental factors. A comprehensive review of the application of Raman spectroscopy as a detection method for liquid-separation techniques has recently been published [3], which also encompasses thin-layer chromatography as well as capillary electrophoresis.

4.5.1 Flow Cells

The great advantage of flow cells when employing Raman detection compared to IR spectroscopy is that the absorbance of the window material is much less of a design consideration because the excitation wavelength of the laser sources that are generally available range between 244 and 1064 nm, which enables glass and quartz to be used.

The first reported attempt that combined Raman spectroscopy detection with HPLC involved an application that demonstrated the NP separation of the three structural isomers of nitroaniline by Chapput and coworkers [105] in 1980. Single wavelength detection was used and it was demonstrated that changing this wavelength by only $10\,cm^{-1}$ dramatically altered the observed chromatogram because of the narrow bandwidths of the Raman peaks. The separation of benzene and anthracene required monitoring at two different wavelengths.

Raman spectra for the individual solutes were obtained using a multichannel detector for the NP separation of azoxybenzenes directly monitored in a flow cell [106]. The identity of two separated isomers was established by comparing the spectra of the isomers using two different excitation wavelengths, one that was "off-resonance" and another that was near resonance. Slight spectral changes due to conjugational interactions enabled the isomers to be differentiated. Millimolar detection limits for the injected sample were reported.

A flow cell with quartz windows, where the inner surface of the measuring chamber was highly polished to ensure maximum collection of the Raman scattered photons, was used for the detection of toluene under RP conditions [107]. The use of a single channel, triple monochromator limited the detection to one wavelength during the chromatographic run and spectra could only be obtained for stationary liquids. The limit of detection (LOD) for toluene was measured to be $480\,mg\,L^{-1}$ for a $20\,\mu L$ injection.

Raman microspectroscopy was used for the detection of a group of nitro-compounds eluting under RP chromatographic conditions using deuterated solvents [108]. Fiber optic coupling and a multichannel CCD detector with a dispersive grating enabled real-time normal Raman spectra to be continuously recorded as shown in Figure 4.17. A micro-LC system using a packed capillary column with a flow rate of $2\,\mu L\,min^{-1}$ enabled the use of expensive deuterated solvents to minimize the spectral overlap between the mobile phase solvents and the solutes. Good detector linearity for the chromatographic peak area was observed for injected masses between 75 and 7000 ng of nitrobenzene with a sample loop of 60 nL being employed.

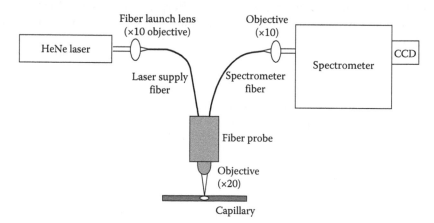

FIGURE 4.17 A schematic showing the fiber-optic probe and exposed capillary flow cell. (From Cooper, S.D., Robson, M.M., Batchelder, D.N., and Bartle, K.D., *Chromatographia*, 44, 257, 1997.)

4.5.2 WAVEGUIDES

Improving sensitivity by increasing the effective optical pathlength by using liquid core waveguides (LCW) has been demonstrated to be effective in circumstances where the solutes do not exhibit resonance-enhanced responses [109–112]. The most common application for the use a liquid core waveguide for Raman detection in LC separations involves the use of Teflon AF tubing and predominately aqueous mobile phases. The RI of the Teflon tube is lower than the RI of the mostly aqueous mobile phase and any light with an incident angle greater than or equal to the critical angle will be totally internally reflected along the length of the inner core of the tube. The light is then collected as it leaves the tube with appropriately matched optics and passes into a spectrometer. Improvement in detection limits by a factor of 1000 can be obtained [111]. A range of internal diameter tubes are available where the length of the tube is optimized so that the total cell volume is compatible with microbore and conventional LC systems to avoid peak band broadening but ensuring the maximum possible pathlength.

After an initial investigation where UV absorption was used in conjunction with a plastic LCW to produce a 30–50 fold improvement in detectability for conventional RP HPLC of some pesticides [113], a preliminary study of Raman active nitroaromatics was undertaken [109]. A 30 cm LCW was used as the flow cell and some spectral responses were observed for sample concentrations of 8 mg mL^{-1}. To further improve the detection limit, a large volume injection method based on on-column focusing [114] was employed and spectra were obtained for all components for concentrations of 1–3 mg mL^{-1} and a LOD of 10 ug mL^{-1} was determined for 4-nitroaniline. A subsequent more detailed study demonstrated that the use of organic modifiers such as methanol, acetonitrile, and acetone all reduced the free spectral window due to significant scattering but no scattering interference from water occurs, the implication is that separation techniques with predominately aqueous mobile phases (e.g., IE and SE) should benefit from LCW detection [112]. It was further demonstrated that for greater efficiency with longer wavelength lasers, it is necessary to use deuterated water due to the increased reabsorption of laser light by water as the excitation wavelength increases. Nucleotide concentration detection limits of 10 ug mL^{-1} were determined but this was only possible because on-column sample focusing and stopped-flow detection were used to enhance the signal intensity. To further enhance detectability, a sophisticated background subtraction method has been developed [115] and was applied to overcome the spectral contribution of the mobile phase to the normal Raman spectroscopic detection using a LCW flow cell [115]. It was demonstrated that the eluent background subtraction (EBS) method was able to accommodate slight variations in peak position and intensity when identifying the spectral features of each component and this improved the accuracy of eluent spectra subtraction from the analytes.

Marquardt and coworkers developed an LCW system which was interfaced to a Raman microscope and diagrams that show the instrument configuration and the design of the flow cell interface are given in Figure 4.18 [110,111]. The system was comprised of a 785 nm diode laser for excitation, a Teflon AF® LCW that was 1 m long with a 50 µm core diameter, and the microscope incorporated a stainless steel cell

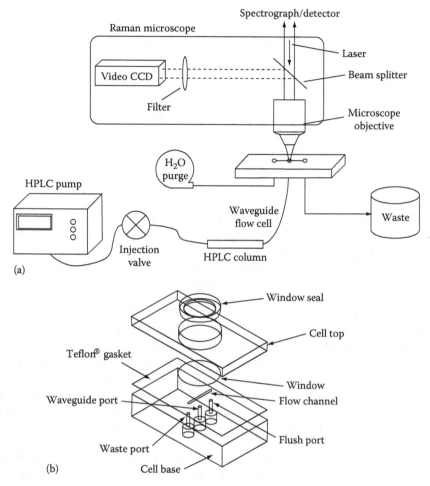

FIGURE 4.18   (a) A diagram of the LCW-RS detection system and (b) flow cell. (Reprinted with permission from Marquardt, B.J., Vahey, P.G., Synovec, R.E., and Burgess, L.W., *Anal. Chem.*, 71, 4808, 1999. Copyright 1999 American Chemical Society.)

block with a 250 µm thick fused silica window. In one application, they demonstrated a LOD of 6 µg mL^{-1} for benzene using flow injection analysis as well as the separation of a BTEX test mixture comprising benzene, toluene, ethyl benzene, *o*-xylene, *m*-xylene, and *p*-xylene using 50:50 water:acetonitrile on a C8 minibore column [110]. Although complete chromatographic separation of the xylenes was not achieved, they were able to be distinguished based on the individual spectral features, which demonstrated the greatest perceived advantage from the coupling of spectroscopic techniques with chromatography. In another application, this was further demonstrated for RP-HPLC of alcohols where an information-rich multivariate data matrix was obtained and individual analytes could be distinguished based on their spectral properties even though less than optimal chromatographic separations were achieved [111]. Further resolution was obtained using a piecewise multivariate curve resolution technique [116].

4.5.3 ENHANCED METHODS

As described earlier, significant signal enhancement can be achieved when resonance Raman effects occur and access to a laser excitation wavelength that matches the electronic absorption band of a chromophore has been a major restriction. Generally, the most widely available lasers have operated in the visible region but these will only be useful for molecules such as long-chain polyenes, porphyrins, and transition metal complexes that have strong visible charge transfer transitions [5]. The use of UV wavelengths for excitation has the advantage that most organic molecules absorb in this region and fluorescence processes are greatly diminished. Alternatively, surface-enhanced resonance can be obtained when an analyte interacts with an appropriate substrate and applications with detection in flowing cells have been developed.

An early example where a flow cell was used and a RR signal was measured involved the detection of two hemoproteins using the 488.0 nm line from an argon laser [117]. Pre-column derivatization of aliphatic amines with dabsyl chloride produces compounds that exhibit absorption maxima at about 440 nm with large molar absorptivities and excitation at 488.0 nm generates RR scattering [118]. Separation of the derivatized amines under RP conditions on a microbore column with single wavelength detection at the RR signal maxima generated a chromatogram where a LOD of 5 ug mL^{-1} was obtained for the amides. A spectrum was obtained by scanning the monochromator when the flow had been stopped.

With an understanding that the use of UV-RR has the potential to greatly increase the variety of molecules that could be directly analyzed using a flow cell configuration, a RR study of two azo dyes was undertaken in conjunction with an argon ion laser [119]. A fiber optic probe connected to a dispersive multichannel array detector enabled a spectrum to be collected every 4 s with an integration time of 1 s and the spectra were corrected for wavelength-dependent absorption that distorts the spectra. The effect of the changing solute concentration on the effective pathlength of the cell also alters the relationship between the scattering intensity and solute concentration with severe nonlinearity being observed. The use of an in-line absorption spectrograph that measured the absorption spectra between the acquisition of each Raman spectrum allowed for the direct correction of each spectrum. To test the system under flowing conditions, an inert packed column containing sand was used and therefore no separation of injected solutes was possible. Each azo dye was investigated separately with a detection limit of 2 µg mL^{-1} obtained. A mixed dye sample was also injected and it was possible to differentiate the spectral components based on fitting reference spectra obtained on the individual solutes.

The use of UV-RR detection coupled to the RP separation of environmentally important polyaromatic hydrocarbons (PAHs) using a tunable laser at 230 nm was described in a review by Asher and coworkers [5]. The advantage of using a tunable UV excitation source for RR to increase sensitivity was demonstrated and greater selectivity because enhancement only occurs for the Raman bands of the chromophore at resonance was discussed.

A further UV-RR study of a group of PAHs was undertaken in a conventional Z-shaped flow cell [120]. The effect of laser power was investigated because photodegradation can be problematic due to possible chemical transformation of the solutes when exposed to shorter wavelength UV light. It was observed that using higher powers and shorter exposure times was optimal because increasing the residence time in the detection cell resulted in significant decomposition. Compared to the usual practice where the S/N ratio is improved by longer exposure times obtained with slower flow rates or stopped flow techniques, the flowing eluent was considered an advantage in minimizing decomposition. Resonance enhancement and on-column focusing resulted in a LOD of 50–200 ng mL^{-1} but quantitation using this method was discouraged because the Raman signal intensity only varied slightly with changes in concentration. It was recommended that an in-line UV absorbance detector should be used for quantitation with the Raman spectra used for identification.

The first attempt at obtaining SERS enhancement was made by Freeman and coworkers who demonstrated that the post-column mixing of a silver sol containing metal-anion cluster particles less than 200 nm with the effluent from an LC column generated detectable Raman spectra in a flow cell [121]. A single channel detector restricted detection to a single wavelength and spectra were obtained under stopped flow conditions. The triphenyl methane dye derivative, pararosaniline hydrochloride, was detected with good reproducibility (~1% variation for 5 μg mL^{-1}) and demonstrated linearity over three orders of magnitude (0.1–50 μg mL^{-1}) with a LOD of 0.1 μg mL^{-1} for measurements where the column was bypassed. With the column in-line, spectra were obtained for 10 μg mL^{-1} of sample in both aqueous and methanol:water solutions with obvious bands due to methanol being detected in the later solvent.

The post-column mixing of the eluent with a silver sol was also used in a preliminary study based on flow injection analysis for the real-time determination of purine bases and demonstrated the viability of developing a chromatographic technique with increased sensitivity due to SERS [122]. An advantage of using a silver sol is that a fresh silver surface is continuously available and temperature control was used to obtain a high degree of stability. The purine bases were separated using an aqueous buffer mobile phase under RP conditions on an analytical C18 column with excellent three-dimensional chromatograms obtained for a solution containing 14–140 μg mL^{-1} of the four bases [123]. The flow cell was a glass capillary but it was found that the surfaces required cleaning with 6 M HNO$_3$ to remove the silver sol that deposited during each chromatographic run.

To overcome the memory effect observed due to silver deposition on surface of the windows when using a flow cell [123], a windowless flow cell for SERS detection of some drugs banned in sport was investigated with post-column mixing of the eluent with a silver colloid [124,125]. The windowless flow cell is shown in Figure 4.19 and is constructed from the alignment of two stainless steel tubes vertically mounted in a aluminum block with a spacing of 1.6 mm so that the liquid flows from the top tube into the bottom tube. The column of liquid that forms in the gap is supported by surface tension. Raman detection was based on a single-channel monochromator, which restricted detection to a single wavenumber and because Raman bands are much narrower, not all the separated solutes were detected, whereas they all could be detected based on their UV absorbances. The LOD for two drugs were determined as 1.6 μg mL^{-1} for pemoline and 5 μg mL^{-1} for 2-mercaptopyridine, which are much higher than the limits obtained using UV absorbance [124]. The follow-up study

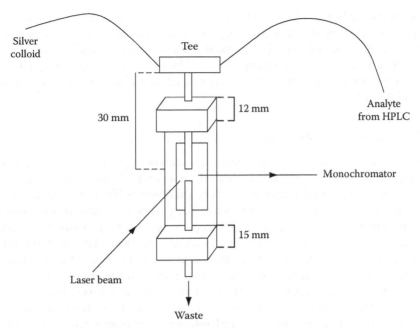

FIGURE 4.19 A diagram of the windowless flow cell. (From *Talanta*, 40, Cabalin, L.M., Ruperez, A., and Laserna, J.J., 1741, Copyright 1993, with permission from Elsevier.)

that investigated various experimental parameters such as spectrometer bandpass, integration time, and the flow rates of Ag sol and eluent did not report directly how the LODs were influenced [125].

A silver-doped sol-gel packed column has been used as both the stationary phase and the SERS substrate for the separation and online detection of *p*-aminobenzoic acid and phenyl acetylene [126]. This work extended previous studies directed at developing SERS substrates using sol-gels and was based on awareness that sol-gels could also be used as chromatographic packings. The possibility of using a detection point at any distance along the packed glass column enables mapping of a solute's concentration profile to obtain more detailed information about the separation characteristics of the sol-gel.

4.5.4 SOLVENT ELIMINATION METHODS

The at-line coupling of Raman spectroscopy for detection has many of the same advantages of the solvent elimination method for IR detection but has the additional advantage that the deposition substrate can be engineered so that improved sensitivity can be obtained due to SERS processes. It was often demonstrated that the use of an FT-IR microscope focused to the same size as a deposit greatly increased the mass sensitivity of FT-IR detection. Two factors are responsible for the greater spatial resolution of confocal Raman microscopy compared to FT-IR microscopy. The use of a laser for Raman spectroscopy compared to the diffuse blackbody radiation source used for IR spectroscopy enables greater diffraction-limited focusing of the more intense, collimated laser light. Also, the shorter wavelengths of light used for Raman spectroscopy also enables the light to be passing through the microscope to be more tightly focused. Based on geometric arguments, it would be expected that the relative area of a sample could be decreased by factor of 10^4 when using a Raman microscope but deposit inhomogeneities can affect reproducibility when examining individual particles.

4.5.5   EVAPORATIVE METHODS

Normal Raman detection of proteomic analytes has been demonstrated using a drop coating deposition Raman (DCDR) method on a range of substrates [127]. It was found that stainless steel coated with a very thin layer (<50 nm) of Teflon was the best surface for the deposition of the protein due to its low wettability and therefore less spreading of the solution when the drops are placed on this substrate. Eluent collection from the HPLC separation of a mixture of human and bovine insulin with subsequent deposition followed by Raman analysis demonstrated the first normal Raman detection of proteins where distinctive individual spectra were obtained. It was observed that the deposits resembled solution spectra and were more homogeneous than microcrystalline samples where crystal orientation effects can decrease reproducibility. Microprinting of a 3 μM solution of lysozyme where 40 deposits of 0.2 nL for a total of 8 nL or ~25 fmol resulted in a 15 μm spot that was readily detected using normal Raman detection and it was suggested that further optimization would enable smaller quantities of protein to be analyzed using normal Raman spectroscopy. In a continuation of this method, three insulin variants (human, bovine, and porcine) were identified after partial least squares classification of normalized second-derivative spectra obtained using the DCDR technique [128].

Another microdispensing technique used for both FT-IR and Raman analysis, described in more detail in the evaporative interfaces of the FT-IR section, showed that Raman analysis was much more sensitive than FT-IR analysis [50]. However, the high spatial resolution of confocal Raman microspectroscopy instrument resulted in less reproducibility compared to FT-IR because of microcrystal phase orientation effects due to the rapid evaporation of the solvent during deposition that were not present in the spectra of the prepared standards.

4.5.6   ENHANCED METHODS

SERS of a group of nitrophenols was demonstrated where fractions of the eluent were collected on the surface of an electrochemically roughened silver electrode and spectra were then obtained [129]. In addition, the pH of the eluent was adjusted to favor the phenolate anions that showed greater enhancement possibly due to specific adsorption on the silver surface.

Post-column mixing of a silver sol with the eluent from an LC system operating with flow rates of 10–200 μL min^{-1} usually used with microbore columns enabled SERS detection of spots deposited on a thin layer chromatography (TLC) plate [130]. A LOD of 750 fmol for pararosaniline was determined under FIA conditions. It was found that aggregation of the colloidal particles occurred when mixed with the dye in the deposition tube, which led to blockages and apparent peak broadening on the TLC plate. To overcome this problem, the dye was directly deposited on the TLC plate and the silver sol was added after the solvent had evaporated by tracking along the same path. Significant improvement in peak shape with minimal loss of intensity demonstrated that this approach is just as effective as direct mixing in solution. Further experiments where different solvents were used showed that as long as the solvent completely evaporated, there were no spectral interferences due to the mobile phase composition. However, no effort was made to enhance the rate of solvent removal and a waiting time of 1 h occurred before addition of the silver sol.

The use of TLC substrates for the at-line collection of the eluent in LC separations prompted Istvan and coworkers [131] to investigated a range of TLC substrates and the influence of different excitation wavelengths on the LOD of a group of amino acids, which are weak Raman scatterers. In addition, the application of a silver sol to promote a SERS response was also explored. It was observed that simple silica plates were the best substrate and the longer wavelength laser gave the best S/N ratio for normal Raman scatterers. An increase in signal intensity of one to three orders of magnitude was observed when the silver sol was applied but the shorter wavelength laser was found to be the most effective. The usual advantage of using a microscope for high spatial resolution was offset by the inhomogeneity of the silver sol deposit and the reproducibility of the SERS-active intensities was relatively poor compared to normal Raman spectra using the unfocussed longer wavelength laser. For improved reproducibility, the importance of matching the laser spot size to the size of the deposited analyte so that maximum averaging of the SERS interactions occurred was again highlighted.

A group of illicit drugs were investigated using SERS detection at-line in the wells of a microtiter plate [132]. The silver halide dispersions were stabilized in a gelatin matrix in the wells of the microtiter plates and selected drops of the eluent containing the analytes were then added. A small aliquot of methanol was added as a "refreshing" solvent that minimized thermal evaporation of the liquid from the each well due to heating from the focused laser spot. Quality spectra were obtained and enabled identification of each component when compared with reference spectra. It was observed that the previously developed separation protocol using acetonitrile in the mobile phase interfered with the response of the analytes because the acetonitrile preferentially adsorbs on the silver surface, which then blocks the surface enhancement process that greatly increases the sensitivity. This was rectified by substitution of methanol for acetonitrile in the mobile phase. In a following study, chromatography with a buffered methanol gradient enabled improved LOD to be obtained for the illicit drugs with unequivocal identification of metabolites being possible if they co-eluted with other solutes whose spectra were established based standards [133].

4.5.7 Concentric Flow Interface: Enhanced Methods

The spray-jet interface developed by Somsen and coworkers [80] as a solvent elimination method with FT-IR detection has been previously described; it has also been adapted for the deposition of solutes so that SERS can be exploited for detection after LC separation. It was used to deposit the RP separated cationic dyes on different

TLC substrates and the addition of a silver colloid generated a SERS response, silica plates were found to be the best substrate [134]. In this study, the limit of identification of the dyes are in the 50–250 ng mL^{-1} (injected) range and because the volume of solvent needing vaporization was reduced using a flow splitter, corresponding deposits of 150–750 pg were analyzed using a microscope fitted with a triple monochromator and a multichannel detector. The influence of ion-pair reagents, such as 1-heptanesulfonate and perchlorate, used to improve the isocratic RP separation of cationic dyes on the generation of the SERS response was then examined [135]. The use of the nonvolatile ion-pair reagents did not adversely affect the deposition process and the presence of 1-hepanesufonate actually increased the SERS intensity but the exact mechanism could not be explained. A further study explored the factors important in the SERS detection of some anionic dyes separated using gradient elution ion-pair RP-LC [136]. The electrostatic repulsion anticipated between the anionic dyes and the negatively charged citrate layer on the silver surface was minimized by adding a drop of 1 M nitric acid to the substrate before the silver sol was deposited. It was also demonstrated that the method was unaffected by the gradient or the additional ion-pair reagents but was dependent on the composition of the TLC plates with silica preferred to aluminum oxide because of reduced spreading of the deposits. An identification limit of 10–20 ng of deposit was reported but a much smaller fraction was imaged by the microscope and the possibility of obtaining SERS from only a few picograms of solute was proposed.

4.6 CONCLUSIONS

The multidimensional advantages of developing hyphenated detection methods based on vibrational spectroscopy have been realized using a variety of strategies. However, the desire to simultaneously obtain good spectral recognition and low detection limits has not been satisfied in the majority of cases due to spectral interferences of the matrix and the small absorptivities of the spectral features.

The general advantages associated with flow cell analysis apply with both IR and normal Raman detection; this includes ease of use, versatility, and reproducibility. Limitations due to spectral overlap can be minimized with judicious choice of solvents that enables spectral features of the solutes to be visible within the restricted spectral regions available due to the concentration dominance of the solvent. The use of single-wavelength detection has been considered both an advantage and a disadvantage; if a specific solute is the target analyte, then it could be possible to find an individual wavelength due to the higher resolving power associated with vibrational spectroscopy. The development of dispersive array detectors with fast acquisition rates overcomes the restrictions associated with single wavelength detection and multicomponent detection with off-line data analysis should be considered as the current state-of the-art when using flow cells. In Raman detection, the use of liquid core waveguides provides for increased interaction area that improves volume sensitivity and when combined with microscopy, the ability to detect the normal Raman spectra of solutes has been demonstrated.

The use of solvent elimination interfaces for the deposition of solutes has been widely employed because of a number of benefits; these include the minimization of spectral interference, greater sensitivity due to an increase in the volume concentration, increased acquisition time for off-line spectral analysis, and the suitability of using alternative detection techniques that require deposited solids (i.e., MALDI-TOF mass spectroscopy). Good mass sensitivity can be obtained but because the deposition process is dependent on a number of variables that can be difficult to control and the effect of morphology on the optical properties of a deposited solid, the ability to obtain quantitative information is fraught with danger. Two commercial instruments are currently available for interfacing FT-IR with LC; one is the LC-Transform system developed by Lab Connections that has been described previously and the other is the DiscovIR™ system available from Spectra Analysis, Inc. (Marlborough, MA) [137] that is able to directly obtain real-time IR spectra from a solid that is vacuum deposited on a cryogenically cooled ZnSe disk.

The reproducibility of microdispensing techniques and the use of optics systems based on microscopy give the greatest mass sensitivity as a deposition interface because the solute can be deposited with minimal spreading so that the area is easily matched by the focused optical beam with maximum overlap. Eliminating the solvent from a minute drop of eluent can be achieved with greater efficiency and without the need for excessive vacuum, temperature, or pressure that characterized the early development of deposition interfaces.

The enhanced modes possible with Raman detection enable improved sensitivity but this is generally combined with a decrease in structural information because the enhancement is linked with specific interactions that occur between the light, a substructure of a molecule, and a surface when SERS results. UV-RR detection using flow cells appears to be the most suitable configuration for obtaining reproducible enhanced sensitivity in flowing solutions because it is less complicated than the addition of an extra stream containing a metered suspension of colloidal silver. However, SERS is more suited to at-line analysis where the post-deposition application of a SERS-active substrate enables enhanced sensitivity where the spectroscopy can be optimized isolated from the separation technique.

REFERENCES

1. McCreery, R. L.; *Raman Spectroscopy for Chemical Analysis*, John Wiley & Sons, New York, 2000.
2. Somsen, G. W.; Gooijer, C.; Brinkman, U. A. T.; *Journal of Chromatography*, 1999, 856, 213.
3. Dijkstra, R. J.; Ariese, F.; Gooijer, C.; Brinkman, U. A. T.; *Trac—Trends in Analytical Chemistry*, 2005, 24, 304.
4. Raman, C. V.; Krishnan, K. S.; *Nature*, 1928, 121, 501.
5. Jones, C. M.; Naim, T. A.; Ludwig, M.; Murtaugh, J.; Flaugh, P. L.; Dudik, J. M.; Johnson, C. R.; Asher, S. A.; *Trac—Trends in Analytical Chemistry*, 1985, 4, 75.
6. Fleischmann, M.; Hendra, P. J.; McQuillan, A. J.; *Chemical Physics Letters*, 1974, 26, 163.
7. Kneipp, K.; Wang, Y.; Kneipp, H.; Perelman, L. T.; Itzkan, I.; Dasari, R.; Feld, M. S.; *Physical Review Letters*, 1997, 78, 1667.
8. Nie, S. M.; Emory, S. R.; *Science*, 1997, 275, 1102.
9. Aroca, R. F.; Ross, D. J.; Domingo, C.; *Applied Spectroscopy*, 2004, 58, 324A.

10. Louden, D.; Handley, A.; Lafont, R.; Taylor, S.; Sinclair, I.; Lenz, E.; Orton, T.; Wilson, I. D.; *Analytical Chemistry*, 2002, 74, 288.
11. Louden, D.; Handley, A.; Taylor, S.; Lenz, E.; Miller, S.; Wilson, I. D.; Sage, A.; *Analytical Chemistry*, 2000, 72, 3922.
12. Louden, D.; Handley, A.; Taylor, S.; Lenz, E.; Miller, S.; Wilson, I. D.; Sage, A.; Lafont, R.; *Journal of Chromatography*, 2001, 910, 237.
13. Brown, R. S.; Hausler, D. W.; Taylor, L. T.; *Analytical Chemistry*, 1980, 52, 1511.
14. Smith, S. L.; Wilson, C. E.; *Analytical Chemistry*, 1982, 54, 1439.
15. Brown, R. S.; Hausler, D. W.; Taylor, L. T.; Carter, R. C.; *Analytical Chemistry*, 1981, 53, 197.
16. Vidrine, D. W.; *Journal of Chromatographic Science*, 1979, 17, 477.
17. Vidrine, D. W.; Mattson, D. R.; *Applied Spectroscopy*, 1978, 3, 502.
18. Johnson, C. C.; Taylor, L. T.; *Analytical Chemistry*, 1983, 55, 436.
19. Johnson, C. C.; Taylor, L. T.; *Analytical Chemistry*, 1984, 56, 2642.
20. Brown, R. S.; Taylor, L. T.; *Analytical Chemistry*, 1983, 55, 723.
21. Fujimoto, C.; Uematsu, G.; Jinno, K.; *Chromatographia*, 1985, 20, 112.
22. Brown, R. S.; Taylor, L. T.; *Analytical Chemistry*, 1983, 55, 1492.
23. Jinno, K.; Fujimoto, C.; *Chromatographia*, 1983, 17, 259.
24. Jinno, K.; Fujimoto, C.; Nakanishi, S.; *Chromatographia*, 1985, 20, 279.
25. Gallignani, M.; Brunetto, M. D.; *Talanta*, 2004, 64, 1127.
26. Vonach, R.; Lendl, B.; Kellner, R.; *Analytical Chemistry*, 1997, 69, 4286.
27. Vonach, R.; Lendl, B.; Kellner, R.; *Journal of Chromatography*, 1998, 824, 159.
28. Schulte-Ladbeck, R.; Edelmann, A.; Quintas, G.; Lendl, B.; Karst, U.; *Analytical Chemistry*, 2006, 78, 8150.
29. Daghbouche, Y.; Garrigues, S.; de la Guardia, M.; *Analytica Chimica Acta*, 1997, 354, 97.
30. Patterson, B. M.; Danielson, N. D.; Sommer, A. J.; *Analytical Chemistry*, 2003, 75, 1418.
31. McKittrick, P. T.; Danielson, N. D.; Katon, J. E.; *Journal of Liquid Chromatography*, 1991, 14, 377.
32. Sabo, M.; Gross, J.; Wang, J.; Rosenberg, I. E.; *Analytical Chemistry*, 1985, 57, 1822.
33. Edelmann, A.; Diewok, J.; Baena, J. R.; Lendl, B.; *Analytical and Bioanalytical Chemistry*, 2003, 376, 92.
34. Istvan, K.; Keresztury, G.; Fekete, J.; *Journal of Liquid Chromatography & Related Technologies*, 2005, 28, 407.
35. Johnson, C. C.; Hellgeth, J. W.; Taylor, L. T.; *Analytical Chemistry*, 1985, 57, 610.
36. Hellgeth, J. W.; Taylor, L. T.; *Analytical Chemistry*, 1987, 59, 295.
37. DiNunzio, J. E.; *Journal of Chromatography*, 1992, 626, 97.
38. Gagel, J. J.; Biemann, K.; *Analytical Chemistry*, 1986, 58, 2184.
39. Kuehl, D.; Griffiths, P. R.; *Journal of Chromatographic Science*, 1979, 17, 471.
40. Kuehl, D.; Griffiths, P. R.; *Analytical Chemistry*, 1980, 52, 1394.
41. Conroy, C. M.; Griffiths, P. R.; Duff, P. J.; Azarraga, L. V.; *Analytical Chemistry*, 1984, 56, 2636.
42. Conroy, C. M.; Griffiths, P. R.; Jinno, K.; *Analytical Chemistry*, 1985, 57, 822.
43. Kalasinsky, K. S.; Smith, J. A. S.; Kalasinsky, V. F.; *Analytical Chemistry*, 1985, 57, 1969.
44. Fujimoto, C.; Jinno, K.; Hirata, Y.; *Journal of Chromatography*, 1983, 258, 81.
45. Jinno, K.; Fujimoto, C.; Ishii, D.; *Journal of Chromatography*, 1982, 239, 625.
46. Jinno, K.; Fujimoto, C.; Hirata, Y.; *Applied Spectroscopy*, 1982, 36, 67.
47. Fujimoto, C.; Oosuka, T.; Jinno, K.; *Analytica Chimica Acta*, 1985, 178, 159.
48. Fujimoto, C.; Oosuka, T.; Jinno, K.; Ochiai, S.; *Chromatographia*, 1987, 23, 512.
49. Dekmezian, A. H.; Morioka, T.; *Analytical Chemistry*, 1989, 61, 458.
50. Surowiec, I.; Baena, J. R.; Frank, J.; Laurell, T.; Nilsson, J.; Trojanowicz, M.; Lendl, B.; *Journal of Chromatography A*, 2005, 1080, 132.

51. Raynor, M. W.; Bartle, K. D.; Cook, B. W.; *HRC—Journal of High Resolution Chromatography*, 1992, 15, 361.
52. Jansen, J. A. J.; *Fresenius' Journal of Analytical Chemistry*, 1990, 337, 398.
53. Robertson, A. M.; Wylie, L.; Littlejohn, D.; Watling, R. J.; Dowle, C. J.; *Analytical Proceedings*, 1991, 28, 8.
54. Robertson, A. M.; Littlejohn, D.; Brown, M.; Dowle, C. J.; *Journal of Chromatography A*, 1991, 588, 15.
55. Robertson, A. M.; Farnan, D.; Littlejohn, D.; Brown, M.; Dowle, C. J.; Goodwin, E.; *Analytical Proceedings*, 1993, 30, 268.
56. Mottaleb, M. A.; Cooksey, B. G.; Littlejohn, D.; *Fresenius' Journal of Analytical Chemistry*, 1997, 358, 536.
57. Mottaleb, M. A.; *Analytical Sciences*, 1999, 15, 57.
58. Mottaleb, M. A.; *Analytical Sciences*, 1999, 15, 1137.
59. Mottaleb, M. A.; *Mikrochimica Acta*, 1999, 132, 31.
60. Mottaleb, M. A.; Littlejohn, D.; *Analytical Sciences*, 2001, 17, 429.
61. Mottaleb, M. A.; Kim, H. J.; *Analytical Sciences*, 2002, 18, 579.
62. Gagel, J. J.; Biemann, K.; *Analytical Chemistry*, 1987, 59, 1266.
63. Gagel, J. J.; Biemann, K.; *Microchimica Acta*, 1988, 95, 185.
64. Lab connections, Carrboro, NC, USA. www.labconnections.com
65. Schunk, T. C.; Balke, S. T.; Cheung, P.; *Journal of Chromatography A*, 1994, 661, 227.
66. Kok, S. J.; Arentsen, N. C.; Cools, P.; Hankemeier, T.; Schoenmakers, P. J.; *Journal of Chromatography A*, 2002, 948, 257.
67. Kok, S. J.; Wold, C. A.; Hankemeier, T.; Schoenmakers, P. J.; *Journal of Chromatography A*, 2003, 1017, 83.
68. Coulier, L.; Kaal, E.; Hankemeier, T.; *Journal of Chromatography A*, 2006, 1130, 34.
69. Ludlow, M.; Louden, D.; Handley, A.; Taylor, S.; Wright, B.; Wilson, I. D.; *Journal of Chromatography A*, 1999, 857, 89.
70. Jordan, S. L.; Taylor, L. T.; McPherson, B.; Rasmussen, H. T.; *Journal of Chromatography A*, 1996, 755, 211.
71. Sanchez, F. C.; Vandeginste, B. G. M.; Hancewicz, T. M.; Massart, D. L.; *Analytical Chemistry*, 1997, 69, 1477.
72. Smith, S. H.; Jordan, S. L.; Taylor, L. T.; Dwyer, J.; Willis, J.; *Journal of Chromatography A*, 1997, 764, 295.
73. Bruheim, I.; Molander, P.; Lundanes, E.; Greibrokk, T.; Ommundsen, E.; *HRC—Journal of High Resolution Chromatography*, 2000, 23, 525.
74. Kalasinsky, V. F.; Whitehead, K. G.; Kenton, R. C.; Smith, J. A. S.; Kalasinsky, K. S.; *Journal of Chromatographic Science*, 1987, 25, 273.
75. Lange, A. J.; Griffiths, P. R.; Fraser, D. J. J.; *Analytical Chemistry*, 1991, 63, 782.
76. Griffiths, P. R.; Haefner, A. M.; Norton, K. L.; Fraser, D. J. J.; Pyo, D.; Makishima, H.; *HRC—Journal of High Resolution Chromatography*, 1989, 12, 119.
77. Norton, K. L.; Lange, A. J.; Griffiths, P. R.; *HRC—Journal of High Resolution Chromatography*, 1991, 14, 225.
78. Griffiths, P. R.; Lange, A. J.; *Journal of Chromatographic Science*, 1992, 30, 93.
79. Lange, A. J.; Griffiths, P. R.; *Applied Spectroscopy*, 1993, 47, 403.
80. Somsen, G. W.; van de Nesse, R. J.; Gooijer, C.; Brinkman, U. A. T.; Velthorst, N. H.; Visser, T.; Kootstra, P. R.; de Jong, A. P. J. M.; *Journal of Chromatography*, 1991, 552, 635.
81. Somsen, G. W.; Gooijer, C.; Brinkman, U. A. T.; Velthorst, N. H.; Visser, T.; *Applied Spectroscopy*, 1992, 46, 1514.
82. Somsen, G. W.; Vanstee, L. P. P.; Gooijer, C.; Brinkman, U. A. T.; Velthorst, N. H.; Visser, T.; *Analytica Chimica Acta*, 1994, 290, 269.
83. Somsen, G. W.; Morden, W.; Wilson, I. D.; *Journal of Chromatography A*, 1995, 703, 613.

84. Somsen, G. W.; Rozendom, E. J. E.; Gooijer, C.; Velthorst, N. H.; Brinkman, U. A. T.; *Analyst*, 1996, 121, 1069.

85. Somsen, G. W.; Hooijschuur, E. W. J.; Gooijer, C.; Brinkman, U. A. T.; Velthorst, N. H.; Visser, T.; *Analytical Chemistry*, 1996, 68, 746.

86. Somsen, G. W.; Jagt, I.; Gooijer, C.; Velthorst, N. H.; Brinkman, U. A. T.; Visser, T.; *Journal of Chromatography A*, 1996, 756, 145.

87. Visser, T.; Vredenbregt, M. J.; ten Hove, G. J.; de Jong, A. P. J. M.; Somsen, G. W.; *Analytica Chimica Acta*, 1997, 342, 151.

88. van de Nesse, R. J.; Hoogland, G. J. M.; Demoel, J. J. M.; Gooijer, C.; Brinkman, U. A. T.; Velthorst, N. H.; *Journal of Chromatography*, 1991, 552, 613.

89. Deruiter, C.; Wolf, J. H.; Brinkman, U. A. T.; Frei, R. W.; *Analytica Chimica Acta*, 1987, 192, 267.

90. Geiger, J.; Korte, E. H.; Schrader, W.; *Journal of Chromatography A*, 2001, 922, 99.

91. Castles, M. A.; Azarraga, L. V.; Carreira, L. A.; *Applied Spectroscopy*, 1986, 40, 673.

92. Dekmezian, A. H.; Morioka, T.; Camp, C. E.; *Journal of Polymer Science Part B: Polymer Physics*, 1990, 28, 1903.

93. Torabi, K.; Karami, A.; Balke, S. T.; Schunk, T. C.; *Journal of Chromatography A*, 2001, 910, 19.

94. Karami, A.; Balke, S. T.; Schunk, T. C.; *Journal of Chromatography A*, 2001, 911, 27.

95. Li, Y.; Brown, P. R.; *Journal of Liquid Chromatography & Related Technologies*, 2003, 26, 1769.

96. Willoughby, R. C.; Browner, R. F.; *Analytical Chemistry*, 1984, 56, 2626.

97. Robertson, R. M.; de Haseth, J. A.; Browner, R. F.; *Mikrochimica Acta*, 1988, 2, 199.

98. Robertson, R. M.; de Haseth, J. A.; Kirk, J. D.; Browner, R. F.; *Applied Spectroscopy*, 1988, 42, 1365.

99. de Haseth, J. A.; Robertson, R. M.; *Microchemical Journal*, 1989, 40, 77.

100. Robertson, R. M.; de Haseth, J. A.; Browner, R. F.; *Applied Spectroscopy*, 1990, 44, 8.

101. Turula, V. E.; de Haseth, J. A.; *Applied Spectroscopy*, 1994, 48, 1255.

102. Turula, V. E.; de Haseth, J. A.; *Analytical Chemistry*, 1996, 68, 629.

103. Bishop, R. T.; Turula, V. E.; de Haseth, J. A.; *Analytical Chemistry*, 1996, 68, 4006.

104. Turula, V. E.; Bishop, R. T.; Ricker, R. D.; de Haseth, J. A.; *Journal of Chromatography A*, 1997, 763, 91.

105. Chapput, A.; Roussel, B.; Montastier, J.; *Journal of Raman Spectroscopy*, 1980, 9, 193.

106. Dorazio, M.; Schimpf, U.; *Analytical Chemistry*, 1981, 53, 809.

107. Hong, T. D. N.; Jouan, M.; Dao, N. Q.; Bouraly, M.; Mantisi, F.; *Journal of Chromatography A*, 1996, 743, 323.

108. Cooper, S. D.; Robson, M. M.; Batchelder, D. N.; Bartle, K. D.; *Chromatographia*, 1997, 44, 257.

109. Dijkstra, R. J.; Bader, A. N.; Hoornweg, G. P.; Brinkman, U. A. T.; Gooijer, C.; *Analytical Chemistry*, 1999, 71, 4575.

110. Marquardt, B. J.; Turney, K. P.; Burgess, L. W.; *Proceedings of the Society of Photo-Optical Instrumentation Engineers*, 1999, 3860, 239.

111. Marquardt, B. J.; Vahey, P. G.; Synovec, R. E.; Burgess, L. W.; *Analytical Chemistry*, 1999, 71, 4808.

112. Dijkstra, R. J.; Slooten, G. J.; Stortelder, A.; Buijs, J. B.; Ariese, F.; Brinkman, U. A. T.; Gooijer, G.; *Journal of Chromatography A*, 2001, 918, 25.

113. Gooijer, C.; Hoornweg, G. P.; de Beer, T.; Bader, A.; van Iperen, D. J.; Brinkman, U. A. T.; *Journal of Chromatography A*, 1998, 824, 1.

114. Ling, B. L.; Baeyens, W.; Dewaele, C.; *Journal of Microcolumn Separations*, 1992, 4, 17.

115. Dijkstra, R. J.; Boelens, H. F. M.; Westerhuis, J. A.; Ariese, F.; Brinkman, U. A. T.; Gooijer, C.; *Analytica Chimica Acta*, 2004, 519, 129.

116. Dable, B. K.; Marquardt, B. J.; Booksh, K. S.; *Analytica Chimica Acta*, 2005, 544, 71.
117. Iriyama, K.; Ozaki, Y.; Hibi, K.; Ikeda, T.; *Journal of Chromatography*, 1983, 254, 285.
118. Koizumi, H.; Suzuki, Y.; *Journal of High Resolution Chromatography & Chromatography Communications*, 1987, 10, 173.
119. Chong, C. K.; Mann, C. K.; Vickers, T. J.; *Applied Spectroscopy*, 1992, 46, 249.
120. Dijkstra, R. J.; Martha, C. T.; Ariese, F.; Brinkman, U. A. T.; Gooijer, C.; *Analytical Chemistry*, 2001, 73, 4977.
121. Freeman, R. D.; Hammaker, R. M.; Meloan, C. E.; Fateley, W. G.; *Applied Spectroscopy*, 1988, 42, 456.
122. Ni, F.; Sheng, R. S.; Cotton, T. M.; *Analytical Chemistry*, 1990, 62, 1958.
123. Sheng, R. S.; Ni, F.; Cotton, T. M.; *Analytical Chemistry*, 1991, 63, 437.
124. Cabalin, L. M.; Ruperez, A.; Laserna, J. J.; *Talanta*, 1993, 40, 1741.
125. Cabalin, L. M.; Ruperez, A.; Laserna, J. J.; *Analytica Chimica Acta*, 1996, 318, 203.
126. Farquharson, S.; Maksymiuk, P.; *Applied Spectroscopy*, 2003, 57, 479.
127. Zhang, D. M.; Xie, Y.; Mrozek, M. F.; Ortiz, C.; Davisson, V. J.; Ben-Amotz, D.; *Analytical Chemistry*, 2003, 75, 5703.
128. Ortiz, C.; Zhang, D. M.; Xie, Y.; Davisson, V. J.; Ben-Amotz, D.; *Analytical Biochemistry*, 2004, 332, 245.
129. Ni, F.; Thomas, L.; Cotton, T. M.; *Analytical Chemistry*, 1989, 61, 888.
130. Soper, S. A.; Ratzlaff, K. L.; Kuwana, T.; *Analytical Chemistry*, 1990, 62, 1438.
131. Istvan, K.; Keresztury, G.; Szep, A.; *Spectrochimica Acta Part A—Molecular and Biomolecular Spectroscopy*, 2003, 59, 1709.
132. Sagmuller, B.; Schwarze, B.; Brehm, G.; Trachta, G.; Schneider, S.; *Journal of Molecular Structure*, 2003, 661, 279.
133. Trachta, G.; Schwarze, B.; Sagmuller, B.; Brehm, G.; Schneider, S.; *Journal of Molecular Structure*, 2004, 693, 175.
134. Somsen, G. W.; Coulter, S. K.; Gooijer, C.; Velthorst, N. H.; Brinkman, U. T.; *Analytica Chimica Acta*, 1997, 349, 189.
135. Seifar, R. M.; Dijkstra, R. J.; Brinkman, U. A. T.; Gooijer, C.; *Analytical Communications*, 1999, 36, 273.
136. Seifar, R. M.; Altelaar, M. A. F.; Dijkstra, R. J.; Ariese, F.; Brinkman, U. A. T.; Gooijer, C.; *Analytical Chemistry*, 2000, 72, 5718.
137. Spectra Analysis, Inc., Marlborough, MA, USA. www.spectra-analysis.com

5 Evaporative Light Scattering and Charged Aerosol Detector

Pierre Chaminade

CONTENTS

5.1 Theory of Operations..146
    5.1.1 Nebulization Process 146
    5.1.2 Solvent Evaporation 148
    5.1.3 Light Scattering 149
    5.1.4 ELSD Response versus Solute Concentration 150
5.2 Parameters Influencing ELSD Response...152
    5.2.1 Gas Pressure and Mobile Phase Flow Rate 152
    5.2.2 Drift Tube Temperature 153
    5.2.3 Mobile Phase Composition 153
        5.2.3.1 Solvents 153
        5.2.3.2 Mobile Phase Additives 154
5.3 Applications of ELS Detection..155
    5.3.1 Micro-Chromatography with ELS Detection...................................155
5.4 Corona-CAD Detector...156
References..158

The principle of the evaporative light scattering detector (ELSD) was introduced in 1978 by Charlesworth [1]. The article titled "Evaporative analyser as a mass detector for liquid chromatography" introduced the term of "mass detector" that is no longer used nowadays. Although Charlesworth developed his mass detector for polymer analysis, lipid and carbohydrate science drew a huge benefit from using ELSD. For many users from these fields, the pioneering publication is from Macrae et al. [2] in 1982 that underline the superiority of this detector over the refractive index (RI) detector for carbohydrate and lipids analysis.

Recently, Dixon and Peterson [3] introduced the principle of using particle charging instead of light scattering. The principle of this new detector marketed as "Corona-CAD" is also reviewed in this chapter.

5.1   THEORY OF OPERATIONS

The ELSD detects the light scattered by solutes after the mobile phase is sprayed and partially or totally evaporated. Depending on the solute melting point, solutes appear as a cloud of particles or droplets. The three stages of the ELSD operation are nebulization, solvent evaporation, and particle detection.

5.1.1   Nebulization Process

Although many principles may be used to create a liquid aerosol from the mobile phase leaving the column, all commercial detectors use a pneumatic nebulizer. Pneumatic nebulizers use a compressed gas such as nitrogen or air to spray the mobile phase using the Venturi effect.

The size characteristics of the wet aerosol at the nebulizer outlet can be predicted from the Nukiyama–Tanasawa equation [4]:

$$d_{sv} = \frac{(585\sqrt{\sigma})}{(v_g - v_l)} + 597 \left(\frac{\mu}{\sqrt{\sigma\rho}} \right)^{0.45} \left(\frac{1000Q_l}{Q_g} \right)^{1.5} \tag{5.1}$$

where

σ is the mobile phase surface tension (dyne/cm)
ρ its density
μ its viscosity (μdyne $\cdot$ s/cm^2)
$(v_g - v_l)$ the difference between the nebulizer gas and liquid velocities (m/s)
Q_l/Q_g the ratio of liquid and gas a volumetric flow rates

This equation is sometimes matter of confusion. According to Nukiyama and Tanasawa, d_{sv} is the diameter of a single drop having the same volume–surface ratio as the total sum of the drop. It corresponds to the drop diameter at the upper 32% of the size distribution of the aerosol. This value is also called the "Sauter mean diameter" and sometimes incorrectly referenced as the mean diameter of drops.

This two termed equation shows that the volume–surface ratio of drops will be influenced by mobile phase characteristics together with the mobile phase flow rate and nebulization gas pressure settings. This equation is assumed to be true for $0.8 < \rho < 1.2$, $30 < \sigma < 73$, $0.01 < \mu < 0.3$ but can be extended as far as $\rho = 0.7 - 0.9$, $\sigma = 19 - 37$, and $\mu = 0.003 - 0.5$ [4]. Most solvents used in chromatography can be evaluated using this equation, however the d_{sv} of useful solvents like chloroform cannot be accurately predicted.

Results for commonly used HPLC solvents are presented in Table 5.1. The exact geometry of the nebulizer must be known in order to calculate the liquid and gas volumetric flow rate together with their velocities. Table 5.1 shows the values of d_{sv} and also d_{max}, the maximum size of drops, calculated for two classical column dimensions operated at their corresponding typical flow rates. At 1 mL/min, bigger droplets are obtained compared with 0.2 mL/min. The same trend (reducing the drop size) is obtained when increasing the nebulization gas flow.

TABLE 5.1

Values of d_{sv} and d_{max} for an ELSD Operated with Conventional or Narrow-Bore Columns

Solvent	Surface Tension (dyne/cm)	Viscosity (dyne·s/cm)	Density (g/mL)	Conventional Column 1 Gas Flow 3.5 L/min d_{sv}	d_{max}	Gas Flow 5 L/min d_{sv}	d_{max}	Narrow-Bore Column Gas Flow 3.5 L/min d_{sv}	d_{max}
Acetone	23.32	0.0036	0.79	8.5	18.6	5.5	12.1	5.0	11.1
Acetonitrile	19.1	0.0038	0.7822	8.3	18.3	5.4	11.8	4.6	10.2
Chloroform	*27.16*	*0.0057*	*1.4798*	*7.4*	*16.4*	*4.8*	*10.5*	*4.0*	*8.8*
Heptane	20.3	0.0042	0.6837	9.0	19.8	5.8	12.8	5.1	11.2
Methanol	22.55	0.0059	0.7913	9.3	20.6	6.0	13.2	5.0	11.1
Tetrahydrofuran	26.4	0.0055	0.64	9.0	19.9	5.8	12.8	5.1	11.2
Toluene	28.53	0.0059	0.8623	9.4	20.6	6.0	13.3	5.3	11.8
Water	72.8	0.01	0.9971	11.8	25.9	7.7	17.0	7.8	17.1

It is also noticeable that some solvents (such as acetone or acetonitrile) give significantly lower droplet size than water.

As proposed by Oppenheimer and Mourey [5], the log-normal distribution of Mugele and Evans can be used to model the number and size distribution of the aerosol. This calculation was also used by P. Van der Meeren and coworkers [6] in their article on modeling the response of the ELSD.

The number of droplets with diameter d_m: dd_n/dd_m is calculated as

$$\frac{dd_n}{dd_m} = Y \frac{\delta d_{max}}{\sqrt{\pi} d_m (d_{max} - d_m)} e^{-[\delta \ln(a d_m/(d_{max}-d_m))+1.5/\delta]^2} \tag{5.2}$$

where

Y is a normalization factor

a and d are constants assumed to be 0.6

The upper limit of distribution d_{max} is calculated using the value of d_{sv} according to

$$d_{max} = d_{sv}\left(1 + ae^{\left(1/4\delta^2\right)}\right) \tag{5.3}$$

In addition, the volume distribution (V_i) can be readily calculated as

$$V_i = n_i \frac{\pi d_m^3}{6} \tag{5.4}$$

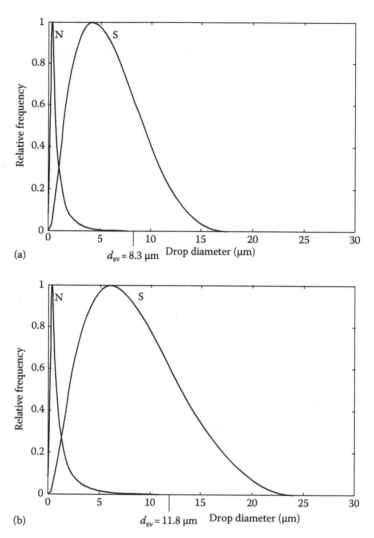

FIGURE 5.1   Number (N) and size (S) distribution for (a) acetonitrile and (b) water.

As shown in Figure 5.1, the pneumatic nebulizer produces a huge number of small (sub-micronic) droplets whatever is the solvent. In the case of water ($d_{sv} = 11.8\,\mu m$), the size distribution is considerably larger than with acetonitrile ($d_{sv} = 8.3\,\mu m$). Thus pneumatic nebulizers produce polydisperse aerosols whose characteristics depend on solvent properties, flow rate, and nebulization gas pressure.

5.1.2   SOLVENT EVAPORATION

Solvent (mobile phase) evaporation arises in a heated tube (or drift tube) located between the nebulizer and the detection cell. Two types can be found in commercial ELSD: either a short direct tube of about 20–30 cm long or a 1–2 m coiled tube. The temperature setting is usually from ambient to about 100°C.

If we suppose a total evaporation of the mobile phase, the diameter of a droplet is given by:

$$d = d_0 \left(\frac{C}{\rho} \right)^{1/3} \tag{5.5}$$

where
  d_0 is the diameter of the starting droplet
  C is the concentration of the solute in the mobile phase
  ρ is the solute density

Thus, starting from a wet aerosol whose characteristics depend on solvent properties and nebulization gas flow rate, the dry aerosol distribution at the drift tube outlet depend also on solute concentration and density.

Most of the publications aimed at modeling the ELSD suppose that the primary aerosol formed by the nebulizer is simply reduced in size by the evaporation process [1,6,7]. The situation is somewhat more complex and it has been shown [8,9] that the distribution of the aerosol is affected during the transport in the drift tube. Impaction of droplets on the tube wall and droplet coalescence may explain the rearrangement of the aerosol. Additionally, a partial volatilization or sublimation of the solute may arise in the drift tube [1]. This phenomenon may undergo a further size reduction of the droplets and it may be thought that, due to a superior surface–volume ratio, little droplets are more affected.

5.1.3 LIGHT SCATTERING

The dry aerosol entering the detection cell is said to be the third aerosol. Depending on the solute melting point, this aerosol is made of liquid droplets or solid particles. In addition, the mobile phase can be totally or partially evaporated depending on operational parameters such as the drift tube length and temperature and mobile phase properties.

The detection cell includes a light source (either a laser or a tungsten-halogen lamp) and a photomultiplier to measure the scattered light. The angle between the light source and the photomultiplier is usually not 90° but 120° or 135° due to a higher intensity of the light diffused at these angles [9].

Three main processes are involved in light diffusion by particles [1]:

1. The Rayleigh scattering occurs when the size of a particle is much less than the wavelength (λ) of the incident light. When the ratio of the radius of the particle (r) to the wavelength r/λ is less than 0.05, the incident light quanta induces dipoles in each particle they hit. These particles behave as point sources and radiate in all directions.
2. When r/λ is 0.05 or more, but $r < \lambda$, the Mie scattering becomes predominant. In this process, the particle is big enough to be hit by several light quanta with differences in amplitudes and phases. The resulting induced oscillating dipoles produces waves which interfere with each other. As illustrated by Stolyhwo [16], the intensity of the light scattered by a particle

has the shape of a daisy. Calculations have shown that Mie scattering is the origin of the angular dependence of the intensity of light collected [5].

3. When the size of the particle is of the same order of magnitude as the incident wavelength, reflection and refraction occur together. The sum of the intensity of the refracted and reflected light equals the total intensity of the incident light.

5.1.4 ELSD Response versus Solute Concentration

The detector response is a complex function which depends on the intensity of light scattered by solute particles from the third aerosol. Depending on particle size, light diffusion is governed by either Rayleigh or Mie scattering or reflection and refraction. With Rayleigh diffusion, the amount of light is proportional to the sixth power of the particle diameter. This amount is proportional to the fourth power with Mie scattering and to the square of the particle diameter with reflection and refraction. The particle size depends on solute concentration and initial droplet size (Equation 5.5), and initial droplet size depends on solvent properties, and liquid and gas flow rates (Equation 5.1). They key relationship is Equation 5.5 as the dependence of particle diameter on concentration is the basis of the quantitative response of the detector.

The light collected by the photomultiplier is thus the reflection of the predominant diffusion process. For this reason, the equation expressing the detector response (y) as a function of the injected amount (m) of solute is expressed as

$$y = Am^b \qquad (5.6)$$

b values vary between 0.67 when the predominant scattering mechanism is reflection–refraction and 2 when Rayleigh scattering prevails. ELSD is thus a non-linear detector and must be carefully calibrated for quantitative analysis.

Figure 5.2a shows the calibration curve obtained for glucose in the 50–500 ppm range. Glucose is conveniently analyzed with an acetonitrile–water mobile phase

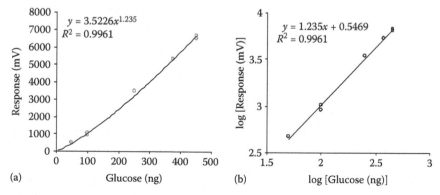

FIGURE 5.2   Calibration curves for glucose. (a) Fitted according to Equation 5.5, (b) log–log transformation. Column Alltech Altima NH$_2$ 100 × 4.6 mm. Mobile phase acetonitrile:water 20:80 @ 1 mL/min. ELSD Eurosep DDL 31. Settings 1 bar air pressure; nebulizer 30°C, drift tube 50°C.

and an amino-bonded silica column [10]. The calibration curve is clearly nonlinear and the *b* coefficient is about 1.2. Such a value suggests that the predominant scattering phenomenon is Mie diffusion. This phenomenon is the most likely to be observed and *b* values are often larger than 1. A characteristic of the ELSD detector is advisable from Figure 5.2: the ELSD response is less sensitive for low amounts than in the rest of the calibration range. This feature becomes obvious if we calculate the sensitivity (*S*) as the signal output per unit mass of solute entering the detector according to the IUPAC recommendations. *S* is thus the derivative of Equation 5.5 with respect to *m* and:

$$S = \frac{dy}{dm} = Abm^{b-1} \tag{5.7}$$

The sensitivity of the ELSD is a function of both *A* and *b* and varies along the calibration curve. For this reason, it is difficult to compare apparatus or experimental conditions on the sensitivity basis [11]. Also, at first glance, a system responding in the Rayleigh region, with the highest *b* values can be thought more sensitive than for Mie or reflection–refraction. Unfortunately, little particles or droplets have less scattering efficiency than big ones [8] and *b* values of 2 are often associated with low *A* values.

In a recent publication, Heron et al. [12] studied the correlation between the *A* and *b* terms of Equation 5.6. A linear relationship can be established between Log *A* and *b* for the same compound with different mobile phase compositions or for different compounds with the same mobile phase composition. Compounds included in this study are lipids analyzed under nonaqueous chromatography or carbohydrates and antibiotics under classical reversed-phase conditions. This finding can have important consequences in quantitative analysis, by simplifying calibration when gradient elution is used or by correcting relative peak areas of a chromatographic profile.

The variation of sensitivity with concentration also explains that the ELSD response varies along the peak profile as originally described by Stolyhwo et al. [13]. During peak elution, the concentration profile evolves as a function of time. When the solute starts leaving the columns, its low concentration leads to small particles. Then the particle size grows according to Equation 5.5 until the peak apex and then diminishes until the peak ends. As sensitivity increases in the ascending part of the peak profile and decreases in the downward part, the peak appears to be narrower than the elution profile. It is thus difficult to compute column efficiency from the ELSD signal [14] as the plate count appears to be overestimated.

A practical advantage is that two overlapping peaks appear to be better separated with an ELSD than with a linear detector. When using an ELSD together with an UV detector to perform a "dual detection" system, it is of common practice to connect the ELSD at the UV outlet. By using this setting, both detectors receive the whole column effluent. The alteration of the elution profile due to the dispersive effect of the UV flow cell is expected to be compensated by the nonlinear response of the ELSD. A recent application of this setting can be found in Ref. [15].

At last, many authors use the "log–log transformation" to "linearize" ELSD response. In Figure 5.2b, the logarithm of the ELSD response is plotted against the logarithm of the amount of glucose injected. Equation 5.6 now becomes:

$$\log(y) = \log(A) + b \cdot \log(m) \tag{5.8}$$

As can be seen from the numerical values of the two calibration curves, the slope of the relationship corresponds to the b coefficient whereas the intercept is logarithm of A. The determination coefficient R^2 is equal to 0.9961 in both cases which demonstrate an equal degree of fit. The "log–log transformation" is just a way to compute the coefficients A and b by linear regression.

5.2 PARAMETERS INFLUENCING ELSD RESPONSE

5.2.1 Gas Pressure and Mobile Phase Flow Rate

The effect of gas pressure and mobile phase flow rate on ELSD response can be understood from the Nukiyama–Tanasawa equation as exemplified in Table 5.1. The higher the gas pressure and thus gas velocity in the nebulizer, the lower the size of droplets. The practical consequence is an increase of the b term of the calibration curve as demonstrated in Figure 5.3. By increasing the air pressure from 1.0 to 1.6 bar, the response shift toward the Rayleigh diffusion with an increase in b and a subsequent decrease in A. This change in A and b with gas pressure is not wide enough to consider it as an effective means to change the response profile of the ELSD.

The same trend can be expected when reducing the mobile phase flow rate. This can be obtained when changing from conventional to narrow-bore columns. From our experience, using 2.1 mm ID columns b values near 2.0 are frequently observed whereas 4.6 mm ID columns lead to lower b values.

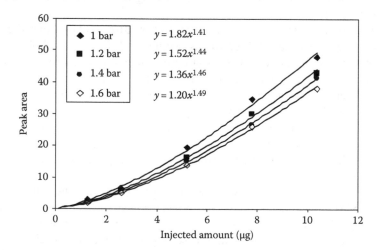

FIGURE 5.3 Effect of gas pressure on ELSD response. Ceramide Type III, ELSD Eurosep DDL 31, nebulizer 30°C, drift tube 40°C. Mobile phase acetonitrile:isopropanol 20:80 0.2 mL/min. Column Kromasil C18 2.1 × 100 mm.

5.2.2 DRIFT TUBE TEMPERATURE

As a rule of thumb, the ELSD response for a particular compound diminishes with increasing drift tube temperature. This trend is linked to the solute vapor pressure [16]. Solutes with low boiling points are more subject to changes in their response than compounds with high boiling points. Also, solids scatter light more efficiently than liquids and solutes with high melting points often give a higher response. As exemplified in Ref. [16], light is more strongly scattered by snow flakes than by rain droplets.

Since ELSD response is also solvent and peak-shape dependent, it is difficult to compare the response of compounds that present wide structural differences. In a publication from our group [17], Deschamps studied the solute response under identical chromatographic conditions. The chromatographic setup used in this study consists in a short size-exclusion column (50 × 7.5 mm PL-Gel from Polymer Laboratories) and pure chloroform as mobile phase. The ELS detector used in this study is a Sedere Sedex 65. This setting was selected as it ensures that all compounds elute with a similar peak shape and also that all of them are adequately dissolved by the mobile phase. With a drift tube temperature set a few degrees higher than the ambient temperature, all solutes exhibit similar responses as depicted in Table 5.2. When setting the drift tube at 80°C, squalene and 1-octadecanol were no longer detectable whereas the other compounds of higher melting and boiling points gave a less sensitive response than at 30°C: the A coefficient is decreased whereas b is close to the maximum value of two for all compounds. This increase in b together with the lower A value is consistent with the hypothesis that the temperature change from 30°C to 80°C diminishes the droplet size and thus the response.

5.2.3 MOBILE PHASE COMPOSITION

5.2.3.1 Solvents

As stated previously, the nebulization process is influenced by the solvent nature. However, the changes in d_{max} and d_{sv} of droplets generated by the pneumatic nebulizer (Table 5.1) cannot be linked easily to the solute response. As reported by

TABLE 5.2
Melting and Boiling Point of Solutes and Response Coefficients for a Drift Tube Setting at 30°C and 80°C

Solute	Melting Point (°C)	Boiling Point (°C)	Drift Tube 30°C			Drift Tube 80°C		
			A	b	R^2	A	b	R^2
Squalene	−75	285	20.87	1.72	0.995			
1-Octadecanol	59.8	210	16.00	1.70	0.985			
Tripalmitin	66	400	18.40	1.57	0.994	8.87	1.81	0.990
1-Triacontanol	87	ND	13.67	1.47	0.991	8.23	1.86	0.992
Ceramide	122	ND	14.04	1.73	0.991	7.87	1.75	0.991
Cholesterol	148.5	360	20.90	1.63	0.995	8.72	1.84	0.991

TABLE 5.3

Response Coefficients for Ceramide III as a Function of Mobile Phase Composition

	Chloroform	Dichloromethane	Methanol	1-Propanol	2-Propanol	Toluene
A	16.45	6	2.36	2.46	3.25	1.95
b	0.73	1.01	1.16	1.28	1.07	0.86
R^2	0.986	0.988	0.997	0.997	0.996	0.984

Charlesworth [1], apart from the initial droplet size, solvent evaporation plays a major role in the solvent dependence of the ELSD response. But, even when taking into account the molar volatility of solvents [2], the change in ELSD response induced by the solvent composition of the mobile phase cannot be predicted.

As an example of the influence of solvents on A and b terms of Equation 5.6, the results in Table 5.3 were obtained by Deschamps [15] with ceramide III in flow injection analysis. Solvents were selected as they allowed an appropriate dissolving of the sample. Chloroform gave the higher response and toluene the lowest. Here also, the relationship between solvent properties and solute response is not elucidated.

From a practical point of view, this dependence of the response on solvent composition is of particular importance when gradient elution is used. In many cases, the shape of the calibration curve varies depending on the position of the solute in the chromatogram (and thus the composition of the mobile phase at the detector inlet). A solvent compensation can be introduced either experimentally, by adding a solvent at the column outlet [18,19] or mathematically by exploiting the $\log A$ vs. b relationship as recently introduced by Heron et al. [20].

5.2.3.2 Mobile Phase Additives

Our group has introduced the use of the sensitivity improvement of the ELSD by the addition of triethylamine (TEA) and formic acid (HCOOH) for lipid analysis. This phenomenon was first reported by Gaudin et al. [21] and further investigated by Deschamps [11,17]. As demonstrated in a later study, an equimolar amount of TEA and HCOOH added to the mobile phase increased the ELSD sensitivity, i.e., increased the a term of Equation 5.5, but little or no effect was noticed on the b term of the equation. When applying an experimental setup consisting in flow injection or size exclusion chromatography (SEC), an enhancement in ELSD response with TEA/HCOOH was found to be solvent, solute, flow rate, and temperature dependant. Tripalmitin was among the solutes with the lowest increase in response by this addition (1.6-fold), whereas ceramide III was one of the most affected (4.7-fold). These values were recorded with SEC with pure chloroform as a mobile phase. Higher values (up to 30×) were obtained with other solvents and ceramide III as test solute. The increase in sensitivity was assumed to be due to a supramolecular assembly between the ion pair TEA:HCOOH and solutes. This explains the decrease of this effect with increasing drift tube temperatures.

5.3   APPLICATIONS OF ELS DETECTION

As outlined in the introductive part of this chapter, the pioneering applications of ELSD concern lipids and carbohydrates [2]. At the present time, it would be nearly impossible to draw a comprehensive review of these application fields. As outlined by Christie [22], HPLC methodology evolves to suit the detection techniques available. Thus, by its compatibility with organic solvents, regardless of their UV absorptivity, ELSD has allowed to develop gradient systems able to resolve complex separations such as lipid classes and molecular species analysis. This historical field is still growing and most of the recent applications of ELSD concern lipids and carbohydrates. Amino acids and peptides can also be detected with ELSD. Using volatile counterions, 20 underivatized amino acids could be separated [23]. In this application, ELSD compares favorably with conductivity detector, RI, UV [24], and its response is more consistent than in electrospray mass spectrometry. ELSD was also used in dual detection for peptide mapping [25].

A growing application field of ELSD is the profiling of plant extracts and, more widely, natural products. As outlined by Ganzera and Stuppner [26] in a recent review, herbal drugs may contain active compounds possessing no chromophore and UV detection is not sufficient to ensure their identification and standardization. As already indicated, ELSD is often used to complete the UV or UV-DAD detection [15,27]. The chromatographic profile is then subjected to principal component-based methods of data analysis to distinguish different groups of substances or between productions.

ELSD is also gaining acceptance in the pharmaceutical industry. In a 10-year-old review [28], the application of ELSD was nearly the classical application field of lipids and carbohydrates plus ginkgolides. A recent application showed that a chromatographic method for gentamycin sulfate and related compounds assay using ELSD can be fully validated according to the ICH standards [29]. The detector provided a sufficient linearity in the range studied and the intermediate precision of less than 5% was acceptable. Other applications of ELS detection for the analysis of antibiotics in pharmaceutical can be found, such as amikacin [30], neomycin [31], gentamicin [32], kanamycin [33], tobramycin [34], bacitracin [34], and streptomycin [35]. Methods for Ibendronate [36], bisphosphonates [37], or piperazine [38] were also developed and validated for quality control of pharmaceuticals. A method able to distinguish six types of cellulose ether and ester derivatives was validated for the assay of hypromellose acetate succinate polymer in pharmaceutical formulations [39]. Likewise, an assay of simethicone [40], polysorbate 80 [41], mannitol [42], or even chloride [43] in pharmaceutical formulations was also reported.

5.3.1   MICRO-CHROMATOGRAPHY WITH ELS DETECTION

According to Stolyhwo et al. [16], the residence time of solutes in the detector was only a few ms, which corresponded mainly to the travel time of the solutes across the drift tube. Thus, though a fully miniaturized detector for micro-LC has already been reported [44], several adaptations of commercially existing detectors for micro-LC were realized by modifying the nebulizer. Hoffmann et al. [45], Trones et al. [46,47],

Molander et al. [48], and Alexander [49] introduced a laboratory-made nebulizer in a Varex detector. Both Heron and Tchapla [50], using a Sedere detector, and Cobb et al. [51], using a Polymer Laboratories detector, modified the detector by sliding a silica tubing of appropriate inner dimension into the standard nebulizer nozzle.

Since the ELSD response also depends on the nebulizing solvents, ELSD parameters, and solutes, it can be difficult to compare performances between the systems described in different publications. Still, the detection limit in micro-LC is generally around 10 ng while for capillary LC, the 1 ng range is reached. Those results were obtained by either using nonaqueous mobiles phases for lipids [44,48,51], Irganox [46,47], or amino acids with an acetonitrile:water mobile phase [51]. A 50 pg detection limit was reported for glucose [49] but it was measured in a flow injection analysis system. Thus this result may be due to the extremely low peak dispersion found with this technique. Some authors reported a linear response shape of the ELSD when used in micro-LC. This trend was found to be statistically relevant [50] and this behavior can be explained by the faster increase of the concentration profile encountered in micro-LC [52]. Details on the consequences of modifying the capillary in the nebulizer can be found in the later reference.

5.4 CORONA-CAD DETECTOR

The Corona-CAD is a recently introduced detector [53]. As ELSD, the operating principle of this detector is to spray the mobile phase into droplets and then to detect them after the mobile phase is (partially) evaporated. The principle is due to Dixon and Peterson [3] who used a TSI particle counter at the outlet of a drift tube used to perform the size reduction of the wet aerosol generated by a pneumatic nebulizer. The principle of this particle counter is based on aerosol charging and is incorporated in the commercial detector Corona-CAD: a stream of nitrogen (or air) is ionized by corona discharge and directed into a chamber at the exhaust of the drift tube. In this chamber, the particles are ionized by impacting the positively charged nitrogen. After particles of less than 10 nm are eliminated by an ion trap, charged particles are detected by an electrometer.

The principle of operation of this detector is thus very close to the ELSD. The same theories concerning the nebulization process and solvent evaporation can be applied with Corona-CAD. The main difference is the response of the detector in relation with particle charging. According to Dixon and Peterson, the sensitivity of the particle counter Sm is given by

$$Sm = \frac{3.01 \times 10^{11}}{\rho_p} dp^{-1.89} \tag{5.9}$$

for particles with a diameter $dp > 10$ nm and in this equation Sm is expressed in units of $fA \cdot m^3 \cdot g^{-1}$. As outlined by these authors, despite the dependence on the volumic mass of the particle (ρ_p), the signal of the counter is not significantly affected by the particle composition, but the prediction of the response of the whole detector is complex to establish.

The change in sensitivity with particle size is thus a characteristic the Corona-CAD shares with ELSD. We can perform a limited extrapolation to predict the response shape of this detector. Here also, the key relationship is Equation 5.5 that links the particle size at the drift tube exhaust to the original droplet size and mobile phase composition. According to this equation, the droplet size increases as the cube root of the solute concentration. Thus, the counter sensitivity is supposed to decrease with increasing solute concentration in the mobile phase and the detector to be more sensitive at low than at high concentration. This would be the exact opposite of the ELSD where big particles scatter light more efficiently than little ones. The situation is somewhat more complicated if we also consider the number and size distribution of the wet aerosol depicted in Figure 5.1. As the particle counter eliminates small particles of less than 10 nm, only the distribution tail is detected. When solute concentration increases, not only the particle size but also the number of detectable particles increases at the drift tube outlet. Thus, the number of particles exceeding 10 nm also increases with solute concentration. As the sensitivity in Equation 5.8 is expressed in fA divided by the particle concentration in the sampled volume, the detector response (but not the sensitivity) is still expected to increase with the particle number and size.

Preliminary results from Dixon and Peterson, obtained in flow injection analysis, showed linear calibration curves over three decades. Under classical chromatographic conditions, results from our laboratory show that the Corona-CAD response can be fitted to the classical ELSD equation $y = A \cdot m^b$ but lead to b values near or less than 1. As shown in Figure 5.4, when compared with ELSD, the Corona-CAD appears to be more sensitive for the low concentrations of the calibration curve. Figure 5.5 illustrates the chromatograms of the extreme points of the calibration curve. A and B chromatograms show the 5 μg/mL injection of ceramide IIIB with Corona-CAD (A) and ELSD (B).

The ELSD is at its limit of detection with a signal-to-noise ratio (SNR) near 3 whereas the Corona-CAD is at its limit of quantification with an SNR of about 10. The chromatogram of the higher concentration also demonstrates differences between Corona-CAD (C) and ELSD (D). Ceramide IIIB corresponds to the molecular species oleyl-phytospingosine and the commercial product is assumed to be 95% pure. The impurities are phytosphingosine acylated by other fatty acids. Figure 5.5c shows

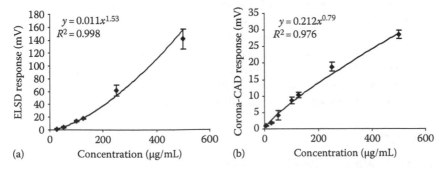

FIGURE 5.4 Calibration curves obtained with ELSD (a) and Corona-CAD (b) for ceramide IIIB. Chromatographic conditions: Hypesil C18 column 50 × 4.6 mm, mobile phase 100% methanol at 1 mL/min, 10 μL 1 injection. (Hazotte, A., EA4041, PhD thesis, University of Paris-Sud.)

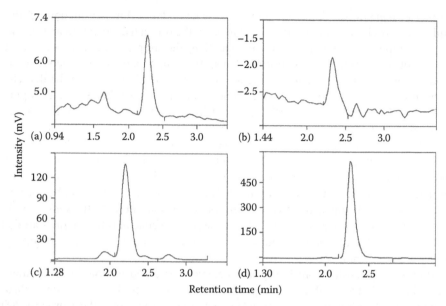

FIGURE 5.5   Zoomed view of the ceramide IIB peak. Left: 5 μg/mL solution of ceramide IIIB with (a) Corona-CAD and (b) ELSD. Right: 250 μg/mL solution of ceramide IIIB with (c) Corona-CAD and (d) ELSD. Chromatographic conditions as in Figure 5.4.

three minor peaks that confirm this statement whereas the chromatographic trace obtained with ELSD only shows the major peak.

Recent publications report similar calibration curves such as Brunelli et al. [54] who recently published the coupling of Corona-CAD with supercritical fluid chromatography in the field of pharmaceutical analysis. This publication also confirms the low influence of solute structure on Corona-CAD response. Górecki et al. [19] using HPLC reported the influence of solvent composition on the response shape of the Corona-CAD and proposed to use a reverse gradient elution at the detector inlet to overcome this problem in gradient elution. Wipf et al. [55] proposed the use of Corona-CAD instead of ELSD for measuring enantiomeric ratios of non-absorbing molecules. Results from this work also show that the Corona-CAD can be used with a linear calibration. Corona-CAD shares the application fields of ELSD as the detection of non-UV absorbing molecules remain an analytical challenge. The first articles on Corona-CAD detection of carbohydrates [56,57] and lipids [58,59] are currently being issued. They confirm the interest of this detector and its possible use as a linear detector on a restricted calibration range.

REFERENCES

1. Charlesworth, J.M.; *Analytical Chemistry*, 1978, 50, 1414.
2. Macrae, R.; Trugo, L.C.; Dick, J.; *Chromatographia*, 1982, 17, 476.
3. Dixon, R.W.; Peterson, D.S.; *Analytical Chemistry*, 2002, 74, 2930.
4. Nukiyama, S.; *Transactions of the Japan Society of Mechanical Engineers*, 1939, 56, S15.
5. Oppenheimer, L.E.; Mourey, T.H.; *Journal of Chromatography A*, 1985, 323, 297.

6. Van der Meeren, P.; Vanderdeelen, J.; Baert, L.; *Analytical Chemistry*, 1992, 64, 1056.
7. Mourey, T.H.; Oppenheimer, L.E.; *Analytical Chemistry*, 1984, 56, 2427.
8. Righezza, M.; Guiochon, G.; *Journal of Liquid Chromatography and Related Technologies*, 1988, 11, 1967.
9. Koropchak, J.A.; Magnusson, L.E.; Heybroek, M.; Sadain, S.; Yang, X.; Anisimov, M.P.; *Advances in Chromatography*, 2000, 40, 275.
10. Slimestad, R.; Vagen, I.M.; *Journal of Chromatography A*, 2006, 1118, 281.
11. Deschamps, F.S.; Gaudin, K.; Lesellier, E.; Tchapla, A.; Ferrier, D.; Baillet, A.; Chaminade, P.; *Chromatographia*, 2001, 54, 607.
12. Heron, S.; Maloumbi, M.; Dreux, M.; Verette, E.; Tchapla, A.; *LCGC Europe*, 2006, 19, 664.
13. Stolyhwo, A.; Martin, M.; Guiochon, G.; *Journal of Liquid Chromatography*, 1987, 10, 1237.
14. Stolyhwo, A.; Colin, H.; Martin, M.; Guiochon, G.; *Journal of Chromatography A*, 1984, 288, 253.
15. van Nederkassel, A.; Vijverman, V.; Massart, D.; Vander Heyden, Y.; *Journal of Chromatography A*, 2005, 1085, 230.
16. Stolyhwo, A.; Colin, H.; Guiochon, G.; *Journal of Chromatography A*, 1983, 265, 1.
17. Frantz, S.; Deschamps, A.B.; Chaminade, P.; *Analyst*, 2002, 127, 35.
18. Héron, S.; *Journal of Chromatography A*, 2004, 1035, 221.
19. Gorécki, T.; Lynen, F.; Szucs, R.; Sandra, P.; *Analytical Chemistry*, 2006, 78, 3186.
20. Heron, S.; Maloumbi, M.; Dreux, M.; Verette, E.; Tchapla, A.; *Journal of Chromatography A*, 2007, 1161, 152.
21. Gaudin, K.; Chaminade, P.; Baillet, A.; Ferrier, D.; Bleton, J.; Goursaud, S.; Tchapla, A.; *Journal of Liquid Chromatography and Related Technologies*, 1999, 22, 379.
22. Christie, W.W.; *Analusis*, 1998, 26, 34.
23. Chaimbault, P.; Petritis, K.; Elfakir, C.; Dreux, M.; *Journal of Chromatography A*, 2000, 870, 254.
24. Petritis, K.; Elfakir, C.; Dreux, M.; *Journal of Chromatography A*, 2002, 961, 21.
25. Bongers, J.; Chen, T.; *Journal of Liquid Chromatography and Related Technologies*, 2000, 23, 933.
26. Ganzera, M.; Stuppner, H.; *Current Pharmaceutical Analysis*, 2005, 1, 144.
27. Lu, Y.; Yu, K.; Qu, H.; Cheng, Y.; *Chromatographia*, 2007, 65, 24.
28. Kohler, M.; Haerdi, W.; Christen, P.; Veuthey, J.; *Trac-Trends in Analytical Chemistry*, 1997, 16, 484.
29. Clarot, I.; Chaimbault, P.; Hasdenteufel, F.; Netter, P.; Nicolas, A.; *Journal of Chromatography A*, 2004, 103, 287.
30. Galanakis, E.; Megoulas, N.; Solich, P.; Koupparis, M.; *Journal of Pharmaceutical and Biomedical Analysis*, 2006, 40, 1120.
31. Megoulas, N.; Koupparis, M.; *Journal of Chromatography A*, 2004, 1057, 131.
32. Megoulas, N.; Koupparis, M.; *Journal of Pharmaceutical and Biomedical Analysis*, 2004, 36, 79.
33. Megoulas, N.; Koupparis, M.; *Analytica Chimica Acta*, 2005, 547, 72.
34. Megoulas, N.; Koupparis, M.; *Analytical and Bioanalytical Chemistry*, 2005, 382, 296.
35. Sarri, A.; Megoulas, N.; Koupparis, M.; *Analytica Chimica Acta*, 2006, 573, 257.
36. Jiang, Y.; Xie, Z.; *Chromatographia*, 2005, 62, 261.
37. Xie, Z.; Jiang, Y.; Zhang, D.; *Journal of Chromatography A*, 2006, 1104, 178.
38. McClintic, C.; Remick, D.; Peterson, J.; Risley, D.; *Journal of Liquid Chromatography and Related Technologies*, 2003, 26, 3104.
39. Rashan, J.; Chen, R.; *Journal of Pharmaceutical and Biomedical Analysis*, 2007, 44, 28.
40. Moore, D.; Liu, T.; Miao, W.; Edwards, A.; Elliss, R.; *Journal of Pharmaceutical and Biomedical Analysis*, 2002, 30, 278.

41. Nair, L.; Stephens, N.; Vincent, S.; Raghavan, N.; Sand, P.; *Journal of Chromatography A*, 2003, 1012, 86.
42. Risley, D.; Yang, W.; Peterson, J.; *Journal of Separation Science*, 2006, 29, 264.
43. Risley, D.; Peterson, J.; Griffiths, K.; McCarthy, S.; *LCGC*, 1996, 14, 1040.
44. Magnus, B.O.; Andersson, L.; Blomberg, G.; *Journal of Microcolumn Separations*, 1998, 10, 249.
45. Hoffmann, S.; Norli, H.R.; Greibrokk, T.; *Journal of High Resolution Chromatography*, 1989, 12, 260.
46. Trones, R.; Andersen, T.; Hunnes, I.; Greibrokk, T.; *Journal of Chromatography A*, 1998, 814, 55.
47. Trones, R.; Andersen, T.; Greibrokk, T.; *Journal of High Resolution Chromatography*, 1999, 22, 283.
48. Molander, P.; Holm, A.; Lundanes, E.; Greibrokk, T.; Ommundsen, E.; *Journal of High Resolution Chromatography*, 2000, 23, 653.
49. Alexander, J.N.; *Journal of Microcolumn Separations*, 1998, 10, 491.
50. Heron, S.; Tchapla, A.; *Journal of Chromatography A*, 1999, 848, 95.
51. Cobb, Z.; Shaw, P.N.; Lloyd, L.L.; Wrench, N.; Barrett, D.A.; *Journal of Microcolumn Separations*, 2001, 13, 169.
52. Gaudin, K.; Baillet, A.; Chaminade, P.; *Journal of Chromatography A*, 2004, 1051, 43.
53. Gamache, P.H.; McCarthy, R.S.; Freeto, S.M.; Asa, D.J.; Woodcock, M.J.; Laws, K.; Cole, R.O.; *LCGC Europe*, 2005, 18, 345.
54. Brunelli, C.; Gorecki, T.; Zhao, Y.; Sandra, P.; *Analytical Chemistry*, 2007, 79, 2472.
55. Wipf, P.; Werner, S.; Twining, L.A.; Kendall, C.; *Chirality*, 2007, 19, 5.
56. Inagaki, S.; Min, J.Z.; Toyo'oka, T.; *Biomedical Chromatography*, 2007, 21, 338.
57. Dixon, R.W.; Baltzell, G.; *Journal of Chromatography A*, 2006, 1109, 214.
58. Moreau, R.A.; *Lipids*, 2006, 41, 727.
59. Cascone, A.; Eerola, S.; Ritieni, A.; Rizzo, A.; *Journal of Chromatography A*, 2006, 1120, 220.

6 Detection and Determination of Heteroatom-Containing Molecules by HPLC: Inductively Coupled Plasma Mass Spectrometry

Sandra Mounicou, Kasia Bierla, and Joanna Szpunar

CONTENTS

6.1 Introduction 162
6.2 Principles of ICP MS 163
    6.2.1 Introduction of Liquid Samples into ICP MS 163
    6.2.2 Ionization 164
    6.2.3 Interface and Mass Analyzers 164
        6.2.3.1 ICP MS Using Quadrupole Analyzers 165
        6.2.3.2 ICP Time-of-Flight Mass Spectrometry 166
        6.2.3.3 ICP with Sector Field Mass Spectrometry 166
    6.2.4 Interferences and Ways of Their Removal 167
6.3 Coupling of HPLC to ICP MS 169
    6.3.1 Size-Exclusion LC–ICP MS 171
    6.3.2 Ion-Exchange HPLC–ICP MS 171
    6.3.3 Reversed-Phase HPLC–ICP MS 172
    6.3.4 Hydrophilic Interaction Chromatography (HILIC)–ICP MS 173
    6.3.5 HPLC–ICP MS Interfacing via Post-Column Volatilization 174
6.4 Coupling of Capillary and Nanoflow HPLC to ICP MS 174
6.5 Quantification of Metallobiomolecules by Isotope Dilution Methods 176

6.6   Application Areas 177
        6.6.1   Analysis of Redox Species 177
        6.6.2   Speciation of Organoarsenic Compounds in Biological Materials 178
        6.6.3   Speciation of Organoselenium Compounds
                  in Biological Materials 178
        6.6.4   Speciation of Metal Complexes in Microorganisms, Plants,
                  and Food of Plant Origin 179
        6.6.5   Speciation of Metal Complexes with Metallothioneins 179
        6.6.6   Speciation of Metal Complexes in Human Body Fluids and Tissues 180
        6.6.7   Metal Speciation in Pharmacology: Metallodrugs 180
        6.6.8   Detection of Chemical Warfare Agents 181
6.7   Conclusions 181
References 181

6.1   INTRODUCTION

Electrospray mass spectrometry (MS) is by far the most popular compound-specific detector in HPLC of environmental contaminants, pharmaceutical residues, and various kinds of biomolecules [1,2]. The detection specificity is achieved by the accurate mass measurement (by TOF MS or FT ICR MS) or by monitoring specific reactions in the collision cell (multiple reaction monitoring [MRM]) modes. Electrospray ionization is, however, known to be critically dependent on the molecular structure of an analyte and to be vulnerable to matrix effects. Total ion current (TIC) HPLC–ES–MS chromatograms are mainly composed of signals of the major, easily ionizable compounds while information on low-abundant species is often lost. If a minor species arrives at the ionization source accompanied by an easily ionizable major one, the ionization of the former is likely to be suppressed, leading to the absence of a relevant peak in the mass spectrum.

An alternative to the "soft" electrospray ionization is the use of "hard" ionization in a high-temperature plasma. Inductively coupled plasma mass spectrometry (ICP MS) allows the production and analysis of the atomic ions of virtually of the elements of the Periodic Table. Consequently, a molecule can be selectively detected by ICP MS in a chromatographic effluent provided that it contains in its structure a heteroatom, such as, e.g., S, P, Se, or a metal. The principal advantage of ICP MS is that its high (femto- or even attomolar sensitivity) is virtually not affected either by the structure of the analyte's molecules, or by the composition of the mobile phase and the co-eluting sample matrix components. The analytical response is linear over a very wide range (5–6 orders of magnitude) and the isotopic specificity offers the potential of isotope dilution quantification.

Obviously, ICP MS does not provide the molecular specificity as a standalone technique; it does, however, as an HPLC detector. The coupling of HPLC and ICP MS has become an established tool for the detection and determination of metallospecies in environmental studies, plant and animal biochemistry, nutrition and clinical chemistry [2,3]. Owing to the number of separation mechanisms available HPLC–ICP MS is also an efficient and versatile technique to monitor purification of trace quantities of metal complexes with biomolecules prior to their structural analysis by electrospray tandem MS. HPLC–ICP MS with enriched stable isotopes is a unique analytical method by which speciation of both endogenous elements and external tracers can be achieved in a single experiment [4–6]. In spite of the elevated running costs (high argon consumption), ICP MS is currently the sole element-specific detector in elemental speciation analysis of real-world samples by HPLC.

The increasing use of ICP MS during the recent years is linked with the progress in three major areas including (1) advances in the detection of nonmetals, (2) decrease in absolute detection limits of a ICP mass spectrometers, and (3) democratization of the access to stable metal isotopes, including isotopically labeled biomolecules. The chapter discusses the basics of ICP MS detection in HPLC, the recent developments, especially in terms of interfacing it with capillary and nanoHPLC, and overviews the application areas.

6.2 PRINCIPLES OF ICP MS

A general scheme of an ICP MS instrument is presented in Figure 6.1. It consists of a sample introduction system, an inductively coupled plasma torch which serves as ionization source, an interface, a mass analyzer, and a detector. A liquid sample is introduced by means of a nebulizer of which the role is to create a fine and uniform aerosol. The sample components are desolvated, vaporized, atomized, and ionized in the highly energetic high-temperature plasma and the resulting elemental ions are transferred to a mass analyzer where they are separated according to their mass-to-charge ratio before the arrival at the detector.

6.2.1 INTRODUCTION OF LIQUID SAMPLES INTO ICP MS

The most commonly used device is a pneumatic nebulizer, which uses mechanical forces of a gas flow (normally argon at a pressure of 20–30 psi) to generate the sample aerosol. The most popular designs of pneumatic nebulizers, usually made of glass, include concentric, microconcentric, microflow, and crossflow.

In the most popular configuration, a nebulizer is accompanied by a spray chamber of which the function is to reject the larger aerosol droplets. Two main types

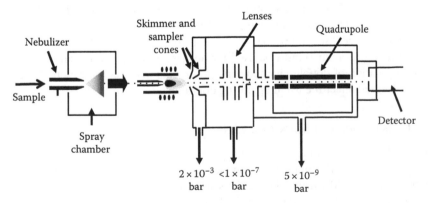

FIGURE 6.1   A general scheme of an ICP MS instrument.

of spray chambers are used in ICP MS instrumentation: double pass and cyclonic spray chambers. The former selects the small droplets by directing the aerosol into a central tube while the larger droplets emerge from the tube and, by gravity, exit the spray chamber via a drain. The cyclonic spray chamber operates by centrifugal force. Droplets are discriminated according to their size by means of a vortex produced by the tangential flow of the sample aerosol and argon gas inside the chamber. Smaller droplets are carried with the gas stream into the ICP MS, while the larger droplets impinge on the walls and fall out through the drain.

Flow rates typical of capillary and nanocapillary columns demand microflow nebulizers discussed in detail in Section 6.4.

6.2.2 Ionization

An inductively coupled plasma (ICP) is usually an argon plasma formed in a quartz torch by coupling radio frequency energy at 27.1 or 40 MHz through a load coil to form an oscillating magnetic field. The torch consists of a set of three concentric tubes. The nebulizer gas, which carries the analytes into the plasma, flows into the central tube, called the injector. The auxiliary gas flows around the injector tube and adjusts the horizontal position of the axial plasma relative to the torch. A third flow: coolant gas flows tangentially through the outer tube, serves as the primary plasma gas, to cool the inside walls of the torch and to center and stabilize the plasma. The temperature in the plasma can attain 6,000–10,000 K and the energy is sufficient to ionize all but a few elements from the Periodic Table. A general advantage of the ICP as an ion source is its capability to generate primarily monoatomic positive ions from most elements. The ionization energy or ionization potential (the energy necessary to remove an electron from the neutral atom) is a minimum for the alkali metals, which have a single electron outside a closed shell and generally increases across a row on the periodic maximum for the noble gases that have closed shells. Most elements are ionized at 90% in the 6000°C of an ICP. The exceptions are As (52%), Se (33%), S (14%), F (9×10^{-4}%).

The ions formed by the ICP discharge are typically positive (M^+ or M^{+2}). Consequently, electronegative elements, such as Cl, I, F, and Br are more difficult to determine by ICP MS. For elements such as S, Se, P, K, and Ca, isobaric and molecular interferences from either the sample matrix or plasma species interfere with the primary isotope (cf. Section 6.2.4). This means that less abundant isotopes with less interference (if available) must be used for determination of these elements. Only a small population of double-charged ions (main example: Ba for which 1%–2% of Ba^{++}) is formed under the usual operating conditions.

6.2.3 Interface and Mass Analyzers

The ICP MS torch operates at the atmospheric pressure. The ions formed must then be introduced into the mass spectrometer via the interface cones. The interface region in the ICP MS transmits the ions travelling in the argon sample stream at atmospheric pressure (1–2 torr) into the low pressure region of the mass spectrometer ($<1 \times 10^{-5}$ torr). This is done through the intermediate vacuum region created

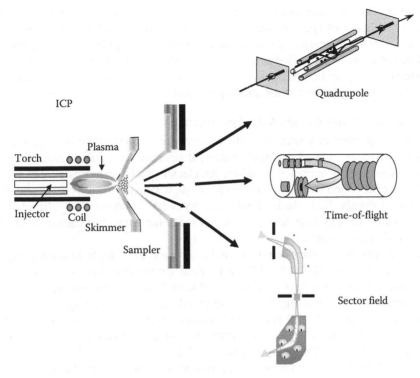

FIGURE 6.2   Mass analyzers used in ICP MS instruments.

by the two interface cones, the sampler and the skimmer. They are usually made of nickel but some applications, notably associated with the addition of oxygen to the plasma gas (necessary an organic modifier present in the mobile phase), required cones made of platinum.

Mass analyzers typically employed for the separation of ions generated in an ICP typically include quadrupole, time-of-flight (TOF), and sector-field instruments (Figure 6.2). The simplicity of the elemental mass spectra makes the use of a quadrupole analyzer sufficient for most applications.

6.2.3.1   ICP MS Using Quadrupole Analyzers

The most popular mass analyzer, a quadrupole mass filter consists of four parallel metal rods arranged in a way that two opposite rods have an applied potential of $(U + V\cos(\omega t))$ and the other two rods have potential of $-(U + V\cos(\omega t))$, where U is a dc voltage and $V\cos(\omega t)$ is an ac voltage. The applied voltages affect the trajectory of ions travelling down the flight path centered between the four rods. For given dc and ac voltages, only ions of a certain mass-to-charge ratio pass through the quadrupole filter and all other ions are thrown out of their original path.

The major advantages of a quadrupole analyzer include its simplicity, relatively low cost, and thus availability. The absolute detection limits are in the sub-picogram range although latest generation instruments allow values down to 1 fg be obtained for some elements. The unit resolution makes the effect of polyatomic interferences prevent the determination of some elements in some matrices, e.g., ^{75}As in sea water (^{40}Ar^{35}Cl). The overall performance is remarkably good, especially that in chromatography where the interfering ions are often separated from the analyte species.

The detection of some elements (e.g., Fe, V, Cr, As, Se) plagued by polyatomic interferences (cf. Section 6.2.4) is the use of multipole collision/dynamic reaction cells between the plasma and the quadrupole analyzer [7,8].

6.2.3.2   ICP Time-of-Flight Mass Spectrometry

TOF mass spectrometry (MS) uses the differences in transit time through a drift region to separate ions of different masses. It operates in a pulsed mode so ions must be produced or extracted in pulses. An electric field accelerates all ions into a field-free drift region with a kinetic energy of qV, where q is the ion charge and V is the applied voltage. Since the ion kinetic energy is expressed as $0.5 mv^2$, lighter ions have a higher velocity than heavier ones and reach the detector installed at the end of the drift region sooner.

The ability to produce complete mass spectra at a high frequency (typically >20,000 s^{-1}) makes TOF MS nearly ideal for the detection of transient signals produced by high-speed chromatographic techniques. The measurement of a time-dependent, transient signal by a sequential scanning using a quadrupole or a sector-field (single collector) mass spectrometer results in two major types of difficulties. The first one is the limited number of isotope intensity measurements that can be carried out within the time-span of a chromatographic peak. The other one is the quantification error known as spectral skew, which arises during the measurement of adjacent mass-spectral peaks at different times along a transient signal. Alleviating these difficulties requires increasing the number of measurement points per time unit and the simultaneous measurement of the isotopes of which the ratio is investigated.

The simultaneous extraction of all m/z ions for mass analysis in TOF MS eliminates the quantification errors of spectral skew, reduces multiplicative noise, and makes TOF MS a valuable tool for determining multiple transient isotopic ratios [9,10]. ICP TOF MS suffers, however, from the lower sensitivity in the monoelemental mode in comparison with the last generation of ICP quadrupole mass spectrometers. This loss of sensitivity is compensated by the fact that the number of isotopes determined during one chromatographic run is no longer limited by peak definition (like in ICP MS employing a quadrupole analyzer). The number of data points per chromatographic peak is independent of the number of measured isotopes.

TOF MS was used in combination HPLC [11–13] but its practical advantages over quadrupole MS for speciation analysis in real samples are far from being convincingly demonstrated, mainly due to its relatively low sensitivity. The need for the simultaneous detection of more than eight isotopes in HPLC is still scarce.

6.2.3.3   ICP with Sector Field Mass Spectrometry

A high-resolution instrument is based on the sequential focusing of an elemental ion using a magnetic sector and electrostatic sector analyzers. Consequently, resolution can be increased up to 10,000; already a resolution of 3,000 allows the elimination of many polyatomic interferences. The increase in resolution brings a decrease in sensitivity but allows the easy determination of elements suffering from polyatomic interferences. Another advantage is the reduction of the background count (owing to preventing the photons formed in the plasma to reach the detector) and a considerable decrease of the detection limits. Consequently, the detection limits (in the low [ca. 300] resolution mode!) are about one order of magnitude lower than those of a quadrupole spectrometer. This gap is shrinking following the introduction of quadrupole instruments equipped with a collision/reaction cell. A wider expansion of high-resolution ICP mass spectrometers (with potentially lower detection limit and larger freedom of interferences) is hampered by the prohibitive cost of instrumentation and the high maintenance costs.

A number of applications of SEC–HR ICP MS have been shown but the advantages in comparison with the use of quadrupole analyzer were unconvincing, the detection limits being controlled by contamination and the irreproducibility of chromatography of sub-picogram amounts of trace element complexes. The most interesting application for HPLC detection of sector field MS is the determination of sulfur and phosphorus, which is of interest in the detection of impurities of drugs [14]. DNA complexes [15,16], and, increasingly, in heteroatom-tagged proteomic applications [17].

Multicollector instruments offer a higher precision for the determination of isotope ratios than those equipped with a single collector when used in the standalone mode. However, their performance in the measurement of isotopic ratios in transient signals is still controversial [18–23].

6.2.4 Interferences and Ways of Their Removal

The most common interferences are summarized in Table 6.1. They are generated by an isotope from a different element (isobaric interference) or by the association of two or more isotopes (polyatomic interference) with a sufficiently close mass/charge ratio to overlap with the analyte signal. They are usually associated with the plasma and nebulizer gas used, matrix components in the sample and solvent, other analyte elements, or entrained oxygen or nitrogen from the surrounding air. Oxides, hydroxides, and hydrides are produced by elements in the sample combining with H, ^{16}O, or ^{16}OH (either from water or air) to form hydride (^{1}H), oxide (^{16}O), and hydroxide ($^{16}O^{1}H$) ions, which occur at 1, 16, and 17 mass units higher than its mass. These interferences are typically produced in the cooler zones of the plasma, immediately before the interface region. Spectral interferences can also be formed from doubly charged ions which produce a peak at half its mass. The formation of oxides and the level of doubly charged species are related to the ionization conditions in the plasma and can usually be minimized by optimization of the nebulizer gas flow, rf power, and sampling position within the plasma.

The ways to compensate for spectral interferences include (1) mathematical correction equations, (2) use of low-temperature plasma, (3) use of collision/reaction cells, and (4) use of high-resolution mass analyzers.

The principle of mathematical correction is based on measuring the intensity of the interfering isotope or of the interfering species at another mass, which, ideally, is free of any interference. A correction is then applied by knowing the ratio of the intensity of the interfering species at the analyte mass to its intensity at the alternate mass.

TABLE 6.1
The Most Common Interferences
Encountered in ICP MS

Analyte Ion	Interference
^{39}K	^{38}ArH
^{40}Ca	^{40}Ar
^{56}Fe	$^{40}Ar^{16}O$
^{80}Se	$^{40}Ar^{40}Ar$
^{51}V	$^{35}Cl^{16}O$
^{75}As	$^{40}Ar^{35}Ci$
^{28}Si	$^{14}N^{14}N$
^{44}Ca	$^{14}N^{14}N^{16}O$
^{55}Mn	$^{40}Ar^{15}N$
^{48}Ti	$^{32}S^{16}O$
^{52}Cr	$^{34}S^{18}O$
^{64}Zn	$^{32}S^{16}O^{16}O$
^{63}Cu	$^{31}P^{16}O^{16}O$
^{24}Mg	$^{12}C^{12}C$
^{52}Cr	$^{40}Ar^{12}C$
^{65}Cu	$^{48}Ca^{16}OH$
^{64}Zn	$^{48}Ca^{16}O$
^{63}Cu	$^{40}Ar^{23}Na$
^{56}Fe	$^{40}Ca^{16}O$
^{64}Zn	$^{48}Ti^{16}O$
^{114}Cd	$^{98}Mo^{16}O$
$^{154}Sm, ^{154}Gd$	$^{138}Ba^{16}O$
^{155}Gd	$^{139}La^{16}O$
$^{156}Gd, ^{156}Dy$	$^{140}Ce^{16}O$
^{57}Fe	$^{40}Ca^{16}OH$
^{66}Zn	$^{31}P^{18}O^{16}OH$
^{80}Se	^{79}BrH
^{64}Zn	$^{31}P^{16}O_2H$
^{69}Ga	$^{138}Ba^{2+}$
^{69}Ga	$^{139}La^{2+}$
$^{70}Ge, ^{70}Zn$	$^{140}Ce^{2+}$

The use a low-temperature plasma (so-called *cold plasma*) can minimize the formation of certain argon-based polyatomic species. Under normal plasma conditions (1000–1400 W rf power), Ar^+ ions combine with matrix and solvent components to generate spectral interferences such as $^{38}Ar^1H$, ^{40}Ar, and $^{40}Ar^{16}O$, which impact the detection limits of some elements, e.g., ^{39}K, ^{40}Ca, and ^{56}Fe. Under cool plasma conditions (500–800 W rf power), many of these interferences are dramatically reduced, which results in significant enhancement of detection limits for this group of elements [24,25].

The use of collision or dynamic reaction cells in front of the quadrupole instruments is becoming more and more widespread in order to eliminate the polyatomic

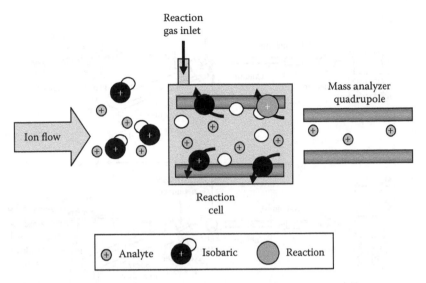

FIGURE 6.3   The principle of operation of a collision cell.

interferences [7,8]. A collision cell is a quadrupole placed between the ion lenses and the quadrupole mass filter, which is pressurized with a reactive gas (e.g., H_2, He). It offers two possible approaches to the chemical resolution of interferences: (1) by collision/reaction of the interfering polyatomic ions with gases such as H_2, He, or Xe, and (2) by reaction of the ion of interest with oxygen. In the first case, a polyatomic ion is destroyed and thus removed from the m/z range of interest whereas in the second case a new product ion is formed and detected at a new, non-interfered m/z value. The principle of operation of a collision cell is presented in Figure 6.3.

An efficient but costly way of interference removal is the use of high-resolution double focusing instruments (cf. Section 6.2.3.3) [26]. They can be used in a low-resolution (R = 300) mode for the analysis of non-interfered isotopes. Medium resolution (R = 4000) guarantees interference-free analysis for most elements in the majority of sample matrices. For example, transition elements are routinely measured in medium resolution due to the formation of many interfering polyatomic species in the mass range 24–70 u. High resolution (R = 10,000) is used for the analysis of elements in the most challenging sample matrices. For example, high resolution is used to separate ^{80}Se and ^{75}As from argon dimer ($^{40}Ar_2$) and argon chloride ($^{40}Ar^{35}Cl$) interferences in chlorine-rich matrices. An increase in resolution, however, entails a loss of transmission and hence of sensitivity.

6.3   COUPLING OF HPLC TO ICP MS

In the simplest case, the coupling of HPLC to ICP MS is achieved by connecting the exit of an HPLC column (i.d. 4.6–10 mm) to a pneumatic or crossflow nebulizer (Figure 6.4a). This interface is principally used when the sample solution is available in sufficient volume (>0.5 mL). The peak width is sufficiently large to allow the use of a quadrupole mass analyzer to simultaneously monitor 12 isotopes. The use of

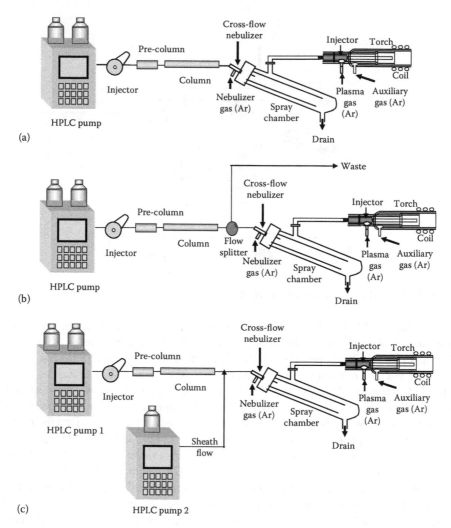

FIGURE 6.4 The HPLC–ICP MS coupling: (a) standard configuration, (b) post-column eluent splitting, (c) post-column dilution.

microbore (i.d. 1.0 mm) HPLC systems, which are becoming popular especially for reversed-phase chromatography, requires the use of micronebulizers, either direct injection (DIN, DIHEN), or micronebulizers (e.g., Micromist) fitted with a small-volume cyclonic nebulization chamber. The optimization of the interface is limited to the choice of the nebulizer matching the flow rate from the column and assuring the stability of the plasma in the presence of the mobile phase. In specific applications, the use of post-column eluent splitting (Figure 6.4b) or a sheath flow (Figure 6.4c) may be necessary.

Several elements can be detected in the same chromatographic peak by scanning or hopping the mass analyzer. The advantage of ICP MS is the independence of the signal intensity of the molecular environment of the determined element and thus the possibility of using inorganic elemental standards for the quantification of high molecular species such as proteins [27]. Indeed, as long as the molecule concentration was sufficiently low, the matrix interference from the surrounding molecule was found to be almost negligible and inorganic quantification standards could be used for quantification with an accuracy of 10% or better [27].

The principal HPLC separation mechanisms used in element speciation analysis include size-exclusion, ion-exchange, and reversed-phase chromatography. The choice of the separation mechanism depends on the physiochemical properties of the analyte species. Separations of trace (picogram) quantities of metal complexes with biomolecules are still under development and many problems, such as unspecific adsorption, species decomposition, and unpredictable memory effects occur.

6.3.1 SIZE-EXCLUSION LC–ICP MS

The number of theoretical plates in SE HPLC is small; this technique is not only insufficient for the discrimination of the small amino acid heterogeneities in the chromatography of metallopeptides but also lacks the resolution in frequently encountered problems, such as, e.g., the separation of blood selenoproteins [28,29] or separation of the human albumin and transferring [30]. Each fraction eluted from a size-exclusion column may still contain hundreds of compounds. In most cases, further signal characterization by orthogonal (complementary) chromatographic techniques is necessary. Mobile phases typically used in size-exclusion separation are generally well tolerated by ICP MS unless high concentrations of salts are required, up to 50 mM Tris-HCl was found to be well tolerated by ICP MS. Size-exclusion LC–ICP MS is a convenient way for the preliminary screening of biological extracts for the presence of metallospecies as the separation conditions are fairly non-denaturating and the columns tolerate fairly concentrated and matrix-rich samples. An example of the application of SEC–ICP MS for the detection of Pb-containing biomolecules in plant cytosol is shown in Figure 6.5.

6.3.2 ION-EXCHANGE HPLC–ICP MS

Ion-exchange chromatography is commonly applied to the speciation of redox forms of As, Se, Sb, Cr, and Fe. Biochemical applications, such as fractionation of metallothionein [11,31–33] and serum proteins [34–36] are becoming popular. Separations on ion-exchange resins require eluents containing high salt concentrations in order to elute multicharged anions and cations in a reasonable time. Buffers concentrations used often exceed 0.1 M. They are likely to affect negatively the sensitivity of the system on the long term because of the clogging of the nebulizer and sampler and skimmer cones. For example, the high potential of separation of the MT-1 and MT-2 isoforms by anion exchange is not fully exploited in the coupled systems because the common end concentration of 0.25 M of buffer is difficult to be tolerated by ICP MS. Much lower buffer concentrations were used in AE chromatography of organometalloid species. Buffer concentrations of 3–20 mM seem to be sufficient [33–40].

Cation-exchange HPLC has been less popular than anion-exchange HPLC. Separations can be carried out in acidic media using several mM pyridine-formate buffer [37] or simply few mM HNO_3 [38].

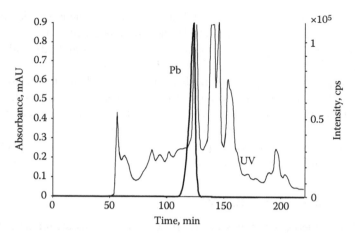

FIGURE 6.5 Size-exclusion LC–ICP MS screening of a plant (*Biscutella leavigata*) cytosol for the presence of Pb-containing species (Peptide Superdex, elution with 10 mM ammonium acetate buffer, pH 7.5).

6.3.3 REVERSED-PHASE HPLC–ICP MS

Reversed-phase HPLC is the preferred separation technique for weakly polar compounds in application areas such as peptide mapping, and amino acid, phospholipid, and protein separations. The chromatography often requires the use of mobile phases, which contain up to 95% of methanol or acetonitrile. The introduction of an organic solvent above a certain concentration level (usually 20%–30% methanol and 10% acetonitrile at 1 mL min^{-1}) into the ICP MS is known to affect negatively the ICP stability, to lead to a decrease in signal intensity and to the deposition of carbon on the cones. The removal of the solvent vapor, using a cooled spray chamber or a membrane desolvator accompanied by the addition of oxygen to the plasma gas and, consequently, the use of platinum cones, are the usual measures taken.

Because an organic solvent modifies the plasma ionization conditions, its concentration has a significant effect on the signal intensity in ICP MS. For elements with high first-ionization energies, the enhancement of the sensitivity in ICP MS by organic modifiers was reported [3] and explained by a charge-transfer reaction from the ionized carbon in the plasma to the incompletely ionized analyte [3]. The influence of an organic modifier in the analyte solvent on the detection sensitivity was tested for sulfur, phosphorus, and iodine [39]. Phosphorus and iodine show an increased signal proportional to the concentration of organic modifier with an intensity maximum at 40%–60%, which corresponds to the metal-like characteristics. For the detection of sulfur, this characteristics are set off by a signal-suppression effect, proportional to the amount of organic modifier [39].

The use of organics-rich mobile phases in HPLC often requires a post-column eluent splitting (Figure 6.4b) or dilution (Figure 6.4c), and/or the reduction of the flow rate. In an application to phospholipid analysis, the solvent load to the plasma was reduced by a fivefold splitting of the mobile phase prior to reaching the nebulizer, by chilling the spray chamber to −5°C, and by an optimization of carrier-gas flow for maximum condensation of organic vapors [40]. In a study of the levothyroxine degradation products, the mobile phase that contain >20% acetonitrile at 300 μL min⁻¹ was diluted (1:3.3) post-column online with a 2% (v/v) nitric acid solution [41]. Narrow peaks can be obtained by HPLC–ICP MS interfaced via a direct injection nebulizer, as shown for the determination of cyanocobalamin and its analogs [42].

The use of ion-pairing reagents in the mobile phase (ion-interaction chromatography) allows the extension of applications of RP HPLC to ionic analytes which would otherwise have been retained on the column. In particular, the latter mode is employed for speciation of organoarsenic [43] and organoselenium [44] compounds. The separation of metal complexes with MTs by RP–HPLC was reviewed [45].

6.3.4 HYDROPHILIC INTERACTION CHROMATOGRAPHY (HILIC)–ICP MS

The separation mechanism is opposite to reversed-phase chromatography. Columns used for HILIC will retain solutes solely through hydrophilic interactions when using mobile phase concentrations in the range of 40%–85% acetonitrile. The least polar compounds elute first whereas the polar ones are retained which makes HILIC the mechanism of choice for the analysis of compounds which usually elutes in the void of RPLC (like most metal complexes do). Buffer concentrations must be selected carefully to avoid reduction in retention (too high concentration) or excessive equilibration times (too low concentrations) [46]. An example of HILIC–ICP MS for detection of Ni complexes in a metal hyperaccumulating plant is presented in Figure 6.6.

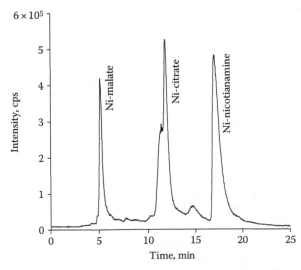

FIGURE 6.6 Hydrophilic interaction chromatography (HILIC)–ICP MS separation of low molecular nickel complexes in a hyperaccumulating plant (*Thlaspi caerulescens*); column Ultra IBD, elution with an increasing concentration of ammonium acetate in methanol–water. (From Gonzalez LaFuente, J.M. et al., *Talanta*, 50, 207, 1999.)

6.3.5 HPLC–ICP MS Interfacing via Post-Column Volatilization

The need for a nebulizer between an HPLC column and an ICP MS spectrometer can be eliminated by the post-column conversion of the elemental species into volatile species (usually hydrides) that can be swept into the plasma. In comparison with the pneumatic nebulization, the online microwave-assisted digestion–hydride generation interface offers 20–100 fold increase in sensitivity and elimination of interferences from the sample matrix or mobile phase components. When a sensitive detector, such as ICP MS is used, the reagent blank values increase the background noise and thus reduce the gain in detection limit to a factor of 2–10 [47].

6.4 COUPLING OF CAPILLARY AND NANOFLOW HPLC TO ICP MS

Problems related to the introduction of organic solvents used in RP HPLC–ICP MS can be alleviated owing to the availability of RP columns with different dimensions. In this way, the nebulizer flow rate can be readily reduced to the level tolerated by the detection system. The flow rate depends strongly on the column geometry and varying the column inner diameter from 8 to 0.075 mm allows changing the flow rate from 10 mL min^{-1} to 300 nL min^{-1} according to what is necessary. Capillary and nanoflow HPLC techniques are becoming increasingly popular to analyze microsamples, such as, e.g., digests of spots in 2D electrophoresis, compartments of individual cells, or human biopsy extracts. They are robust, provide high resolution, especially when used in the gradient mode, and can be used in single or multidimensional separations. Their use is stimulated by the development of systems with the parallel ICP MS and ESI MS/MS detection from heteroatom-tagged proteomics and metallometabollomics [48].

The coupling of capillary and especially of nanoflow HPLC, with ICP MS requires interfaces capable of overcoming the discrepancy (100–1000 times) between the column flow rates and those required by conventional nebulizers. Also, the large dead volume (40–100 cm^3) of the double-pass Scott spray chamber results in long washout times and peak broadening. Recently, a number of dedicated interfaces between capillary HPLC and ICP MS were reported. They were based on different micronebulizers: the MCN 6000 [49], the PFA 100 [49–52], and the direct injection high efficiency nebulizer (DIHEN) [39]. The disadvantages of the former two included the sample uptake rates of 10–100 µL, which were still higher than those demanded by capillary HPLC (4 µL min^{-1}), and the large dead volume of the desolvation system of the MCN 6000. Even with the modified DIHEN, the peaks were relatively broad (15 s peak width at half height) [39].

The potential of capillary and nanoflow HPLC–ICP MS coupling was reviewed [53]. The most promising seem to be interfaces based on a total consumption micronebulizer (DS-5) that operates at flow rates in the range 0.5–7.5 µL min^{-1} [54]. The interfaces between ICP MS and capillary HPLC (300 µm) [55] and nano-HPLC (75 µm) [48] were described. The sheathless interface between capillary HPLC and ICPMS [55] allowed the efficient nebulization and transport into the plasma of mobile phases containing up to 100% organic solvent without either cooling the spray chamber or oxygen addition. The minimal peak broadening (5 s at the half height) allowed baseline resolution of a mixture containing more than 30 selenopeptides, many of which could not be separated using the conventional HPLC–ICP MS coupling [55]. The coupling of nanoflow HPLC to ICP MS, using a sheath flow made possible the nonspecific isotope dilution [56]. The peak width at half-height was reduced to 3 s for a 11 nL sample volume injection. Such small sample injection volumes are often not practical because ICP MS is sensitive to the mass of an analyte and not to its concentration. Therefore, nanoHPLC–ICP MS should be combined with a preconcentration step, which can be easily automated using commercial 2D HPLC systems.

The use of capillary HPLC (flow rate of 4 µL min⁻¹) and nanoHPLC (flow rate of 200 nL min⁻¹) is likely to alleviate problems with organic-rich mobile phases. At such low flow rates, the introduction of up to 100% of organic solvent becomes possible without either cooling the spray chamber or oxygen addition [55,57]. A considerable loss of signal intensity was nevertheless observed at acetonitrile concentrations >80% introduced into and ICP at 4 µL min⁻¹ [55,57]. The addition of a post-column sheath flow to couple nanoHPLC with ICP MS demonstrated that a sheath flow could buffer the influence of the organic-rich mobile phase, and thus the signal intensity remained stable during an organic solvent gradient in reversed-phase HPLC [56].

A novel nebulizer working at sample uptake rates of less than 500 nL min⁻¹ was developed for a sheathless interfacing of nanoHPLC (75 µm column i.d.) with ICP MS [48]. It is schematically shown in Figure 6.7.

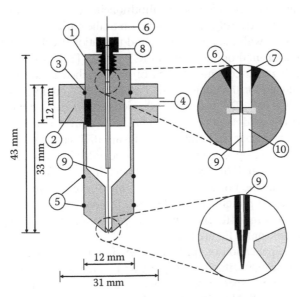

FIGURE 6.7 A scheme of the nebulizer allowing a sheathless interfacing of nanoHPLC and ICP MS; (1), nebulizer body (male part); (2), nebulizer body (female part); (3), *o*-ring; (4), argon inlet; (5), *o*-rings; (6), fused silica capillary; (7), Teflon holder; (8), PEEK screw; (9), fused silica needle; (o.d. 150, i.d. 20, orifice 10 µm), (10), Teflon holder.

6.5 QUANTIFICATION OF METALLOBIOMOLECULES BY ISOTOPE DILUTION METHODS

The two principal approaches to improve the precision and accuracy of the quantification of biomolecules and the related metal complexes by isotope dilution analysis (IDA) include [58–60]

1. Non-speciated IDA (when the isotopic spike ignores the speciation of the analyte compounds). This technique is used for biomolecules, for which isotopically labeled calibration standards are unavailable. It consists of the continuous introduction of isotopically enriched, species-unspecific spike solution after the separation step. Quantification by external calibration gives rise to problems, which result from matrix-induced differences in detector sensitivity between standard and sample.
2. Speciated IDA (when a species-specific spike is used). An isotopically labeled analyte species is added to the sample and is supposed to co-elute with the analyte species after an entire analytical procedure. Because the basis of quantification is the measurement of the isotope ratio of the speciated element in the mixture that produces a chromatographic peak, incomplete recoveries and matrix effects can be corrected for. The use of this approach is limited by the availability of an isotopically labeled analyte molecule and the equilibration of the spike with the analyte species.

A typical application of non-speciated IDA concerned the determination of the metal stoichiometry and quantification of metallothionein complexes in eel liver cytosol fractionated by SEC–AE FPLC [32]. A solution of the enriched isotopes ^{111}Cd, ^{65}Cu, and ^{67}Zn was mixed with the AE effluent, and the ratios $^{114}Cd/^{111}Cd$, $^{63}Cu/^{65}Cu$, and $^{64}Zn/^{67}Zn$ were measured online using an ICP (Q)MS [32]. An online isotope dilution method, using post-column addition of the enriched isotopes ^{65}Cu, ^{67}Zn, and ^{106}Cd was developed for the quantification of metal-MT complexes by HPLC–ICP TOF MS [11]. ICP MS IDA was also proposed for the determination of oxidized metallothioneins by a Cd-saturation method in rat liver cytosol [61], and for selenium species in yeast and flour [62] and blood plasma [63] samples. Nanoflow HPLC–ICP MS IDA was developed to quantify selenopeptides in a tryptic digest of a selenium-containing protein [56]. ^{76}Se was introduced in the makeup flow. This method enabled the determination of tryptic peptides, miscleaved and/or oxidized peptides, incompletely digested protein, and undigested protein in one run, and allowed the precise evaluation of the efficiency and quality of tryptic digestion by using several nanoliters of sample only (Figure 6.8).

The principal application of speciated IDA, in the here-reviewed area, was the determination of selenomethionine in yeast and gluten [64–66] and in serum [57]. A method based on the species-specific IDA was developed for the accurate determination of Asp-Tyr-SeMet-Gly-Ala-Ala-Lys in a tryptic digest of an aqueous extract of selenized yeast [67]. For this purpose, a ^{77}Se-labeled peptide standard had been purified from yeast grown on $^{77}SeO_4$-rich culture by 2D LC and quantified by reversed IDA. The sample mixed with the ^{77}Se-labeled peptide spike was analyzed

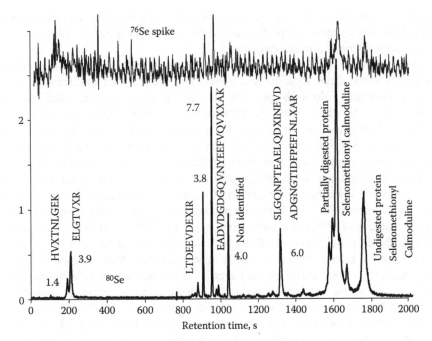

FIGURE 6.8 Quantification of selenium species in tryptic digestate of selenized calmoduline by nonspecific isotope dilution nanoRP HPLC–ICP MS; column: C18 PepMap, elution: gradient 5%–80% water–CH₃CN (0.1% TFA). The numbers denote the Se content in individual peaks in pg.

with capillary HPLC–ICP collision cell MS. The use of a labeled peptide allowed the correction for retention time shift and possible peak distortion due to the injection of a complex salt-rich matrix onto a capillary column. The isotope ratio of selenium ($^{78}Se/^{80}Se$) was measured in the peak corresponding to the peptide of interest, allowing its accurate quantification. The determined concentration of the peptide, which was quantitatively formed from a selenized 12 kDa heat shock protein, allowed the quantification of the latter by cHPLC–ICP MS directly in a yeast extract, without any additional purification [67].

6.6 APPLICATION AREAS

6.6.1 ANALYSIS OF REDOX SPECIES

Differentiation of redox elements species is of primary concern in occupational health and environmental studies. The species of interest can be separated either in native states or one or both can be derivatized to produce chromatographable species. Another important consideration is the tendency of some elements to hydrolyze, which requires the presence of a complexing agent in the mobile phase. The problems to be solved during the method development include the need for maintaining the species stability on the timescale of the chromatographic or electrophoretic run, and controlling polyatomic interferences in ICP MS, especially in the analysis of carbon or chlorine-rich matrices, such as humus-rich samples or seawater, respectively. The analysis should be performed as soon as possible after sampling to avoid species degradation and interconversion.

Anion-exchange HPLC–ICP MS is a versatile analytical technique for speciation analysis of redox states. Efficient multielement separations of metal states in environmental matrices followed by the online ICP MS detection were reported [68]. The most widely analyzed have been the redox states of As [69–71], Cr (66) and Se [68].

Arsenic, highly toxic as As(III), should be differentiated from As(V) and organoarsenic compounds that show a by far lesser toxicity. Inorganic arsenic is predominant in ground water whereas monomethylarsinic acid and dimethylarsonic acid are the most commonly detected species in marine water sand sediment and soil extracts The different bioavailability and toxicity of chromium species [Cr(VI) is more toxic than Cr(III)] has both occupational and health implications. The analytical technique of choice is usually anion-exchange HPLC with ICP MS detection [72–74]. The essential Se(IV) needs to be differentiated from Se(VI); trimethylselenonium cation, selenomethionine, and selenocysteine are often included in separations, especially in extracts biological materials [44]. The trace analysis of bromate is becoming increasingly important as this ion is believed to be carcinogenic and may be formed during the oxidative processes of drinking water [75–78]. The inorganic species of Sb are more toxic than the methylated ones, and Sb(III) is 10 times more toxic than Sb(V). Sb(V), a number of environmental studies on Sb speciation have been reported [79,80].

6.6.2 Speciation of Organoarsenic Compounds in Biological Materials

The toxicity of arsenic is critically dependent on the chemical form in which it occurs. Speciation of arsenic in food is of particular interest due to the potential accumulation of arsenic in the food chain and the risk to man. The metabolism of inorganic arsenic by marine and terrestrial plants and animals leads to the formation of a range of organic arsenic species that may be considered as naturally occurring compounds. The most widely referred to of this group is arsenobetaine, which is the major organoarsenic compound in marine animals, and arsinoyl-ribosides (arsenosugars), which are products of the As metabolism in marine plants and some bivalves. Another field of interest includes studies of the metabolism of arsenic following its administration to humans and experimental animals by means of the determination of arsenic speciation in urine. Analytical methods used for speciation of arsenic have been recently reviewed [43].

6.6.3 Speciation of Organoselenium Compounds in Biological Materials

Speciation of organoselenium compounds in biological materials has become one of major topics in species selective chemical analysis. Selenium has been shown both to be essential for life and to be toxic at levels little above those required for health. Dietary levels of the desired amount of Se are in a very narrow range: dietary levels lower than $0.1\,mg\,kg^{-1}$ of this element will result in its deficiency whereas consumption of food containing more than $1\,mg\,kg^{-1}$ will lead to toxic manifestations. Selenium exists in biological systems in the form of inorganic species such as Se(IV) (SeO_3^{2-}—selenite), Se(VI) (SeO_4^{2-}—selenate) or selenides (e.g., HgSe), or in the form of organic species having a range of molecular masses and charges, starting from the simplest MeSeH and ending at complex selenoproteins [81]. Chemical forms of selenium reported in the context of speciation studies are summarized in recent reviews [44,82–84]. The major fields of interest include (1) speciation of volatile Se species in the environment carried out by gas chromatography-based techniques; (2) speciation of redox states in natural waters, soils, and sediments [85–87]; (3) speciation of metabolite products (amino acids and peptides) in microorganisms, plants, nutritional supplements and human [88,89] and animal urine [90] (4) characterization of selenoproteins in yeast [91], plants [92,93], and mammals [28,29,37].

6.6.4 SPECIATION OF METAL COMPLEXES IN MICROORGANISMS, PLANTS, AND FOOD OF PLANT ORIGIN

Various analytical approaches have been proposed to study metal speciation in plants. The species studied most often are phytochelatins (PCs)—a class of oligopeptides composed only of three amino acids: cysteine (Cys), glutamic acid (Glu), and glycine (Gly) with a general formula: $(GluCys)_n Gly$ where $n = 2 - 11$. PCs can detoxify metals by forming a metal–PC complex in which the metal is bound to the thiol group of the cystein unit. Other species investigated by HPLC–ICP MS include metal complexes with polysaccharides, metallophores (hydroxy acids, e.g., citric or malic acid, and non-proteinaceous amino acids), and water-soluble proteins. There have also been a number of reports of the screening of plant extracts for the presence of stable metallospecies without the unambiguous identification of the latter.

Size-exclusion LC–ICP MS is a convenient technique for screening for the presence of stable metal complexes with PCs [31,94], amino acids [46], polysaccharides [79,95], or protein [17] in plants. The use of other separation mechanisms has been scarce probably due to the difficulties with preservation of metal–ligand complexes in other than physiological conditions, HILIC–ICP MS was recently proposed to study hydrophilic metal complexes [46].

6.6.5 SPECIATION OF METAL COMPLEXES WITH METALLOTHIONEINS

HPLC–ICP MS shows a considerable potential for studying metal–metallothionein complexes [45,96]. Metallothioneins (MTs) are a group of nonenzymatic low molecular mass (6–7 kDa), cysteine-rich metal-binding proteins, resistant to thermocoagulation and acid precipitation. They are considered to intervene in the metabolism, homeostatic control, and detoxification of a number of essential (Zn, Cu) and toxic (Cd, Hg, As) trace elements. Challenges related to speciation analysis of the metal–MT include the identification of metals involved in the complex, calculation of the stoichiometry of the complex, and the identification of amino acid sequence of the ligand present. As a result, the requirements for a suitable analytical technique concern (1) selectivity with regard to the different MT isoforms and sub-isoforms, (2) selectivity with regards to metals complexed by an individual MT isoform, (3) sensitivity to cope with baseline (non-induced) MT levels in real-world samples, (4) accurate quantification.

Size-exclusion LC–ICP MS has become a routine technique for the screening of a raw sample extract for the MT fraction at concentrations down to the non-induced levels. Different separation mechanisms have been employed for speciation of MT–metal complexes, each with a different purpose. Size-exclusion chromatography tolerates raw sample extracts and is a useful technique to indicate the concentration of MT in the sample and to isolate the MT fraction prior to a finer characterization by HPLC using other separation mechanisms. The SEC separation efficiency is rather poor, the MT-1 and MT-2 cannot be separated from each other but they both can be separated from the MT-3 fraction [97]. Anion-exchange LC offers at least the separation between the MT-1 and MT-2 isoform classes [11,31]. A finer separation within each of the classes is preferably carried out by reversed-phase HPLC in standard [98], microbore [95,99], or capillary [46] format.

6.6.6 SPECIATION OF METAL COMPLEXES IN HUMAN BODY FLUIDS AND TISSUES

Progress in the understanding of metal functions in metalloproteins, enzymes, and nucleic acids is determined by the availability of the information on metal species in a complex bioligand environment (e.g., blood serum). Speciation studies of trace elements associated with proteins are necessary to elucidate the roles that trace elements play in the structures and functions of biological macromolecules [100].

The largest interest in the field of the biomedical speciation analysis has been enjoyed by some essential metals, such as, e.g., Fe, Cu, and Zn, and toxic metals, such as, e.g., Al, Cr, Pb, Cd, and Hg. The samples include blood (subdivided by centrifugation into plasma [serum] and red cells [erythrocytes]) and breast milk [2]. Liver and kidney have been the most widely studied organs because of their crucial function in the metabolism of toxic metals [2]. The speciation of arsenic and selenium metabolites in urine has been discussed in previous sections.

Size-exclusion LC with ICP MS detection is a convenient method for the fractionation of the metal complexing proteins in serum [101–103] and in milk [65,104–107]. The multielement detection is readily feasible but this technique was also used for the determination of species of a single element. SEC–ICP HR MS was used for the analysis of trace element species in bovine and human serum [101] and in liver extracts [102]. Anion-exchange HPLC–ICP MS was proposed for studies involving trace metal complexes with albumin and transferrin isoforms [34–36,38,108]. Analysis of nickel species in cytosols of normal and malignant human colonie tissues using 2D liquid chromatography with ICP-sector field MS detection was reported [109].

6.6.7 METAL SPECIATION IN PHARMACOLOGY: METALLODRUGS

Platinum (cisplatin, carboplatin), ruthenium, and gold (auranofin) compounds are well known in cancer therapy whereas some other gold compounds (aurithiomalate, aurothioglucose) are important antiarthritic drugs. A wide range of Tc compounds are used for diagnostic imaging of renal, cardiac and cerebral functions, and of various forms of cancer. Gadolinium (III) polyaminopolycarboxylic crown complexes are employed as magnetic resonance imaging contrast reagents. Vanadium and tungsten complexes have been considered for use as insulin mimetics [81].

HPLC–ICP MS plays an important role in studies of metallodrugs including: (1) studies of the drug purity, stability, and chemical transformations that occur under physiological conditions [17,110,111], (2) studies of interactions of drugs and their metabolites with biologically relevant molecules (amino acids, proteins, nucleotides, DNA fragments) [30,112,113], (3) studies of the kinetics of metal binding with blood plasma [114,115].

6.6.8 DETECTION OF CHEMICAL WARFARE AGENTS

Detection of chemical warfare agents and products of their degradation is an emerging field of species selective chemical analysis.

The species of interest include 2-chlorovinylarsenous acid (CVAA) and 2-chlorovinylarsonic acid (CVAOA), which are degradation compounds of the chemical warfare agent lewisite [116], diphenylarsinic acid (DPAA) and phenylarsonic acid (PAA) [117,118], and organophosphorus chemical warfare degradation products such as ethyl methylphosphonic acid (EMPA, the major hydrolysis product of VX), isopropyl methylphosphonic acid (IMPA, the major hydrolysis product of Sarin (GB)), and methylphosphonic acid (MPA, the final hydrolysis product of both) [119].

6.7 CONCLUSIONS

HPLC–ICP MS is a sensitive and robust technique for species-specific detection and determination of compounds containing in their structure a heteroelement, such as, e.g., sulfur, phosphorous, arsenic, selenium, or metal. ICP MS offers the unparalleled detection sensitivity, especially in the presence of a variety of mobile phases and co-elution of matrix components from the sample. HPLC–ICP is the primary technique for screening biological extracts for the presence of such molecules and for the species-specific determination of toxic metal redox forms in the environment. It starts being widely employed for studies of drug metabolites, provided the latter contain a heteroelement. ICP MS is a convenient detection to monitor quantitatively the elution of heteroatom-containing species during purification protocols for electrospray MS. The development trends include the miniaturization of the chromatographic sample introduction and the automations of setups for metallometabollomics and heteroatom-tagged proteomics.

REFERENCES

1. Wang, H.; Hanash, S.; *Journal of Chromatography B: Analytical Technologies in the Biomedical and Life Sciences*, 2003, 787, 11.
2. Szpunar, J.; *Analyst*, 2005, 130, 442.
3. Montes-Bayon, M.; DeNicola, K.; Caruso, J.A.; *Journal of Chromatography A*, 2003, 1000, 457.
4. Suzuki, K.T.; Ogra, Y.; *Food Additives and Contaminants*, 2002, 19, 974.

5. Kobayashi, Y.; Ogra, Y.; Suzuki, K.T.; *Journal of Chromatography B: Biomedical Sciences and Applications*, 2001,760, 73.
6. Shiobara, Y.; Ogra, Y.; Suzuki, K.T.; *Life Sciences*, 2000, 67, 3041.
7. Koppenaal, D.W.; Eiden, G.C.; Barinaga, C.J.; *Journal of Analytical Atomic Spectrometry*, 2004, 19, 561.
8. Tanner, S.D.; Baranov, V.I.; Bandura, D.R.; *Spectrochimica Acta, Part B: Atomic Spectroscopy*, 2002, 57, 1361.
9. Balcerzak, M.; *Analytical Sciences*, 2003, 19, 979.
10. Ferrarello, C.N.; Fernandez de la Campa, M.R.; Sanz-Medel, A.; *Analytical and Bioanalytical Chemistry*, 2002, 373, 412.
11. Infante, H.G.; Van Campenhout, K.; Blust, R.; Adams, F.C.; *Journal of Chromatography A*, 2006, 1121, 184.
12. Ferrarello, C.N.; Ruiz Encinar, J.; Centineo, G.; Garcia Alonso, J.I.; Fernandez de la Campa, M.R.; Sanz-Medel, A.; *Journal of Analytical Atomic Spectrometry*, 2002, 17, 1024.
13. Vazquez Pelaez, M.; Costa-Fernandez, J.M.; Sanz-Medel, A.; *Journal of Analytical Atomic Spectrometry*, 2002, 17, 950.
14. Evans, E.H.; Wolff, J.-C.; Eckers, C.; *Analytical Chemistry*, 2001, 73, 4722.
15. Cartwright, A.J.; Jones, P.; Wolff, J.-C.; Evans, E.H.; *Journal of Analytical Atomic Spectrometry*, 2005, 20, 75.
16. Siethoff, C.; Feldmann, I.; Jakubowski, N.; Linscheid, M.; *Journal of Mass Spectrometry*, 1999, 34, 421.
17. Nageswara Rao, R.; Kumar Talluri, M.V.N.; *Journal of Pharmaceutical and Biomedical Analysis*, 2007, 43, 1.
18. Becker, J.S.; Dietze, H.-J.; *Fresenius' Journal of Analytical Chemistry*, 2000, 368, 23.
19. Clough, R.; Belt, S.T.; Evans, E.H.; Fairman, B.; Catterick, T.; *Analytica Chimica Acta*, 2003, 500, 155.
20. Clough, R.; Belt, S.T.; Evans, E.H.; Fairman, B.; Catterick, T.; *Journal of Analytical Atomic Spectrometry*, 2003, 18, 1039.
21. Clough, R.; Belt, S.T.; Fairman, B.; Catterick, T.; Evans, E.H.; *Journal of Analytical Atomic Spectrometry*, 2005, 20, 1072.
22. Gunther-Leopold, I.; Wernli, B.; Kopajtic, Z.; Gunther, D.; *Analytical and Bioanalytical Chemistry*, 2004, 378, 241.
23. Gunther-Leopold, I.; Waldis, J.K.; Wernli, B.; Kopajtic, Z.; *International Journal of Mass Spectrometry*, 2005, 242, 197.
24. Wollenweber, D.; Straßburg, S.; Wunsch, G.; *Fresenius' Journal of Analytical Chemistry*, 1999, 364, 433.
25. Murphy, K.E.; Long, S.E.; Rearick, M.S.; Ertas, O.S.; *Journal of Analytical Atomic Spectrometry*, 2002, 17, 469.
26. Stuewer, D.; Jakubowski, N.; *Journal of Mass Spectrometry*, 1998, 33, 579.
27. Svantesson, E.; Pettersson, J.; Markides, K.E.; *Journal of Analytical Atomic Spectrometry*, 2002, 17, 491.
28. Palacios, O.; Encinar, J.R.; Bertin, G.; Lobinski, R.; *Analytical and Bioanalytical Chemistry*, 2005, 383, 516.
29. Palacios, O.; Ruiz Encinar, J.; Schaumloffel, D.; Lobinski, R.; *Analytical and Bioanalytical Chemistry*, 2006, 384, 1276.
30. Sulyok, M.; Hann, S.; Hartinger, C.G.; Keppler, B.K.; Stingeder, G.; Koellensperger, G.; *Journal of Analytical Atomic Spectrometry*, 2005, 20, 856.
31. Ferrarello, C.N.; Bayon, M.M.; De La Campa, R.F.; Sanz-Medel, A.; *Journal of Analytical Atomic Spectrometry*, 2000, 15, 1558.
32. Rodriguez-Cea, A.; De La Campa, M.D.R.F.; Gonzalez, E.B.; Fernandez, B.A.; Sanz-Medel, A.; *Journal of Analytical Atomic Spectrometry*, 2003, 18, 1357.

33. Rodriguez-Cea, A.; Linde Arias, A.R.; Fernandez de la Campa, M.R.; Costa Moreira, J.; Sanz-Medel, A.; *Talanta*, 2006, 69, 963.
34. Nagaoka, M.H.; Yamazaki, T.; Maitani, T.; *Biochemical and Biophysical Research Communications*, 2002, 296, 1207.
35. Nagaoka, M.H.; Akiyama, H.; Maitani, T.; *Analyst*, 2004, 129, 51.
36. Belen Soldado Cabezuelo, A.; Montes Bayon, M.; Blanco Gonzalez, E.; Garcia Alonso, J.I.; Sanz-Medel, A.; *Analyst*, 1998, 123, 865–869.
37. Moreno, P.; Quijano, M.A.; Gutierrez, A.M.; Perez-Conde, M.C.; Camara, C.; *Analytica Chimica Acta*, 2004, 524, 315.
38. Nagaoka, M.H.; Maitani, T.; *Journal of Inorganic Biochemistry*, 2005, 99, 1887.
39. Wind, M.; Eisenmengerb, A.; Lehmann, W.D.; *Journal of Analytical Atomic Spectrometry*, 2002, 17, 21.
40. Kovacevic, M.; Leber, R.; Kohlwein, S.D.; Goessler, W.; *Journal of Analytical Atomic Spectrometry*, 2004, 19, 80–84.
41. Kannamkumarath, S.S.; Wuilloud, R.G.; Stalcup, A.; Caruso, J.A.; Patel, H.; Sakr, A.; *Journal of Analytical Atomic Spectrometry*, 2004,19, 107.
42. Chassaigne, H.; Lobinski, R.; *Analytica Chimica Acta*, 1998, 359, 227.
43. Francesconi, K.A.; Kuehnelt, D.; *Analyst*, 2004, 129, 373.
44. Poatajko, A.; Jakubowski, N.; Szpunar, J.; *Journal of Analytical Atomic Spectrometry*, 2006, 21, 639.
45. Prange, A.; Schaumloffel, D.; *Analytical and Bioanalytical Chemistry*, 2002, 373, 441.
46. Montes-Bayon, M.; Profrock, D.; Sanz-Medel, A.; Prange, A.; *Journal of Chromatography A*, 2006, 1114, 138.
47. Gonzalez LaFuente, J.M.; Marchante-Gayon, J.M.; Fernandez Sanchez, M.L.; Sanz Medel, A.; *Talanta*, 1999, 50, 207.
48. Giusti, P.; Lobinski, R.; Szpunar, J.; Schaumloffel, D.; *Analytical Chemistry*, 2006, 78, 965.
49. Wind, M.; Wesch, H.; Lehmann, W.D.; *Analytical Chemistry*, 2001, 73, 3006.
50. Wind, M.; Edler, M.; Jakubowski, N.; Linscheid, M.; Wesch, H.; Lehmann, W.D.; *Analytical Chemistry*, 2001, 73, 29.
51. Wind, M.; Wegener, A.; Eisenmenger, A.; Kellner, R.; Lehmann, W.D.; *Angewandte Chemie—International Edition*, 2003, 42, 3425.
52. Wind, M.; Gosenca, D.; Kubler, D.; Lehmann, W.D.; *Analytical Biochemistry*, 2003, 317, 26.
53. Schaumloffel, D.; *Analytical Bioanalytical Chemistry*, 2004, 379, 351.
54. Schaumloffel, D.; Prange, A.; *Fresenius Journal of Analytical Chemistry*, 1999, 364, 452.
55. Schaumloffel, D.; Encinar, J.R.; Lobinski, R.; *Analytical Chemistry*, 2003, 75, 6837.
56. Giusti, P.; Schaumlöffel, D.; Preud'homme, H.; Szpunar, J.; Lobinski, R.; *Journal of Analytical Atomic Spectrometry*, 2006, 25, 255.
57. Encinar, J.R.; Schaumloffel, D.; Ogra, Y.; Lobinski, R.; *Analytical Chemistry*, 2004, 76, 6635.
58. Hill, S.J.; Pitts, L.J.; Fisher, A.S.; *Trac—Trends in Analytical Chemistry*, 2000, 19, 120.
59. Monperrus, M.; Krupp, E.; Amouroux, D.; Donard, O.F.X.; Rodriguez Martin-Doimeadios, R.C.; *Trac—Trends in Analytical Chemistry*, 2004, 23, 261.
60. Schaumloffel, D.; Lobinski, R.; *International Journal of Mass Spectrometry*, 2005, 242, 217.
61. Valles Mota, J.P.; Linde Arias, A.R.; Fernandez De La Campa, M.R.; Garcia Alonso, J.I.; Sanz-Medel, A.; *Analytical Biochemistry*, 2000, 282, 194.
62. Diaz Huerta, V.; Hinojosa Reyes, L.; Marchante-Gayon, J.M.; Fernandez Sanchez, M.L.; Sanz-Medel, A.; *Journal of Analytical Atomic Spectrometry*, 2003, 18, 1243.
63. Hinojosa Reyes, L.; Marchante-Gayon, J.M.; Garcia Alonso, J.I.; Sanz-Medel, A.; *Journal of Analytical Atomic Spectrometry*, 2003, 18, 1210.
64. McSheehy, S.; Kelly, J.; Tessier, L.; Mester, Z.; *Analyst*, 2005, 130, 35.
65. Michalke, B.; Schramel, P.; *Journal of Analytical Atomic Spectrometry*, 2004, 19, 121.
66. Wolf, W.R.; Zainal, H.; *Food Nutrition Bulletin*, 2002, 23, 120.

67. Polatajko, A.; Ruiz Encinar, J.; Schaumlöffel, D.; Szpunar, J.; *Chemia Analityczna*, 2005, 50, 265.
68. Martínez-Bravo, Y.; Roig Navarro, A.F.; López Benet, F.; Hernández Hernández, F.J.; *Journal of Chromatography A*, 2001, 926, 265.
69. Bissen, M.; Frimmel, F.H.; *Fresenius' Journal of Analytical Chemistry*, 2000, 367, 51.
70. Garcia-Manyes, S.; Jimenez, G.; Padro, A.; Rubio, R.; Rauret, G.; *Talanta*, 2002, 58, 97.
71. Nakazato, T.; Taniguchi, T.; Tao, H.; Tominaga, M.; Miyazaki, A.; *Journal of Analytical Atomic Spectrometry*, 2000, 15, 1546.
72. Gürleyük, H.; Wallschläger, D.; *Journal of Analytical Atomic Spectrometry*, 2001, 16, 926.
73. Vanhaecke, F.; Saverwyns, S.; Wannemacker, G.D.; Moens, L.; Dams, R.; *Analytica Chimica Acta*, 2000, 419, 55.
74. Vonderheide, A.P.; Meija, J.; Tepperman, K.; Puga, A.; Pinhas, A.R.; States, J.C.; Caruso, J.A.; *Journal of Chromatography A*, 2004, 1024, 129.
75. Guo, Z.-X.; Cai, Q.; Yu, C.; Yang, Z.; *Journal of Analytical Atomic Spectrometry*, 2003, 18, 1396.
76. Schminke, G.; Seubert, A.; *Fresenius' Journal of Analytical Chemistry*, 2000, 366, 387.
77. Seubert, A.; Nowak, M.; *Fresenius' Journal of Analytical Chemistry*, 1998, 360, 777.
78. Nowak, M.; Seubert, A.; *Analytica Chimica Acta*, 1998, 359, 193.
79. Amereih, S.; Meisel, T.; Kahr, E.; Wegscheider, W.; *Analytical and Bioanalytical Chemistry*, 2005, 383, 1052.
80. Amereih, S.; Meisel, T.; Scholger, R.; Wegscheider, W.; *Journal of Environmental Monitoring*, 2005, 7, 1200.
81. Chassaigne, H.; Mounicou, S.; Casiot, C.; Lobinski, R.; Potin-Gautier, M.; *Analusis*, 2000, 28, 357.
82. Infante, H.G.; Hearn, R.; Catterick, T.; *Analytical and Bioanalytical Chemistry*, 2005, 382, 957.
83. Dumont, E.; Vanhaecke, F.; Cornelis, R.; *Analytical and Bioanalytical Chemistry*, 2006, 385, 1304.
84. B'Hymer, C.; Caruso, J.A.; *Journal of Chromatography A*, 2006, 1114, 1.
85. Orero Iserte, L.; Roig-Navarro, A.F.; Hernandez, F.; *Analytica Chimica Acta*, 2004, 527, 97.
86. Ochsenkuhn-Petropoulou, M.; Michalke, B.; Kavouras, D.; Schramel, P.; *Analytica Chimica Acta*, 2003, 478, 219.
87. Bueno, M.; Potin-Gautier, M.; *Journal of Chromatography A*, 2002, 963, 185.
88. Juresa, D.; Darrouzes, J.; Kienzl, N.; Bueno, M.; Pannier, F.; Potin-Gautier, M.; Francesconi, K.A.; Kuehnelt, D.; *Journal of Analytical Atomic Spectrometry*, 2006, 21, 684.
89. Kuehnelt, D.; Kienzl, N.; Traar, P.; Le, N.H.; Francesconi, K.A.; Ochi, T.; *Analytical and Bioanalytical Chemistry*, 2005, 383, 235.
90. Suzuki, K.T.; Kurasaki, K.; Okazaki, N.; Ogra, Y.; *Toxicology and Applied Pharmacology*, 2005, 206, 1.
91. Encinar, J.R.; Sliwka-Kaszynska, M.; Poatajko, A.; Vacchina, V.; Szpunar, J.; *Analytica Chimica Acta*, 2003, 500, 171.
92. Kannamkumarath, S.S.; Wrobel, K.; Wrobel, K.; Vonderheide, A.; Caruso, J.A.; *Analytical and Bioanalytical Chemistry*, 2002, 373, 454.
93. Bryszewska, M.A.; Ambroziak, W.; Rudzinski, J.; Lewis, D.J.; *Analytical and Bioanalytical Chemistry*, 2005, 382, 1279.
94. Infante, H.G.; Cuyckens, F.; Van Campenhout, K.; Blust, R.; Claeys, M.; Van Vaeck, L.; Adams, F.C.; *Journal of Analytical Atomic Spectrometry*, 2004, 19, 159.
95. Chassaigne, H.; Lobinski, R.; *Analytical Chemistry*, 1998, 70, 2536.
96. Lobinski, R.; Chassaigne, H.; Szpunar, J.; *Talanta*, 1998, 46, 271.
97. Richarz, A.-N.; Bratter, P.; *Analytical and Bioanalytical Chemistry*, 2002, 372, 412.
98. Ferrarello, C.N.; Fernandez de la Campa, M.R.; Carrasco, J.F.; Sanz-Medel, A.; *Spectrochimica Acta, Part B: Atomic Spectroscopy*, 2002, 57, 439.

99. Polec, K.; Perez-Calvo, M.; Garcia-Arribas, O.; Szpunar, J.; Ribas-Ozonas, B.; Lobinski, R.; *Journal of Inorganic Biochemistry*, 2002, 88, 197.
100. Garcia, J.S.; De Magalhaes, C.S.; Arruda, M.A.Z.; *Talanta*, 2006, 69, 1.
101. Wang, J.; Houk, R.S.; Dreessen, D.; Wiederin, D.R.; *Journal of the American Chemical Society*, 1998, 120, 5793.
102. Wang, J.; Dreessen, D.; Wiederin, D.R.; Houk, R.S.; *Analytical Biochemistry*, 2001, 288, 89.
103. Wang, J.; Houk, R.S.; Dreessen, D.; Wiederin, D.R.; *Journal of Biological Inorganic Chemistry*, 1999, 4, 546.
104. De La Flor St Remy, R.R.; Sanchez, M.L.F.; Sastre, J.B.L.; Sanz-Medel, A.; *Journal of Analytical Atomic Spectrometry*, 2004, 19, 1104.
105. Rivero Martino, F.A.; Fernandez Sanchez, M.L.; Sanz Medel, A.; *Journal of Analytical Atomic Spectrometry*, 2002, 17, 1271.
106. Coni, E.; Bocca, B.; Galoppi, B.; Alimonti, A.; Caroli, S.; *Microchemical Journal*, 2000, 67, 187.
107. Sanchez, L.F.; Szpunar, J.; *Journal of Analytical Atomic Spectrometry*, 1999, 14, 1697.
108. Nagaoka, M.H.; Maitani, T.; *Biochimica et Biophysica Acta—General Subjects*, 2000, 1523, 182.
109. Bouyssiere, B.; Knispel, T.; Ruhnau, C.; Denkhaus, E.; Prange, A.; *Journal of Analytical Atomic Spectrometry*, 2004, 19, 196.
110. Hann, S.; Koellensperger, G.; Stefanka, Z.; Stingeder, G.; Furhacker, M.; Buchberger, W.; Mader, R.M.; *Journal of Analytical Atomic Spectrometry*, 2003, 18, 1391.
111. Makarov, A.; Szpunar, J.; *Journal of Analytical Atomic Spectrometry*, 1999, 14, 1323.
112. Galettis, P.; Carr, J.L.; Paxton, J.W.; McKeage, M.J.; *Journal of Analytical Atomic Spectrometry*, 1999, 14, 953.
113. Hartinger, C.G.; Hann, S.; Koellensperger, G.; Sulyok, M.; Groessl, M.; Timerbaev, A.R.; Rudnev, A.V.; Stingeder, G.; Keppler, B.K.; *International Journal of Clinical Pharmacology and Therapeutics*, 2005, 43, 583.
114. Szpunar, J.; Makarov, A.; Pieper, T.; Keppler, B.K.; Lobinski, R.; *Analytica Chimica Acta*, 1999, 387, 135.
115. Vacchina, V.; Torti, L.; Allievi, C.; Lobinski, R.; *Journal of Analytical Atomic Spectrometry*, 2003, 18, 884.
116. Kinoshita, K.; Shikino, O.; Seto, Y.; Kaise, T.; *Applied Organometallic Chemistry*, 2006, 20, 591.
117. Ishizaki, M.; Yanaoka, T.; Nakamura, M.; Hakuta, T.; Ueno, S.; Komuro, M.; Shibata, M.; Kitamura, T.; Honda, A.; Doy, M.; Ishii, K.; Tamaoka, A.; Shimojo, N.; Ogata, T.; Nagasawa, E.; Hanaoka, S.; *Journal of Health Science*, 2005, 51, 130.
118. Kinoshita, K.; Shida, Y.; Sakuma, C.; Ishizaki, M.; Kiso, K.; Shikino, O.; Ito, H.; Morita, M.; Ochi, T.; Kaise, T.; *Applied Organometallic Chemistry*, 2005, 19, 287.
119. Richardson, D.D.; Sadi, B.B.M.; Caruso, J.A.; *Journal of Analytical Atomic Spectrometry*, 2006, 21, 396.

7 HPLC with Electrochemical Detection

Fumiyo Kusu and Akira Kotani

CONTENTS

7.1 Introduction 188
7.2 What Is HPLC-ED 189
   7.2.1 HPLC-ED System 189
   7.2.2 Redox Reaction and Potential 189
   7.2.3 Signal Current 190
7.3 HPLC-ED Instrumentation 191
   7.3.1 Modes of HPLC 191
   7.3.2 System Components 191
      7.3.2.1 Column 191
      7.3.2.2 Mobile Phase 192
      7.3.2.3 Pump 197
      7.3.2.4 Injectors 198
      7.3.2.5 ED Detector 198
   7.3.3 Optimization of the System Based on Noise Analysis 200
      7.3.3.1 Power Spectrum of Chromatographic Baseline Noise 200
      7.3.3.2 Pump Selection 200
      7.3.3.3 Electrochemical Cell and Working Electrode 203
7.4 Strategy to Apply HPLC-ED for an Intended Analyte 204
   7.4.1 Cyclic Voltammetry 204
      7.4.1.1 Reversible Redox Reaction 206
      7.4.1.2 Electrochemical Reaction Followed by a Chemical
           Reaction, Depending on pH 206
   7.4.2 Hydrodynamic Voltammetry 208
   7.4.3 Determination of Flavonoids by HPLC-ED 208
      7.4.3.1 CLC-ED System 208
      7.4.3.2 Column Connection 211
      7.4.3.3 Material of the Pathway 211
      7.4.3.4 Column Temperature 211
      7.4.3.5 Preparation of Mobile Phase 211
      7.4.3.6 Cell Volume in Electrochemical Flow Cell 212

	7.4.3.7	Optimization of CLC-ED Conditions	212
	7.4.3.8	Method Validation	214
	7.4.3.9	Pretreatment of Samples	216
7.5	Conclusion		217
References			218

7.1 INTRODUCTION

High-performance liquid chromatography (HPLC) with electrochemical detection (ED) is used for the determination of electroactive constituents in complex mixtures. Although ED can give high sensitivity and selectivity, an improper system design or operation is likely not to show such intrinsic characteristics of the analytical results by HPLC-ED. Therefore, a good understanding of the principles and operations underlying HPLC-ED is essential. The aim of this chapter is to provide a basic understanding of the principles of HPLC-ED to aid those who are initiating analysis in this area.

Various types of detectors, such as amperometric, coulometric, conductometric, and pulsed-amperometric detectors, are included in ED in a broad sense. Those detectors are utilized for detection based on the electrochemical or electrical properties of components of chromatographic eluates. The features of those detectors are

- Conductometric detectors are sensitive to ionic solutes in eluents with low conductivity.
- Amperometric and coulometric detectors are sensitive to solutes that can be oxidized or reduced. The signal current is the result of an electrochemical conversion of the solute by oxidation or reduction on a working electrode at an applied potential. The difference between coulometric and amperometric detectors lies in the area of the working electrode, thus in the efficiency of the electrochemical conversion. The working electrodes employed in coulometric detectors, which achieve nearly 100% conversion, have a much greater surface area (ca. 5 cm^2) than those used for amperometric detectors. Large electrode surface area results in a lower signal-to-noise ratio (S/N) and an increase in detection limits [1].
- Pulsed-amperometric detectors are also sensitive to oxidizable solutes, such as sugars, amino acids, usually in alkaline eluents. The signal current is the result of an electrochemical conversion of the solute by oxidation on a Au or Pt working electrode at an applied potential held for a few hundred milliseconds using a pulsed waveform [2].

Because the system design in amperometric detector is simpler than coulometric and pulsed-amperometric detectors, the tuning and optimization of HPLC with amperometric detection for high-sensitivity determinations are easily performed, even by a novice analyst. From a practical standpoint, HPLC with amperometric detection, is widely applied to the analysis of electroactive components in biological, pharmaceutical, and agricultural samples. The primary focus in the current text is the analysis of catechins in natural products, and this will be used as an illustrative means to demonstrate utility of HPLC-ED.

7.2 WHAT IS HPLC-ED

7.2.1 HPLC-ED System

A simple HPLC-ED system is shown in Figure 7.1. This system is comprised of a solvent delivery system, injector, column, and the electrochemical detector with data recording (or data handling system). In the most common electrochemical detector, a working electrode (WE), a reference electrode (RE), and a counter electrode (CE) are located in the thin-layer type flow cell filled with eluent.

The electrochemical detector measures the electrical current generated by electroactive analytes in the HPLC eluent on the working electrode in the flow-type cell. The analytes can be oxidized or reduced at the working electrode, which is maintained at a specific applied potential.

7.2.2 Redox Reaction and Potential

The relationship between the concentrations of an oxidant (Ox) and reductant (Red) couple and the potential of an inert electrode at which a reaction occurs (at 25°C) is given by the Nernst equation according to

$$Ox + ne^- \rightleftharpoons Red \tag{7.1}$$

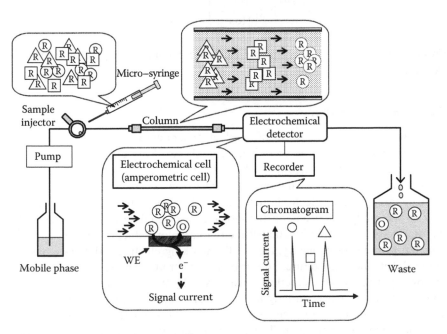

FIGURE 7.1 Schematic diagrams of HPLC-ED system (oxidation mode) R, reductant; O, oxidation product; WE, working electrode.

$$E = E^0 + \frac{0.059}{n} \log\left(\frac{[Ox]_i}{[Red]_i}\right) \tag{7.2}$$

where

 E^0 is the standard electrode potential of the redox couple
 n is the number of electrons gained by the oxidant
 subscript indicates concentrations at the electrode interface

It is obvious from the Nernst equation that if an applied potential to the electrode is positive compared to E^0, the reductant species will be oxidized, and if an applied potential is negative compared to E^0, the oxidant species will be reduced.

The equation for a redox reaction accompanied with proton transfer according to

$$Ox + nH^+ + ne^- \rightleftharpoons Red \tag{7.3}$$

is given by:

$$E = E^0 + \frac{0.059}{n} \log\left(\frac{[Ox]_i}{[Red]_i}\right) + \frac{0.059}{n} \log([H^+]) \tag{7.4}$$

$$= E^0 + \frac{0.059}{n} \log\left(\frac{[Ox]_i}{[Red]_i}\right) - \frac{n}{0.059} pH \tag{7.5}$$

From Equation 7.5, it is clear that the electrode potential is dependent on the electrolyte pH.

7.2.3 SIGNAL CURRENT

The signal current is the result of an electrochemical conversion of the electroactive analyte by oxidation or reduction at the working electrode surface at an applied potential. The magnitude of the signal current is a function of the rate of mass transfer of analyte molecules toward the working electrode and the rate of electron transfer at the electrode surface. In other words, the electron transfer process of the analyte depends on the molar flux of analyte molecules, as well as the thermodynamics and kinetics of the oxidation or reduction. The molar flux is affected by detector geometry, area of the working electrode, flow rate, viscosity of the eluent, and temperature. The current at electrodes having various geometries in flowing streams are theoretically expressed by the equations in Table 7.1. It is noted that the current is directly proportional to the concentration of the analyte and the current sensitivity is dependent on detector geometry, area of the working electrode, flow rate, and so on. The current is amplified and plotted against retention time to yield a chromatogram.

TABLE 7.1
Current Equations for Various Electrode Geometries

Electrode Geometry	Detector Schematic	Current Equation [3]
Planar (parallel flow)		$i = 0.68nFCD^{2/3}v^{-1/6}(A/b)^{1/2}V^{1/2}$
Tubular		$i = 1.61nFC(DA/r)^{2/3}V^{1/3}$
Wall-jet		$i = 0.898nFCD^{2/3}v^{-5/12}a^{-1/2}A^{3/8}V^{3/4}$

Schematic: a, inlet; b, outlet; c, working electrode; d, spacer.

Definition of terms in limiting current equations: a, diameter of inlet; A, electrode area; b, channel height; C, concentration in mM; F, Faraday constant; D, diffusion coefficient; v, kinematic viscosity; r, radius of tubular electrode; V, average volume flow rate; n, number of electrons.

7.3 HPLC-ED INSTRUMENTATION

7.3.1 MODES OF HPLC

Unless a mobile phase is less conductive, various modes of HPLC with ED can be used for the analysis of electroactive substances. Reverse-phase chromatography and ion-exchange chromatography are frequently utilized for determination of small organic molecules. In most popular cases for the analysis of organic compounds by HPLC-ED, reverse-phase chromatography is performed in an isocratic mode so as to maintain constant solvent conditions.

7.3.2 SYSTEM COMPONENTS

7.3.2.1 Column

Various columns, the characteristics of which are shown in Table 7.2, can be used for HPLC-ED. Since the most critical aspect for ED is the analyte concentration in the region of the working electrode surface, micro HPLC and capillary LC yield higher mass sensitivity due to lower sample dilution compared to conventional bore HPLC (Figure 7.2). The sensitivity in detection can be enhanced by using a narrower bore column operated at a higher flow rate as shown by the results presented in Figure 7.3 [4–6]. The narrower bore column afforded the lower detection limits (Table 7.3). Reproducibility was improved by thermostating the column, which also helped reduce baseline fluctuation.

TABLE 7.2

Characteristics of Columns in Popular Use for HPLC-ED

Column Type[a]	Typical i.d.[a] (mm)	Typical Flow Rate[b] (mL/min)	Void Volume[c] (mL)	Typical Sample Loading (μg)
Conventional	3.0–4.6	0.5–1.0	0.7–1.6	40–100
Microbore	1.0–2.0	0.05–0.2	0.075–0.3	5–20
Capillary	0.2–0.5	0.001–0.01	0.001–0.02	0.05–1

Note: Reversed-phase chromatography and ion-exchange chromatography is widely used for HPLC-ED.

The general characteristics of typical HPLC analysis column in use today are:

- 50–250 mm long packed with a 3- or 5-μm octadecylsilica, octadecylpolymer, and ion exchange resin.
- Operating pressure ranges 3.0–18 MPa.
- Has 4,000–20,000 theoretical plates (N).

[a] Designations of these column types are not universally accepted and may vary with manufacturers. Column i.d. is typical.

[b] In the case of reversed-phase chromatography, typical flow rate range using a mobile phase of methanol or acetonitrile in water or aqueous buffer.

[c] Void volumes are based on lengths of 150 mm.

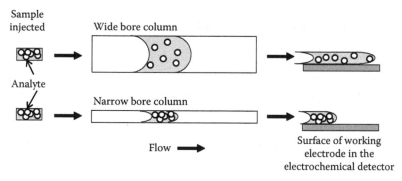

FIGURE 7.2   Comparison of the analyte concentrations at the electrode surface between the elute after a wide-bore column and after a narrow-bore column.

7.3.2.2   Mobile Phase

In reversed-phase HPLC with ED, isocratic elution is the preferred separation mode since baseline variations are enhanced in gradient elution due to the composition change of mobile phase. The change in mobile phase composition causes a change in the double-layer structure at the working electrode surface.

In HPLC-ED, the mobile phase serves a dual purpose; the first to assist selectivity between analytes and the chromatographic stationary phase, and the second, to act as a supporting electrolytic solution within the electrochemical detector. HPLC solvents, therefore should also be compatible with the electrochemical condition of electrolysis of the analyte at the working electrode. The general requirements of

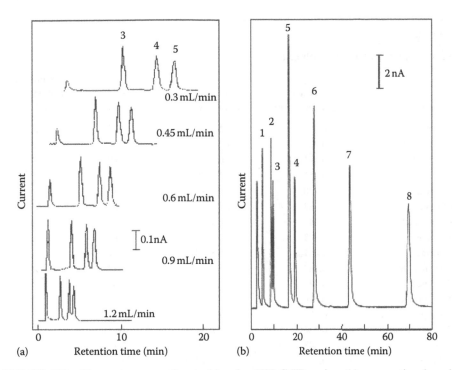

FIGURE 7.3 Chromatograms of catechins by HPLC-ED using (a) conventional and (b) microbore columns: (a) HPLC conditions: column, ODS (LiChrospher 100 RP-18 ODS column, 4.0 × 125 mm, 5 μm, Cica-Merck, Tokyo, Japan); mobile phase, acetonitrile: 0.1 M phosphate buffer, pH 2.5 (15:85, v/v); applied potential, +0.6 V vs. Ag/AgCl; flow rates were indicated in the figure. Each catechin was injected into HPLC-ECD in a 5 pmol amount. (b) HPLC conditions: ODS column (Capcell pak C18 UG120, 1.0 × 150 mm, 3 μm, Shiseido, Tokyo, Japan); mobile phase, methanol–water–phosphoric acid (19:81:0.5, v/v/v) column temperature, 40°C; flow rate, 25 μL/min; applied potential, +0.6 vs. Ag/AgCl. Each catechin was injected into HPLC-ECD in a 1 pmol amount. Peaks: 1, GC; 2, EGC; 3, C; 4, EC; 5, EGCg; 6, GCg; 7, ECg; 8, Cg. (Used with permission from Kotani, A., Hayashi, Y., Matsuda, R., and Kusu, F., Optimization of HPLC-ECD conditions for determination of catechins with precision and efficiency based on the FUMI theory, *Anal. Sci.*, 2003, 19, 865.)

TABLE 7.3
Detection Limits of Epigallocatechin Gallate by Various Type of HPLC-ED

	Column Size			Detection Limit		
Type	i.d. (mm)	Length (mm)	Injection Volume (μL)	Amount (fmol)	Concentration (ng/mL)	Reference
Conventional	4.0	250	5	500	29.0	[7]
Conventional	4.6	250	80	160	0.92	[8]
Semi-micro	1.0	150	5	5	0.46	[6]
Capillary	0.2	150	0.05	0.075	0.46	[9]

TABLE 7.4
Physical Properties of Solvent

Solvent	Viscosity Coefficient[a] (cP)	Electric Conductivity[a] (S/cm)	Relative Permittivity[a]	Solvent for a Mobile Phase in HPLC-ED[b]
Water	0.890	6×10^{-8}	78.39	✓
Formamide	3.30	$<2 \times 10^{-7}$	111_{20}	•
Acetonitrile	0.341_{30}	6×10^{-10}	35.9	✓
Methanol	0.551	1.5×10^{-9}	32.7	✓
Acetic acid	1.130	6×10^{-9}	6.19	•
Ethanol	1.083	1.4×10^{-9}	24.6	✓
Isopropanol	2.044	6×10^{-8}	19.9	•
Acetone	0.303	5×10^{-9}	20.6	•
Dioxane	1.087_{30}	5×10^{-15}	2.21	×
Tetrahydrofuran	0.460		7.58	×
Ethyl acetate	0.426	$<1 \times 10^{-9}$	6.02	×
Nitro methane	0.614	5×10^{-9}	36.7	•
Dichloromethane	0.393_{30}	4×10^{-11}	8.93	×
Dichloroethane	0.73_{30}	4×10^{-11}	10.37	×
Benzene	0.603	4×10^{-17}	2.27	×
Toluene	0.553	8×10^{-16}	2.38	×
Dimethyl formamide	0.802	6×10^{-8}	36.7	•
Dimethyl sulfoxide	1.99	2×10^{-9}	46.5	•
Hexane	0.294	$<1 \times 10^{-16}$	1.88	×
Propylene carbonate	2.53	1×10^{-8}	64.92	•
Ethylene carbonate	1.9_{40}	$5 \times 10^{-8}_{40}$	89.8_{40}	•

[a] These values are cited from Ref. [10]. Subscript values refer to temperature at which viscosity coefficient, electric conductivity, and relative permeability were obtained. No subscript means data collected at 25°C.
[b] The solvent is (✓) suitable, (•) usable, or (×) not suitable.

the solvents are (1) high purity, (2) high solubility of analyte, electrolytes and electrode reaction products, and (3) stability at the applied potential and in the flow system components. The characteristics of solvents used for HPLC-ED are shown in Table 7.4. The mobile phase usually contains between 10^{-3} and 10^{-2} M salt concentration for suitable eluent conductivity in the detector. Since redox potential is also a function of the pH, mobile phases are buffered and this also serves to maintain better retention behavior of the solutes during the chromatographic separation. If possible, surface active chemicals of an ion pair reagent, which may influence the redox reaction of analytes, should be avoided. Additives of antioxidants in mobile phase solvents should be removed prior to use. Mobile phases must also be degassed.

The effect of mobile phase composition on the separation of analytes using reversed-phase HPLC is clearly seen in the chromatograms of 15 flavonoids (Figure 7.4) shown in Figure 7.5. This separation was undertaken on an ODS column

Flavonols

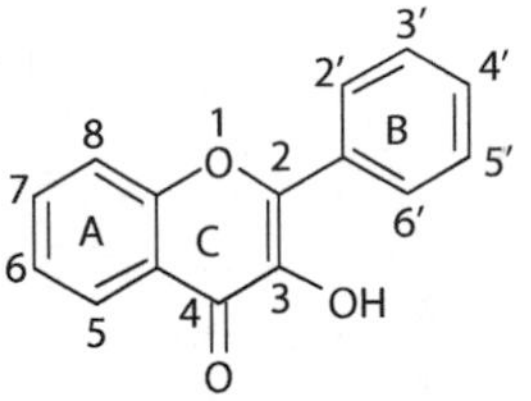

Galangin	3,5,7-Trihydroxyflavone	GL
Fisetin	3,3′,4′,7-Tetrahydroxyflavone	FI
Kaempferol	3,4′,5,7-Tetrahydroxyflavone	KF
Morin	2′,3,4′,5,7-Pentahydroxyflavone	MR
Quercetin	3,3′,4′,5,7-Pentahydroxyflavone	QC
Myricetin	3,3′,4′,5,5′,7-Hexahydroxyflavone	MC

Flavones

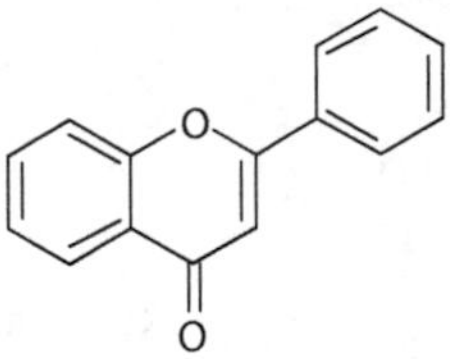

Baicalein	5,6,7-Trihydroxyflavone	B
Baicalin	5,6-Dihydroxyflavone-7-glycoside	BG
Wogonin	5,6-Dihydroxy-8-methoxyflavone	WO
Luteolin	3′,4′,5,7-Tetrahydroxyflavone	LO
Hyperoside	4′,5,5′,7-Tetrahydroxyflavone-3-glycoside	HP

Flavanones

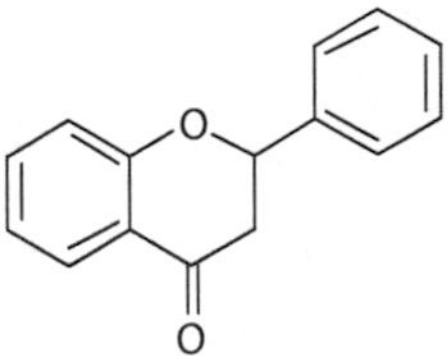

| Fustin | 3,3′,4′,7-Tetrahydroxyflavanone | FU |
| Hesperidin | 5,3′-Dihydroxy-4′-methoxyflavanone-7-rhamnoglucoside | HE |

Isoflavones

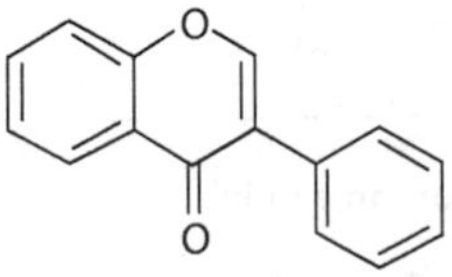

| Daidzein | 4′, 7-Dihydroxyisoflavone | D |
| Daidzein | 4′-Hydroxyisoflavone-7-glycoside | DG |

FIGURE 7.4 Structures of the flavonoids examined by HPLC-ED.

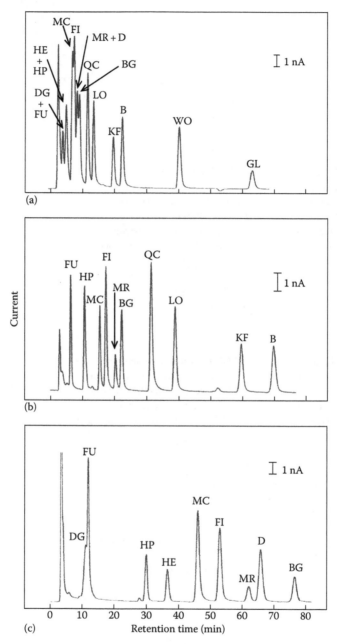

FIGURE 7.5   Chromatographic separation of flavonoids with a methanol–water–phosphoric acid ((a) 50:50:0.5; (b) 40:60:0.5; (c) 30:70:0.5; v/v/v) mixture. HPLC conditions: Column ODS (Capcell pak C18 UG120, 1.0 × 150 mm, 3 μm); column temperature 40°C; flow rate 25 μL/min; applied potential (a) +0.8, (b) +0.6, (c) +0.8 V vs. Ag/AgCl; the commercial available electrochemical cell (radial flow cell, BAS) was constructed from a GC working, Ag/AgCl reference and stainless steel counter electrode was employed. Each flavonoid was injected at an amount of 1 pmol.

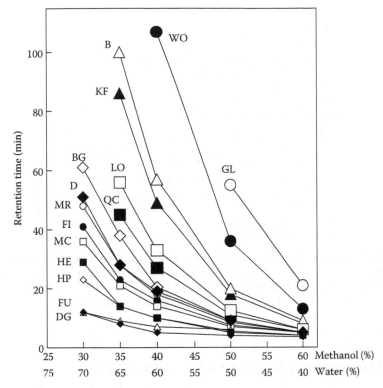

FIGURE 7.6 Effects of methanol in mobile phase on elution of flavonoids. Mobile phase, methanol–water–phosphoric acid ($x{:}y{:}0.5$, v/v/v) mixture. The HPLC-ED conditions used were the same as in Figure 7.5 except for the applied potential, +0.8 V vs. Ag/AgCl and mobile phase.

with a mobile phase composed of methanol–water–phosphoric acid. With increasing methanol concentration, the retention time for all the flavonoids decreased, as did the separation factor (Figure 7.6).

7.3.2.3 Pump

The background current produced at a working electrode consists of faradaic noise current and non-faradaic noise current. The former results from not only unwanted redox reactions of trace impurities in a mobile phase but also unwanted oxidation of the working electrode surface. The latter arises from charging a double layer at the interface between the working electrode and solution of mobile phase (Figure 7.7). The faradaic noise current may be removed by careful purification of solvents and chemicals in the mobile phase and by proper selection of an applied potential where the working electrode is not oxidized. On the other hand, it is inevitable in any HPLC-ED experiment that the non-faradaic noise currents coming from the double layer exist. Since bias current (or potential) is often supplied to original background current at the working electrode to subtract a part of the non-faradaic noise currents in ED detectors, stable noise current is important to minimize noise level.

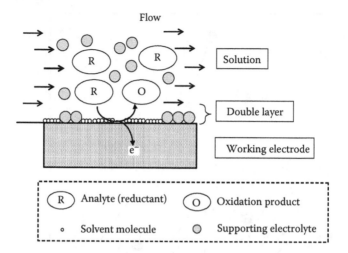

FIGURE 7.7   Diagrams of electrochemical oxidation reaction on working electrode in electrochemical flow cell. Since the noises are especially to be influenced by the characteristics of the interface between an electrode and solution in the electrochemical flow cell, a suitable condition of HPLC-ED should be chosen for keeping the interface stable.

TABLE 7.5
Characteristics of Typical Piston Pumps for HPLC-ED

HPLC System Type	Column i.d. (mm)	Range of Enable Flow Rate	Accuracy of the Flow Rate (%)
Conventional[a]	3.0–4.6	0.01–5.0 mL/min	±0.5
Semi-micro[a]	1.0–2.0	0.001–1.0 mL/min	±1
Capillary (Micro)	0.2–0.5	0.01–50 μL/min	±2

[a]   A small stroke (e.g., 4–6 μL/stroke) pump which is composed of parts of inert materials (e.g., PEEK tube).

The flow rate stability of the HPLC pump is essential in determination by HPLC-ED. The flow rate fluctuation varies with the design of the HPLC pump. The general characteristics of pumps in HPLC-ED are shown in Table 7.5. If a reduction in the fluctuation of flow is required for determination with high sensitivity, pulse damping is achieved with a pulse damper.

7.3.2.4   Injectors

To introduce the sample to the column under high pressure, an injector such as a Rheodyne 7125 or 7725 injector is used.

7.3.2.5   ED Detector

The signal current is the result of an oxidation or reduction of the analyte at the working electrode at constant applied potential in an electrochemical detector, where

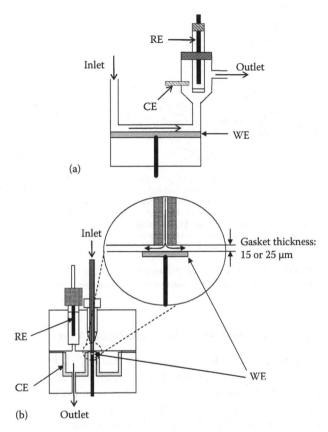

FIGURE 7.8 Schematic diagrams of (a) cross and (b) radial flow electrochemical cells. WE, working electrode; RE, reference electrode; CE counter electrode.

the reference and counter electrodes were located downstream of the working electrode. Several thin-layer cells are now commercially available. Two examples are shown in Figure 7.8. The currents measured are in the pA–mA range.

Although there is a wide range of chemical species that can be oxidized or reduced, the analytes determined by HPLC-ED are limited by the potential range available at the working electrode in contact with the eluent. The most commonly used working electrode is a glassy carbon (GC) electrode at which redox reaction of analytes is carried out in an aqueous eluent at potentials in the −1 to +1.2 V (vs. Ag/AgCl) range. The working electrode surface should be clean and smooth for efficient electrochemical reaction of analytes. The most common method for cleaning the electrode surface is mechanical polishing on a polishing pad using alumina or diamond polishes (Figure 7.9). After the mechanical polishing, the electrode surface must be rinsed thoroughly with an appropriate solvent to remove all traces of the polishing material. During periods of inoperation, the working electrode should be maintained at the applied potential, in contact with the mobile phase at a low flow rate.

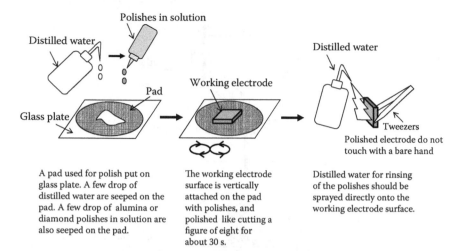

Polishes in solution

Distilled water

Distilled water

Working electrode

Pad

Glass plate

Tweezers

Polished electrode do not
touch with a bare hand

A pad used for polish put on
glass plate. A few drop of
distilled water are seeped on the
pad. A few drop of alumina or
diamond polishes in solution are
also seeped on the pad.

The working electrode
surface is vertically
attached on the pad
with polishes, and
polished like cutting a
figure of eight for
about 30 s.

Distilled water for rinsing
of the polishes should be
sprayed directly onto the
working electrode surface.

FIGURE 7.9 Procedure for GC electrode polishing. For more information about electrode polishing and other methods of electrode pretreatment. (See the literature Bott, A.W., *Curr. Sep.*, 16, 79, 1997.)

7.3.3 OPTIMIZATION OF THE SYSTEM BASED ON NOISE ANALYSIS

In order to get highly reproducible and sensitive results by HPLC-ED, optimization of the operating conditions is necessary. For this purpose, the analyst must often carry out many runs under various conditions, and then evaluate the variation in response according to changes in S/N of the analyte. In recent years, it was noted that power spectral analysis of baseline noise is a useful tool to examine the size of noise and determine its origin [12].

7.3.3.1 Power Spectrum of Chromatographic Baseline Noise

An example of power spectral analysis of baseline noise is shown in Figure 7.10. The baseline noise of the chromatogram (Figure 7.10a) was converted to a power spectrum by Fourier transform (Figure 7.10b). The power density of the low frequency was larger than that of the high frequency, and it was apparent that the baseline noise contained $1/f$ noise. Thus, the reduction of low-frequency noise in $1/f$ fluctuation is essential for improvement of sensitivity and precision on HPLC-ED analysis. Tracing the source of low-frequency noise using the power spectral analysis, allows the selection of the best components of the HPLC-ED, such as pump, electrochemical flow cell, and working electrode material.

7.3.3.2 Pump Selection

The power spectral analysis of baseline noise in chromatographic data obtained using two types of pumps, reciprocating and syringe pumps, is shown in Figure 7.11. In this case, the low-pass filter of the amplifier was set at the cutoff frequency of 10 Hz. Figure 7.12a illustrates the baselines of the chromatograms at the flow rates of 0.36, 0.45, and 0.54 mL/min using a syringe pump. The power spectra for each baseline are shown in Figure 7.12b, and two major bands are clearly apparent in the spectra.

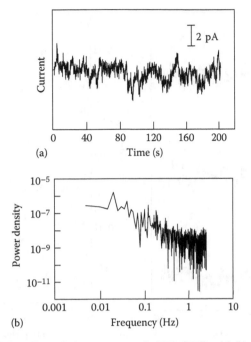

(a)   Time (s)

(b)   Frequency (Hz)

FIGURE 7.10   (a) A baseline of chromatogram in HPLC-ED and (b) its power spectrum. The digital data of 256 data points obtained from chromatographic baseline were converted by Fourier transform [13]. The analog data of chromatogram was converted to digital data by an analog-to-digital (A/D) converter.

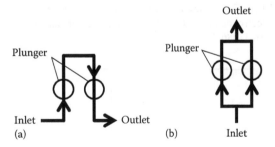

FIGURE 7.11   Flow line in (a) dual series of piston pump (PU-880) and (b) dual parallel piston pump (DP-8020). The arrows indicate the flow of the mobile phase in the pump.

These two bands correspond to the frequencies of the fundamental tone and harmonic noise caused by the reciprocating motion of the pistons in the pump, caused by pressure/flow fluctuations associated with pulsation of the pump. This pulsation effect was not readily distinguished by visual inspection of chromatographic baseline data.

The introduction of a pulse damper between the pump and the injector resulted in the disappearance of the pulsation bands in Figure 7.12b as shown in Figure 7.13a. In Figure 7.13b, the power spectrum obtained using a dual-piston parallel-type, small stroke pump, without a pulse damper is illustrated. The power density in this case

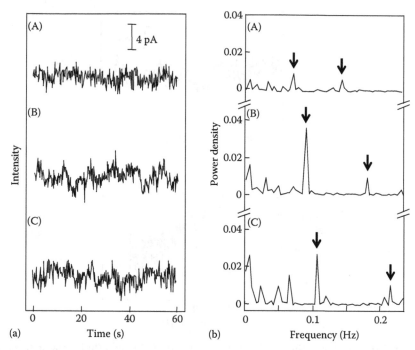

FIGURE 7.12 Baselines of (a) chromatogram and (b) their power spectra. The baselines were obtained using the HPLC-ED system with PU-880 pump and EC-840 detector at the flow rates of (A) 0.36, (B) 0.45, and (C) 0.54 mL/min. The arrows indicate the noises originating from the pulsation of the pump. (Reprinted with permission from Kotani, A., Hayashi, Y., Matsuda, R., and Kusu, F., Prediction of measurement precision of apparatus using a chemometric tool in electrochemical detection of high-performance liquid chromatography, *J. Chromatogr. A.*, 2003, 986, 239.)

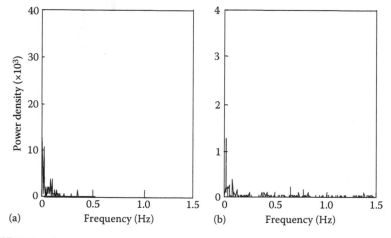

FIGURE 7.13 Power spectra of baselines of chromatograms by the HPLC systems with (a) PU-880 pump, EC-840 detector, and pulse damper system, and (b) DP-8020 pump and EC-840 detector.

is considerably smaller than that for the syringe pump. Thus, a dual-piston, parallel-type, small stroke pump is recommended as a HPLC pump because of flow rate stability and reduction in noise.

7.3.3.3   Electrochemical Cell and Working Electrode

Various types of electrochemical flow cells have been developed and utilized for ED detectors [3]. The baseline noise level of the cells illustrated in Figure 7.14 is assessed in Figure 7.15. The EDP-1 detector has a wall-jet type cell [14] whose inlet line is perpendicular to the working electrode, whereas the inlet line of EC-840 detector is oblique. Figure 7.15 shows the power spectra obtained from the chromatographic baselines. The bands with high power density appeared around 0.09 and 0.18 Hz for the spectra of the EC-840 detector (Figure 7.15a). Contrary to this, the band intensities for the EDP-1 detector (Figure 7.15b) are about one tenth of the band intensities of the EC-840 detector. The wall-jet type cell thus seems to resist the noise created by the pump pulsation.

Carbon is the most common working electrode material for the electrochemical detectors. One factor that influences performance for electrolysis at the working electrode is the electrochemical characteristics of working electrode material. Thus, the electrochemical properties that lead to low background currents, low noise, and high efficiency must be carefully selected. Based on power spectral analysis, comparison of noise levels was made between two chromatograms obtained using

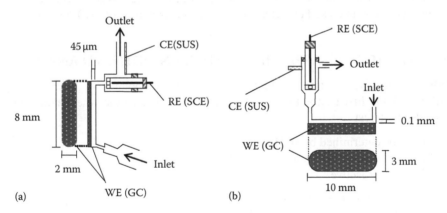

FIGURE 7.14   Composition and dimensions of electrochemical flow cells in (a) Jasco EC-840 and (b) Kotaki EDP-1 electrochemical detectors. WE, working electrode; RE, reference electrode; CE, counter electrode; GC, glassy carbon; PFC, plastic formed carbon; SCE, saturated calomel electrode; SUS, stainless steel. The electrochemical cell of the EC-840 detector is made from a GC working electrode, SCE, and SUS counter electrode. That of the EDP-1 detector is made from a GC or a PFC working electrode, SCE, and SUS counter electrode. GC electrodes were purchased form Tokai Carbon (Tokyo, Japan), and PFC electrodes that is made of graphite/carbon composite has favorable electrochemical properties arising from the edge of neutral crystalline graphite, were purchased from Tsukuba Materials Information Laboratory (TMIL, Ibaraki, Japan). (Used with permission from Kotani, A., Hayashi, Y., Matsuda, R., and Kusu, F., Prediction of measurement precision of apparatus using a chemometric tool in electrochemical detection of high-performance liquid chromatography, *J. Chromatogr. A.*, 2003, 986, 239.)

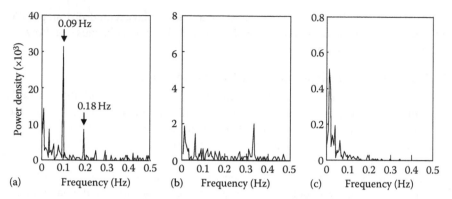

FIGURE 7.15 Power spectra of baselines of chromatograms by the HPLC systems with (a) PU-880 pump and EC-840 detector, (b) PU-880 pump and EDP-1 detector using GC working electrode, and (c) PU-880 pump and EDP-1 detector using PFC working electrode.

GC and plastic formed carbon (PFC) working electrodes. With the PFC working electrode in HPLC, the power density of noise is found to be smaller, as seen in Figure 7.15c, than that with the GC working electrode (Figure 7.15b).

The noise level of a chromatogram can be assessed from a section of baseline even if not referring to a total chromatogram. The power spectrum can be useful for regular check of the performance of an HPLC-ED apparatus, as well as the optimization of experimental condition and the selection of components in a HPLC-ED system.

7.4   STRATEGY TO APPLY HPLC-ED FOR AN INTENDED ANALYTE

When attempting to start out in the analysis of compounds of interest using HPLC-ED, the first task is to refer to literature, and determine whether the compounds of interest

> #1 can be determined by an HPLC-ED
> #2 can be determined by an HPLC with any detection except ED
> #3 are electroactive

Following from a literature search, an experimental plan, such as that presented in Figure 7.16, may be adopted. In the absence of any information regarding the redox chemistry of the compound of interest or related species, cyclic voltammograms of the compound in an electrolytic solution of a candidate of mobile phase may be examined. When some guidance regarding HPLC-ED of the compounds of interest is available, reported HPLC-ED conditions can usually be reproduced, at least as a starting point, but fine-tuning of the method is recommended (Figure 7.17).

7.4.1   Cyclic Voltammetry

Cyclic voltammetry is useful for observing redox behavior of an interesting analyte over a wide potential range using simple electrochemical equipment shown in Figure 7.18. In cyclic voltammetry, the potential is linearly changed in the positive

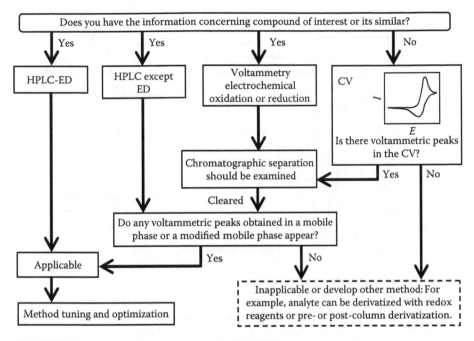

FIGURE 7.16   Scheme of strategy to apply HPLC-ED for an intended analyte.

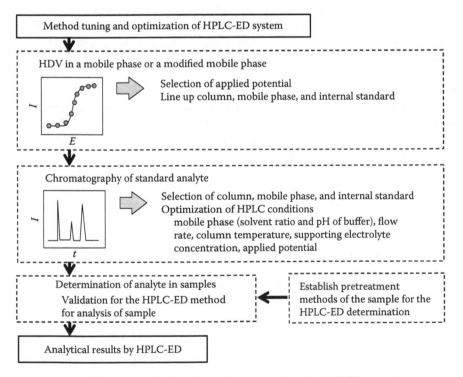

FIGURE 7.17   Scheme of method tuning and optimization of HPLC-ED system.

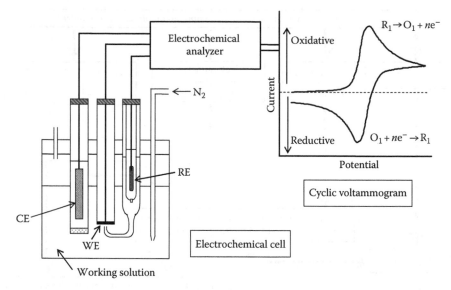

FIGURE 7.18 Diagram of instrument for measuring cyclic voltammetry and typical beaker-type electrochemical cell. WE, working electrode; RE, reference electrode; CE, counter electrode. Working solution were contained analyte and supporting electrolyte.

(or negative) direction with a scan rate of 10–100 mV/s. Then, the potential is linearly changed in the negative (or positive) direction and the current resulting from the redox electrode reactions is recorded as a cyclic voltammogram. A typical cyclic voltammogram for an electrochemically reversible redox couple is shown in Figure 7.19.

7.4.1.1   Reversible Redox Reaction (Figure 7.19)

Reductant R_1 can be oxidized and its oxidation product O_1 is stable in the electrolyte solution and easily reduced at the potential more negative than $E^0(O_1/R_1)$. The cyclic voltammograms of R_1 are shown in Figures 7.19 and 7.20. Provided the standard redox potential of R_1 is $E^0(O_1/R_1)$, a redox peak couple, of which the mid potential is $E^0(O_1/R_1)$, appears on the cyclic voltammogram. When the starting potentials were more negative than $E^0(O_1/R_1)$, no faradaic current is observed. If the electrolyte solution was stirred by bubbling inert gas, only oxidation current was recorded, since R_1 was supplied by the convection and O_1 removed from the surface of the working electrode (Figure 7.20). In a diagnostic test for HPLC-ED, it is noted that the analyte R_1 is oxidizable at the potential more positive than $E^0(O_1/R_1)$.

7.4.1.2   Electrochemical Reaction Followed by a Chemical Reaction, Depending on pH (Figure 7.21)

The cyclic voltammograms of reductant R_2 are shown in Figure 7.21. The oxidation of R_2 is accompanied by the proton transfer and the electron transfer is followed by a chemical reaction of its oxidation product O_2. The chemical reaction product R_3 is also oxidized. The R_2 generates O_2 on oxidation, which subsequently undergoes a chemical reaction to form R_3 at pH higher than 1.

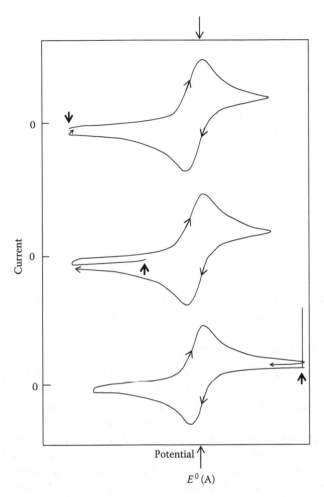

FIGURE 7.19 Typical cyclic voltammogram for an electrochemically reversible redox couple (O_1/R_1). Started potential for positive or negative scan of cyclic voltammogram was indicated by arrows.

The oxidation peak (R_2 to O_2), caused by the electron transfer from R_2 to O_2, and is shifted to negative potential with pH. After the first positive potential scan, two or three peaks due to the redox reactions of the products were observed on the cyclic voltammograms at pH 4 and pH 7. In a diagnostic test for HPLC-ED, it is primarily noted that the analyte R_2 is oxidizable at the potential positive than the oxidation peak (R_2 to O_2).

If the interesting analyte gave an oxidation (or a reduction) peak on the cyclic voltammogram, the analyte is expected to be detected by an HPLC-ED at an applied potential positive (or negative) than the mid-potential or half-peak potential. A list of the mid-potential and half-peak potential of flavonoids at pH 7.5 is shown in Table 7.6, as a reference. However, the detection potential should be fixed using a hydrodynamic voltammogram of the analyte.

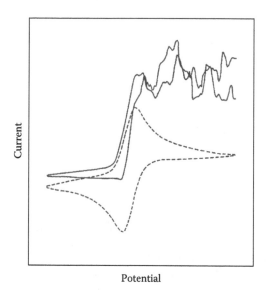

Potential

FIGURE 7.20   Cyclic voltammogram for an electrochemically reversible redox couple (O_1/R_1) in a stationary solution (----) and stirred the solution (—).

7.4.2   HYDRODYNAMIC VOLTAMMETRY

A hydrodynamic voltammogram shows the dependence of the chromatographic peak height on the applied potential in HPLC-ED. A voltammogram is readily obtained by reinjecting the analyte at a suitable concentration at different applied potentials and plotting the peak height vs. the potentials. Since the voltammograms may vary with construction of the detector, composition of mobile phase, the working electrode material, and the characteristics of the reference electrode, as well as the electrochemical reaction of the analyte, it is thus important to ascertain the optimum detection potential for the analyte using the HPLC system for the determination.

The determination of catechins by capillary LC with ED will be used to the method of tuning and optimization of the HPLC system.

7.4.3   DETERMINATION OF FLAVONOIDS BY HPLC-ED

The system tuning and optimization of CLC-ED with a packed 0.2 mm inner diameter column was carried out for determining attomole quantities of catechins. These catechin structures are shown in Figure 7.22.

7.4.3.1   CLC-ED System
The CLC-ED system (Figure 7.23) comprised of the following:

- Separation mode: Reverse-phase chromatography in a isocratic mode
- Column: Inertsil ODS-3 Capillary EX-Nano column (150 × 0.2 mm i.d.; 3 μm, GL Science)

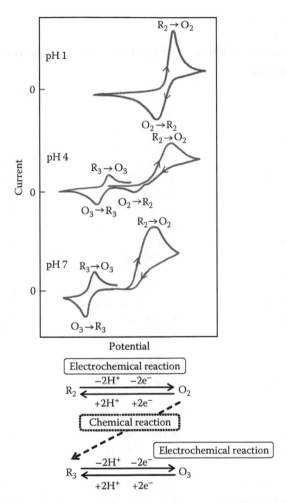

FIGURE 7.21 Effect of pH on cyclic voltammograms and electrochemical reaction followed by a chemical reaction (O_2/R_2 and O_3/R_3) at pH 1, 4, and 7.

- Column temperature: 50°C
- Mobile phase: Phosphoric acid (85%)–methanol–water (0.5:27.5:72.5, v/v/v)
- Pump: MP710i pump (polyetheretherketone [PEEK] is used in the pathway; GL Science)
- Injector: C4-0004-05 microbore internal injector was fitted with 50 nL sample loop (Valco, Houston, TX, USA)
- Electrochemical detector: Decade electrochemical detector with column oven (Antec Leyden, Zoeterwoude, the Netherlands)
- Electrochemical cell: μVT-03 (Antec Leyden) was constructed from a glassy carbon working electrode (φ 0.7 mm), a Ag/AgCl reference electrode, and a stainless steel counter electrode

TABLE 7.6
Half-Wave Potentials ($E_{1/2}$) of the First Oxidation Waves of Flavonoids

Flavonoids	Substituents	$E_{1/2}$ (V vs. Ag/AgCl)[a]
Flavonols		
Myricetin	3,5,7,3',4',5'-OH	−0.030[r]
Quercetin	3,5,7,3',4'-OH	0.020[r]
Fisetin	3,7,3',4'-OH	0.030[r]
Kaempferol	3,5,7,4'-OH	0.080
Galangin	3,5,7-OH	0.280
Morin	3,5,7,2',4'-OH	0.105
Flavones		
Baicalein	5,6,7-OH	−0.060[r]
Baicalin	5,6-OH-7-glucose	0.080[r]
Luteolin	5,7,3',4'-OH	0.180[r]
Rutin	5,7,3',4'-OH-3-rutinose	0.180[r]
Hyperoside	5,7,4',5'-OH-3-glucose	0.185[r]
Wogonin	5,7-OH-8-OCH$_3$	0.360
Apigenin	5,7,4'-OH	>0.500[b]
Flavanones		
Fustin	3,7,3',4'-OH	0.132[r]
Naringenin	5,7,4'-OH	0.590
Isoflavones		
Daidzein	7,4'-OH	0.500
Daidzin	4'-OH-7-glucose	0.538
Puerarin	7,4'-OH-8-C-glucose	0.540
Catechins		
(−)-Gallocatechin	3,5,7,3',4',5'-OH	−0.030
(−)-Epigallocatechin	3,5,7,3',4',5'-OH	−0.035
(−)-Epigallocatechin gallate	5,7,3',4',5'-OH-3-gallate	−0.020
(−)-Gallocatechin gallate	5,7,3',4',5'-OH-3-gallate	−0.010
(−)-Catechin gallate	5,7,3',4'-OH-3-gallate	0.040
(−)-Epicatechin gallate	5,7,3',4'-OH-3-gallate	0.080
(+)-Epicatechin	3,5,7,3',4'-OH	0.082[r]
(−)-Epicatechin	3,5,7,3',4'-OH	0.082[r]
(−)-Catechin	3,5,7,3',4'-OH	0.092[r]
(+)-Catechin	3,5,7,3',4'-OH	0.102[r]
Biflavonoid		
Silibinin		0.450

[a] These values are cited from literatures [15,16].
[b] No obvious plateau was observed. Apigenin was oxidized at potentials more positive than 0.5 V vs. Ag/AgCl.
[r] Stands for reversible or quasi-reversible electrode reaction.

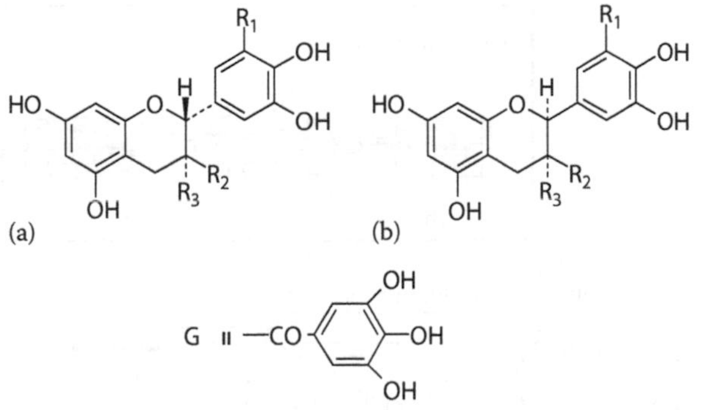

a	R_1	R_2	R_3	MW[a]
EC	H	H	OH	290.27
ECg	H	H	OG	442.37
EGC	OH	H	OH	306.27
EGCg	OH	H	OG	458.37

b	R_1	R_2	R_3	MW[a]
C	H	H	OH	290.27
Cg	H	H	OG	442.37
GC	OH	H	OH	306.27
GCg	OH	H	OG	458.37

FIGURE 7.22 Catechin structures. Catechins: C, catechin; EC, epicatechin; GC, gallocatechin; EGC, epigallocatechin, Cg, catechin gallate; ECg, epicatechin gallate; GCg, gallocatechin gallate; EGCg, epigallocatechin gallate. [a]MWs were calculated using Advanced Chemistry Development (ACD/Labs) Software V8.14 for Solaris. (Used with permission from Kotani, A., Takahashi, K., Hakamata, H., Kojima, S., and Kusu, F., Attomole catechins determination by capillary liquid chromatography with electrochemical detection, *Anal. Sci.*, 2007, 23, 157.)

7.4.3.2 Column Connection

In general, a capillary tube is connected between an injector and a capillary column, using compression nuts and ferrules. However, to minimize void volume, the Inertsil ODS-3 Capillary EX-Nano column was directly connected to the injector, and the capillary fused silica tube was directly connected to the electrochemical cell (Figure 7.23).

7.4.3.3 Material of the Pathway

A comparison was made of the effects of the pathway including pump and tubes on chromatographic baseline noise and the detection limits of catechins using MP710i and MP711 pumps, that use PEEK and stainless in the pathway in the pump, respectively. As shown in Table 7.7, chromatographic baseline noise of CLC-ED using MP710i pump was smaller than that using the MP711 pump. The pump with PEEK tubing in the pathway was suitable for the CLC-ED.

7.4.3.4 Column Temperature

The column temperature was 50°C in the column oven.

7.4.3.5 Preparation of Mobile Phase

An examination was made of how the ratio of water to methanol in the mobile phase influenced the separation for the determinations of catechins. Separation increased as the water content of the mobile phase increased. The most appropriate mobile phase was selected to be methanol–water (27.5:72.5).

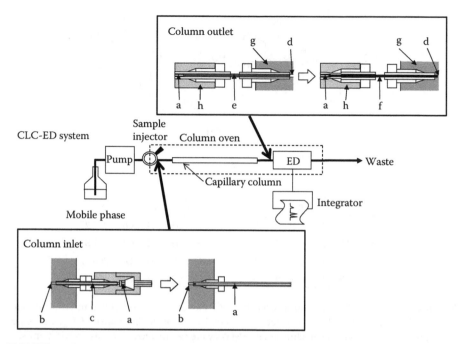

FIGURE 7.23 Capillary LC with ED (CLC-ED) system and schematic diagrams of the improvement on connection in column inlet and outlet of CLC-ED system. a, capillary column; b, outlet port of injector to column (0.15 mm i.d.); c, connection tube; d, inlet port to ED cell (0.3 mm i.d.); e, fused silica capillary tube (0.05 mm i.d.; 0.375 mm o.d.); f, fused silica capillary tube (0.05 mm i.d.; 0.3 mm o.d.); g, flow cell; h, union.

TABLE 7.7
Effects of Wetted Material on CLC-ECD Pump on Baseline Noise Level and Detection Limit of Epicatechin

Pump (Material in the Pathway)	Width of Baseline Noise (pA)	Detection Limit of EC (amol)
MP 711 (stainless)	0.42	260
MP 710i (PEEK)	0.34	180

The thickness of gasket in the flow cell was 25 μm.

7.4.3.6   Cell Volume in Electrochemical Flow Cell
The effects of the thickness of a gasket in the flow cell on the chromatographic baseline noise level and on the detection limit of catechins were examined. As shown in Table 7.8, the detection limit with 15 μm-gasket showed a lower value than that with 25 μm-gasket.

7.4.3.7   Optimization of CLC-ED Conditions
Applied potential: Hydrodynamic voltammograms (Figure 7.24) were measured in order to select the optimal detection potential for determination of catechins. Catechins were oxidized at potentials more positive than +0.3 V vs. Ag/AgCl.

TABLE 7.8
**Effect of Gasket Thickness in Electrochemical
Flow Cell on Chromatographic Peak Height
of Catechins and Baseline Noise Width**

	Catechin	Gasket Thickness	
		15 μm	25 μm
Peak height (pA)	GC	516	467
	C	355	298
	EC	309	246
	GCg	413	378
	ECg	271	254
	Cg	237	210
Baseline noise (pA)		0.29	0.44

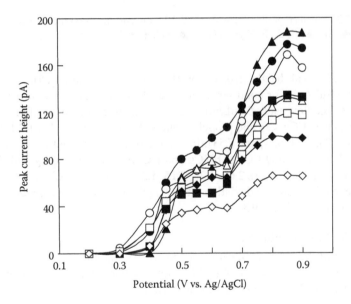

FIGURE 7.24 Hydrodynamic voltammogram of catechins (○, EGC; ●, GC; ▲, C; △, EGCg; □, GCg; ■, EC; ◆, ECg; ◇, Cg). (Used with permission from Kotani, A., Tatahashi, K., Hakamata, H., Kojima, S., and Kusu, F., Attomole catechins determination by capillary liquid chromatography with electrochemical detection, *Anal. Sci.*, 2007, 23, 157.)

Two oxidation waves, one at +0.5 to 0.6 V vs. Ag/AgCl and the other at +0.85 V vs. Ag/AgCl, were observed in the hydrodynamic voltammogram. For potentials more positive than +0.9 V vs. Ag/AgCl, sensitivity was increased, but reproducibility was decreased, possibly due to contamination of the electrode surface by oxidation products. Therefore, for highly sensitive determination without loss of selectivity and reproducibility, the value +0.85 V vs. Ag/AgCl was adopted for the present study.

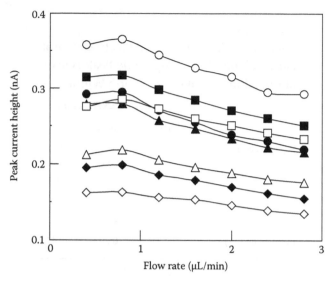

FIGURE 7.25 Effect of flow rate in CLC-ED on chromatographic peaks of catechins (○, EGC; ●, GC; ▲, C; △, EGCg; □, GCg; ■, EC; ◆, ECg; ◇, Cg).

Flow rate: The peak height depends on the flow rate (Figure 7.25). The flow rate was set at 4.0 μL/min for good separation and the efficiency of the measurement in this case. At the flow rate of 4 μL/min, the chromatographic peaks of GC, EGC, C, EGCg, EC, GCg, ECg, Cg, and propyl gallate (I.S.) were obtained at 1.20, 1.53, 1.71, 2.25, 2.50, 3.03, 4.24, 5.77, and 7.53 min, respectively. The resolution (R_S) of GC and EGC, EGC and C, C and EGCg, EGCg and EC, EC and GCg, GCg and ECg, ECg and Cg, and Cg and I.S. were 2.68, 1.26, 3.15, 1.33, 2.53, 4.71, 4.67, and 8.26, respectively. The relative standard deviations (RSDs) of the retention time for catechins and I.S. peaks were less than 1.9%.

7.4.3.8 Method Validation

Linearity: Peak height was linearly related to the amount of catechins present in standard solution (from 200 amol to 500 fmol ($r > 0.998$)), i.e., the concentration of catechins from 4 nM to 10 μM, as shown in Table 7.9.

Accuracy: The accuracy, as relative error (RE) at 25 fmol, was within the range of −2.8% to 4.4% (Table 7.9). Each catechin, 25 fmol, was detected with a RSD of 2.2% ($n = 5$). The RSDs of the inter-day precisions for 25 fmol catechins were less than 3.0%.

Detection limit: The amount of detection limit (S/N = 3) for GC, EGC, C, EGCg, EC, GCg, ECg, and Cg were 54, 61, 61, 75, 75, 67, 75, and 89 amol, respectively, i.e., the concentration of detection limit (S/N = 3) for GC, EGC, C, EGCg, EC, GCg, ECg, and Cg were 331, 374, 354, 459, 435, 614, 664, and 787 pg/mL, respectively.

TABLE 7.9

Intraday Accuracy, Precision, Calibration Curve Linearity, Detection Limit, and Quantitation Limit for the Measurement of Catechins Peaks

Catechin	Nominal Amount (fmol)	Determined Amount (fmol)	Accuracy (%, RE)	RSD (%, n = 5)	Linearity			LOD (amol)	LOQ (amol)
					Regression Equation[a,b]	r[b]	fmol		
GC	25.0	26.1	4.4	2.2	$y = 6.59x + 39.6$	0.998	0.2–500	54	180
EGC	25.0	25.0	0.1	1.5	$y = 7.26x + 42.1$	0.998	0.2–500	61	203
C	25.0	24.3	−2.8	1.8	$y = 6.10x + 33.5$	0.999	0.2–500	61	203
EGCg	25.0	24.5	−2.0	1.6	$y = 6.34x + 37.2$	0.998	0.2–500	75	250
EC	25.0	25.7	2.8	1.9	$y = 5.84x + 31.7$	0.999	0.2–500	75	250
GCg	25.0	25.5	2.2	1.9	$y = 7.57x + 49.7$	0.998	0.2–500	67	223
ECg	25.0	25.4	1.4	1.8	$y = 5.23x + 35.2$	0.998	0.2–500	75	250
Cg	25.0	26.1	4.4	1.8	$y = 3.80x + 24.2$	0.998	0.2–500	89	297

Source: Used with permission from Kotani, A., Takahashi, K., Hakamata, H., Kojima, S., Kusu, F., Attomole catechins determination by capillary liquid chromatography with electrochemical detection, *Anal. Sci.*, 2007, 23, 157.

[a] x and y indicate catechin amount (fmol) and peak current height (pA), respectively.

[b] The number of plots used to show linearity in the concentration range was 14.

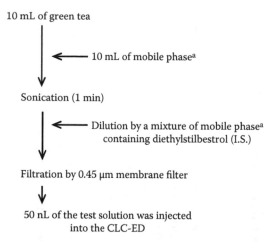

10 mL of green tea

⟵ 10 mL of mobile phase[a]

Sonication (1 min)

⟵ Dilution by a mixture of mobile phase[a]
containing diethylstilbestrol (I.S.)

Filtration by 0.45 μm membrane filter

50 nL of the test solution was injected
into the CLC-ED

FIGURE 7.26   Preparation of test solution in the present method for determining catechins in green tea. [a]The dry residue was dissolved by methanol–water–phosphoric acid mixture (27.5:72.5:0.5, v/v/v).

The amount of quantitation limit (S/N = 10) for GC, EGC, C, EGCg, EC, GCg, ECg, and Cg were 180, 203, 203, 250, 250, 223, 250, and 297 amol, respectively, i.e., the concentration of quantitation limit (S/N = 10) for GC, EGC, C, EGCg, EC, GCg, ECg, and Cg were 1.10, 1.25, 1.18, 1.53, 1.45, 2.05, 2.21, and 2.62 ng/mL, respectively.

7.4.3.9   Pretreatment of Samples

Pretreatment of samples prior to analysis may be necessary, and examples of pretreatment protocol are summarized in the following and outlined in Figures 7.26 and 7.27.

Pretreatment of a tea sample in the determination catechins—Protocol: Ten mL of a sample of tea and 10 mL of a mobile phase are mixed by the sonication for 1 min. This mixture is diluted 500 times with mobile phase and then filtered through a 0.45 μm membrane filter. The test solutions can be injected into the CLC-ED system (Figure 7.26). Using this pretreatment, the recovery of the catechins was approximately 100% (based on 50 μg/mL) with an RSD (n = 5) of less than 3% [9].

Pretreatment of plasma sample for determining catechins—Protocol: The plasma obtained from 20 μL of blood is used for determining the free and conjugated form catechins in plasma. Ten microliters of plasma are mixed with 10 μL 0.4 M phosphate buffer (pH 3.6) containing 2% ascorbic acid and 0.1% ethylenediaminetetraacetic acid disodium salt, and 30 μL 0.1 M phosphate buffer (pH 6.8) containing 500 U β-glucuronidase, 40 U sulfatase, and 5.0 pmol propyl gallate as the internal standard. The mixture is incubated at 37°C for 45 min to digest conjugated forms of catechins completely. Catechins in the enzyme-hydrolyzed solution are extracted three times with 50 μL of aliquots of ethyl acetate. The collected ethyl acetate is evaporated to dryness using a nitrogen stream. The dry residue is dissolved in 10 μL phosphoric acid (85%)–methanol–water (0.5:27.5:72.5, v/v/v) solution to obtain a test solution for total amounts of free and conjugated form catechins. A test solution for free form catechins is prepared in a similar way without the enzyme digestion: 10 μL of

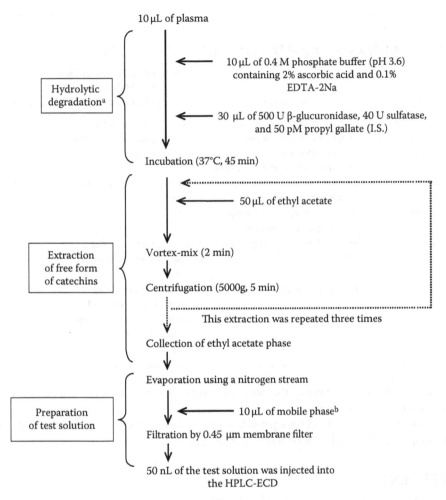

FIGURE 7.27 Preparation of test solution in the present method for determining catechins in human plasma. [a]For determination of free catechins, 10 μL of plasma were mixed with the 0.4 mol/L phosphate buffer (pH 3.6) containing 2% ascorbic acid and 0.1% EDTA-2Na. [b]The dry residue was dissolved by methanol–water–phosphoric acid mixture (27.5:72.5:0.5, v/v/v).

plasma are mixed with the aforementioned solution without enzyme, catechins in the mixture are extracted with ethyl acetate, and the collected ethyl acetate is evaporated to dryness using a nitrogen stream. Then, the test solutions can be injected into the CLC-ED system (Figure 7.27). Using this pretreatment, the recovery of the catechins was 88% (based on 10 ng/mL), with an RSD ($n = 5$) of less than 5% [9].

7.5 CONCLUSION

This chapter provides an overview of practical HPLC-ED for determining electro-active analytes. Typical electroactive natural products analyzed by HPLC-ED are shown in Table 7.10. Using a properly constructed system and operating the system

TABLE 7.10

Typical Analytes of Natural Products Analyzed by HPLC-ED

Group	Analyte	Typical HPLC-ED Conditions (Column, Mobile Phase, Applied Potential)	References
Catechol amine and its metabolite	Norepinephrine epinephrine, dopamine, methanephrine, homovanillic acid, vanillylmandelic acid, synephrine	RP (C_{18}), methanol-citric and phosphoric buffer mixture, +0.5 to +1.0	[17–21]
Flavone	Baicalin, baicalein, wogonin, luteolin, hyperoside	RP (C_{18}), water–methanol mixture	[22,23]
Flavonol	Rutin, galangin, fisetin, kaempferol, morin, quercetin, myricetin		[23–25]
Phenol	Magnolol, honokiol, hesperidin, fustin	RP (C_{18}), water-methanol-phosphoric acid mixture, +0.8 V	[26,27]
Estrogens	Estrone, estriol, β-estradiol	RP (C_{18}), water-methanol-acetic acid mixture, +1.0 V	[28]
Hydroxychromans	Tocopherol	RP (C_{18} or C_{30}), methanol-acetate buffer mixture, +0.7 to +1.0	[29–32]

with the optimum condition, the analytes can be determined with high selectivity, sensitivity, and precision.

Details of the historical background, the basic theory of electrochemistry, and the principle of HPLC-ED can be found in Refs. [33–37].

REFERENCES

1. Kissinger, P.T.; *Analytical Chemistry*, 1977, 49, 447A.
2. LaCourse, W.R.; *Pulsed Electrochemical Detection in High Performance Liquid Chromatography*, Wiley, New York, 1997.
3. Elbicki, J.M.; Morgan, D.M.; Weber, S.G.; *Analytical Chemistry*, 1984, 56, 978.
4. Kotani, A.; Hayashi, Y.; Matsuda, R.; Kusu, F.; *Analytical Science*, 2003, 19, 865.
5. Kotani, A.; Iwagami, T.; Ueda, T.; Kimura, Y.; Hayashi, Y.; Matsuda, R.; Kusu, F.; *ITE Letters*, 2003, 4, 783.
6. Kotani, A.; Miyashita, N.; Kusu, F.; *Journal of Chromatography B*, 2003, 788, 269.
7. Yang, B.; Arai, K.; Kusu, F.; *Analytical Biochemistry*, 2000, 283, 77.
8. Umegaki, K.; Sugisawa, A.; Yamada, K.; Higuchi, M.; *Journal of Nutritional Science and Vitaminology*, 2001, 47, 402.
9. Kotani, A.; Takahashi, K.; Hakamata, H.; Kojima, S.; Kusu, F.; *Analytical Science*, 2007, 23, 157.
10. Riddick, J.A.; Bunger, W.B.; Sakano, T.K.; *Organic Solvents, Physical Properties and Methods of Purification*, 4th edn., Wiley, New York, 1986.
11. Bott, A.W.; *Current Separations*, 1997, 16, 79.
12. Kotani, A.; Hayashi, Y.; Matsuda, R.; Kusu, F.; *Journal of Chromatography A*, 2003, 986, 239.

13. Hayashi, Y.; Matsuda, R.; *Analytical Chemistry*, 1994, 66, 2874.
14. Caudill, W.L.; Howell, J.O.; Wightman, R.M.; *Analytical Chemistry*, 1982, 54, 2532.
15. Yang, B.; Kotani, A.; Arai, K.; Kusu, F.; *Chemical and Pharmaceutical Bulletin*, 2001, 49, 747.
16. Yang, B.; Kotani, A.; Arai, K.; Kusu, F.; *Analytical Science*, 2001, 17, 599.
17. Unceta, N.; Rodriguez, E.; de Balugera, Z.G.; Sampedro, C.; Goicolea, M.A.; Barrondo, S.; Salles, J.; Barrio, R.J.; *Analytica Chimica Acta*, 2001, 444, 211.
18. Patel, B.A.; Arundell, M.; Parker, K.H.; Yeoman, M.S.; O'Hare, D.; *Journal of Chromatography B*, 2005, 818, 269.
19. Kusu, F.; Li, X.-D.; Takamura, K.; *Chemical and Pharmaceutical Bulletin*, 1992, 40, 3284.
20. Kusu, F.; Matsumoto, K.; Takamura, K.; *Chemical and Pharmaceutical Bulletin*, 1995, 43, 1158.
21. Arai, K.; Jin, D.; Kusu, F.; Takamura, K.; *Journal of Pharmaceutical and Biomedical Analysis*, 1997, 15, 1509.
22. Kotani, A.; Kojima, S.; Hayashi, Y.; Matsuda, R.; Kusu, F.; *Journal of Pharmaceutical and Biomedical Analysis*, 2008, 48, 780.
23. Jin, D.; Hakamata, H.; Takahashi, K.; Kotani, A.; Kusu, F.; *Bulletin of Chemical Society Japan*, 2004, 77, 1147.
24. Danila, A.-M.; Kotani, A.; Hakamata, H.; Kusu, F.; *Journal of Agriculture and Food Chemistry*, 2007, 54, 1139.
25. Jin, D.; Hakamata, H.; Takahashi, K.; Kotani, A.; Kusu, F.; *Biomedical Chromatography*, 2004, 18, 662.
26. Kotani, A.; Kojima, S.; Hakamata, H.; Jin, D.; Kusu, F.; *Chemical and Pharmaceutical Bulletin*, 2005, 53, 319.
27. Xia, J.; Kotani, A.; Hakamata, H.; Kusu, F.; *Journal of Pharmaceutical Biomedical Analysis*, 2006, 41, 1401.
28. Penalver, A.; Pocurull, E.; Borrull, F.; Marce, R.M.; *Journal of Chromatography A*, 2002, 964, 153.
29. Delgado-Zamarreno, M.M.; Bustamante-Rangel, M.; Sanchez-Perez, A.; Carabias-Martinez, R.; *Journal of Chromatography A*, 2004, 1056, 249.
30. Williamson, K.S.; Hensley, K.; Floyd, R.A.; *Methods in Biological Oxidative Stress*, Humana Press, Totowa, NJ, 2003, pp. 67–76.
31. Puspitasari-Nienaber, N.L.; Ferruzzi, M.G.; Schwartz, S.J.; *Journal of the American Oil Chemists' Society*, 2002, 79, 633.
32. Yamauchi, R.; Noro, H.; Shimoyamada, M.; Kato, K.; *Lipids*, 2002, 37, 515.
33. Gunasingham, H.; Fleet, B.; *Electroanalytical Chemistry*, 1989, 16, 89.
34. Wang, J.; *Analytical Electrochemistry*, 3rd edn., Wiley, Hoboken, NJ, 2006.
35. Bard, A.J.; Faulkner, L.R.; *Electrochemical Method, Fundamentals and Applications*, Wiley, New York, 1980.
36. Flanagan, R.J.; Perrett, D.; Whelpton, R.; *Electrochemical Detection in HPLC: Analysis of Drugs and Poisons*, The Royal Society of Chemistry, Cambridge, U.K., 2005.
37. Acworth, I.N.; Naoi, M.; Parvez, H.; Parvez, S.; *Coulometric Electrode Array Detectors for HPLC*, Progress in HPLC-HPCE, vol. 6, VSP, Utrecht, the Netherlands, 1997.

8 Liquid-Phase Chemiluminescence Detection for HPLC

Paul S. Francis and Jacqui L. Adcock

CONTENTS

8.1 Introduction 221
8.2 Chemiluminescence Detection for Liquid Chromatography 223
8.3 Practical Considerations 224
    8.3.1 Instrument Components 224
    8.3.2 Optimization 225
8.4 Commonly Used Reagents and Their Applications 225
    8.4.1 Luminol and Related Reagents 226
    8.4.2 Lucigenin and Acridinium Esters 230
    8.4.3 Peroxyoxalate Reagents 232
    8.4.4 Tris(2,2′-Bipyridine)Ruthenium(III) 238
    8.4.5 Potassium Permanganate 241
    8.4.6 Other Liquid-Phase Chemiluminescence Reagents 242
8.5 Concluding Remarks 244
References 245

8.1 INTRODUCTION

The emission of light from an electronically excited intermediate or product of a chemical reaction is known as *chemiluminescence* [1–4]. The general process is shown in Scheme 8.1, where species A and B react to form C, with a certain proportion in an electronically excited state (C*) that can subsequently relax to the ground state by emitting a photon. Chemiluminescence is commonly observed in the near-ultraviolet, visible and/or near-infrared regions (~350 to 900 nm), which corresponds to excess chemical energy from 340 to 130 kJ mol^{-1}.

$$A + B \rightarrow C^* + \text{other products}$$
$$C^* \rightarrow C + \text{light}$$

$$(8.1)$$

In some cases, the excited intermediate (C*) transfers energy to a suitable fluorophore (Fl), which may then emit its characteristic fluorescence (Scheme 8.2). This phenomenon is referred to as *indirect* or *sensitized* chemiluminescence [1–4]:

$$C^* + Fl \rightarrow C + Fl^*$$

$$Fl^* \rightarrow Fl + \text{light}$$

$$(8.2)$$

Irrespective of the reaction pathway, once the final excited state is reached, the emission process is identical to that of other modes of luminescence, and for practical applications can be considered as instantaneous. However, unlike photoluminescence, the production of the excited state in chemiluminescence depends on the physical processes of solution mixing and the kinetics of the chemical reaction. The transient emission is therefore initiated as soon as the reactants are combined. Most analytically useful chemiluminescent reactions last for seconds or minutes, but some persist for much longer. Commercially available glow sticks, used as novelty items and for emergency lighting, are a well-known application of intense, long-lived chemiluminescence reactions.

Chemiluminescence is sometimes described as *cold light* or *light without heat*. It should be noted that much of the excess energy from these exothermic reactions is released as heat, resulting in a small increase in solution temperature. However, it is far lower than the temperatures required for *incandescent* light, such as that emitted from household light globes containing a filament through which an electric current is passed. As shown in Scheme 8.3, the chemiluminescence quantum yield (the proportion of reacting molecules that result in the emission of a photon) depends on several factors, including the efficiencies of the chemical reaction (ϕ_{cr}), conversion of chemical potential energy into electronic excitation (ϕ_{ec}) and—in the case of sensitized chemiluminescence—energy transfer (ϕ_{et}), and the proportion of excited molecules that emit a photon (ϕ_l):

$$\phi_{cl} = \phi_{cr} \times \phi_{ec} (\times \phi_{et}) \times \phi_l \qquad (8.3)$$

Furthermore, the chemiluminescence intensity at any particular moment is dependent on the quantum yield, the number of reacting molecules, and the rate of the reaction. These relationships can be exploited for quantitative measurement of the reactants, or species that catalyze, enhance, or inhibit the light-producing reaction. Chemiluminescence detection can be extremely sensitive, due in part to the absence of an excitation light source, which provides superior signal-to-background ratios compared to fluorescence, and the ability to detect light emitted from a relatively large volume of solution. Moreover, chemiluminescence detection can be quite selective toward the target analytes, depending on the number of compounds that evoke an intense emission, which can be manipulated by the choice of reagent and reaction conditions.

8.2 CHEMILUMINESCENCE DETECTION
FOR LIQUID CHROMATOGRAPHY

This subject has been discussed in several previous book chapters [5–8] and reviews [9,10]. We would therefore like this chapter to serve as both an update of contemporary practice in HPLC with chemiluminescence detection, and a practical account of the advantages and pitfalls of this approach.

The most commonly used mode of detection for HPLC is the absorption of ultraviolet (UV) light, which provides reasonable sensitivity for many organic compounds, using reliable and fairly low-cost instrumentation. The downside of the wide applicability is relatively poor selectivity, which must therefore be derived almost entirely from the chromatographic separation and preliminary sample preparation. Furthermore, many compounds of interest in biological or environmental samples are present at concentrations below the limits of detection that can be achieved using UV-absorbance, and therefore the analyst must often turn to more sensitive and selective modes of detection, such as fluorescence or electrochemistry.

Laser-induced fluorescence can provide exceedingly low limits of detection for analytes containing a suitable fluorophore, but with a considerable increase in instrument complexity. On the other hand, chemiluminescence detection is generally superior to electrochemistry or standard fluorescence procedures in terms of sensitivity, selectivity, and analytical working range, and can be performed using simple, robust, inexpensive instrumentation. In general, a chemiluminescence detector consists of a transparent flow cell mounted against a photosensitive device (such as a photomultiplier tube). The column eluate is merged with the chemiluminescence reagent prior to entering the flow cell (see the following section).

Like electrochemical detection, chemiluminescence is dependent on oxidation and reduction processes, but the interactions that initiate the light-producing pathways can be considerably more complex than simple electron transfer. Although rarely exploited to date, chemiluminescence detection can therefore provide a means to rapidly assess the chemical reactivity of sample components eluted from a chromatographic column, which can be particularly useful when coupled with other methods of detection such as UV-absorbance.

The previous discussion on the advantages of chemiluminescence detection prompts the question of what has restricted its use in liquid chromatography (LC). First, chemiluminescence detection requires the column eluate to be merged with a reagent within a suitable detector. The assembly of the flow manifold and daily preparation of reagent solutions increases the total labor requirements, and therefore this approach is less attractive in situations where detection based on the native absorbance or fluorescence of the analytes is appropriate. The complexity of chemiluminescence detection could be considered similar to automated, rapid versions of the pre- or post-column derivatizations used to improve the fluorescence character of analytes.

Second, a large majority of HPLC methods with chemiluminescence detection published in the open literature are based on three well-established reagent classes:

(1) luminol and its derivatives, (2) peroxyoxalate reagents, and (3) ruthenium complexes [5–10]. While a range of other reagents have been explored, in many cases the reaction pathways leading to the emission of light and the influence of the chemical environment are yet to be fully established. Therefore, the application of this mode of detection to new analyte classes has often been based on an imperfect, empirical understanding of the relationship between analyte structure and chemiluminescence intensity with any particular reagent.

The perception that chemiluminescence has limited applicability is reflected in the reluctance of the major HPLC manufacturers to develop chemiluminescence detectors as part of their suite of components. In addition, the lack of purpose-built commercial instrumentation limits the promotion of this approach and the advice and support available from HPLC manufacturers for analysts to develop new procedures.

Nevertheless, a wide range of HPLC procedures incorporating chemiluminescence detection have been developed and successfully applied in areas such as industrial process monitoring, pharmaceutical and biomedical analysis, and forensic science.

8.3   PRACTICAL CONSIDERATIONS

8.3.1   INSTRUMENT COMPONENTS

For chemiluminescence detection, the column eluate must be merged with one or more reagent solutions. Because the pressure within the flow system after the separation column (and UV-absorbance detector, if included) is not high, the additional manifold can be constructed from relatively cheap components (peristaltic pumps, low-pressure polymer fittings), such as those employed for flow-injection analysis. A simple T- or Y-piece fitting can be used to combine solutions (Figure 8.1).

It is crucial that the geometry and dimensions of the flow cell maximize the proportion of light that is emitted when the reacting mixture is in front of the detector. For fast chemiluminescence reactions, this means that the column eluate and reagent solutions should merge at (or as close as possible to) the point of detection. For slower chemiluminescence reactions, the length of tubing between the final T-piece and the detector should be optimized so that the most intense portion of the emission is detected.

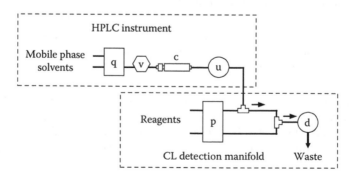

FIGURE 8.1   An example manifold for HPLC with chemiluminescence detection. Key: q = quaternary pump; v = injection valve; c = column; u = UV-absorbance detector; p = peristaltic pump; d = chemiluminescence detector.

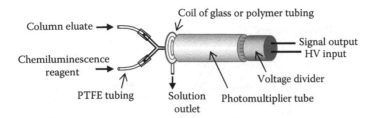

FIGURE 8.2 A flow-through chemiluminescence detector.

In either case, the "dead volume" of the flow cell should be minimized, to ensure reproducible mixing and rapid washing between injections, and to minimize band broadening.

Unlike fluorescence detection, a light source is not required for chemiluminescence. Wavelength discrimination usually offers no advantage (because different analytes generally lead to the same emission) and should be avoided due to the detrimental effect on sensitivity. One design that satisfies these requirements is a coil of glass or transparent polymer tubing with similar internal diameter to the manifold tubing (0.3–1.0 mm) positioned against a photomultiplier tube (Figure 8.2). A light-tight housing is essential. Some chemiluminescence detectors that incorporate this design are now commercially available. Fluorescence detectors (with the excitation source switched off) have also been used for chemiluminescence detection. Although suitable for some applications, they generally provide lower sensitivity, because they are not designed for efficient mixing of column eluate and reagent in front of the photodetector window. Furthermore, the relatively short residence time of the reacting mixture within fluorescence flow cells limits the proportion of the emitted light that can be captured.

8.3.2 OPTIMIZATION

As with any analytical procedure, the optimization of experimental conditions can lead to significant improvements in performance. The optimum conditions (reagent concentration, pH, flow rates, selection of "enhancers") for chemiluminescence detection are often dependent not only on the reagent, but also the class of analyte under investigation. Much of our knowledge on chemiluminescence detection is derived from flow-injection analysis studies using predominantly aqueous conditions, and therefore it is also important to examine the effect of mobile phase solvents on these reactions. For example, acetonitrile quenches the response from chemiluminescence reactions with acidic potassium permanganate, but methanol has been found to be a suitable alternative [11]. Moreover, the optimum pH for separation may not be optimal for detection, but this can be addressed by merging the column eluate with an acid, base, or buffer solution prior to merging with the reagent.

8.4 COMMONLY USED REAGENTS AND THEIR APPLICATIONS

As shown in Figure 8.3, the application of chemiluminescence detection in LC is dominated by a few well-known reagents. The following survey is not exhaustive, but is intended to illustrate the variety of ways that liquid-phase chemiluminescence

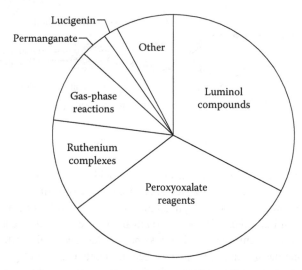

FIGURE 8.3   Relative use of different reagent classes for post-column chemiluminescence detection.

has been applied as a sensitive and selective mode of detection for HPLC. Gas-phase chemiluminescence has also been successfully utilized (e.g., see Refs. [12–18]), but will not be discussed in this chapter.

8.4.1   LUMINOL AND RELATED REAGENTS

The oxidation of luminol (5-amino-2,3-dihydro-1,4-phthalazinedione) in alkaline solution evokes a blue luminescence (λ_{max} = 425 nm) that emanates from electronically excited 3-aminophthalate (Figure 8.4) [19–21].

A variety of oxidants, such as hydrogen peroxide, periodate, hypochlorite, and permanganate, can be used. The reaction with hydrogen peroxide is slow, but can

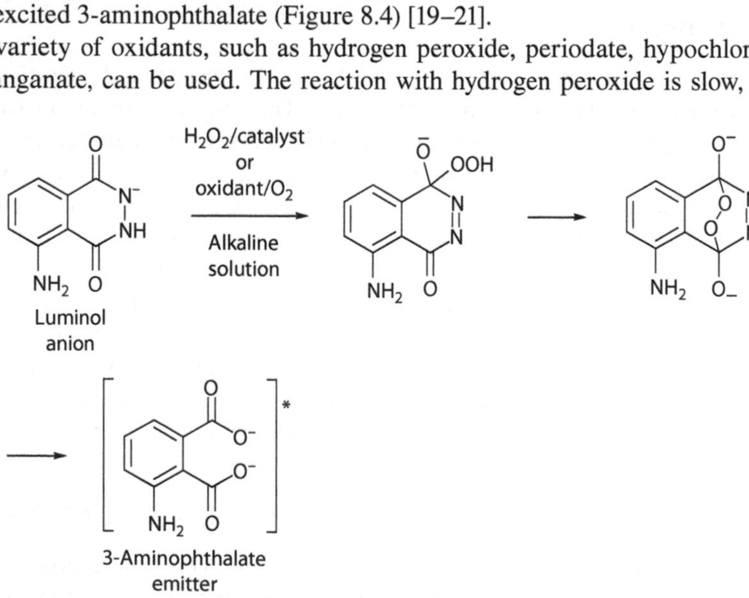

FIGURE 8.4   The chemiluminescent oxidation of luminol in alkaline solution.

be effectively catalyzed by certain transition-metal ions and complexes, such as cobalt(II), copper(II), iron(II) and hexacyanoferrate(III) (ferricyanide), and peroxidase enzymes, which can be used for the highly sensitive detection of these species. For example, cobalt has been detected at picomolar levels after cation-exchange chromatography without pre-concentration [22]. Badocco and coworkers recently described the determination of iron(II), cobalt(II), and manganese(II) using ion chromatography and reported detection limits of 0.50, 0.24, and 375 nM, respectively [23]. This chemistry has also been exploited for the post-column detection of many organic molecules, which mainly fall into four categories: organic hydroperoxides; substrates of enzymatic reactions that produce hydrogen peroxide; compounds derivatized with luminol analogues; and compounds that enhance or inhibit the light-producing pathway.

The determination of organic hydroperoxides using HPLC with luminol chemiluminescence detection has been extensively used to obtain quantitative analytical data on the peroxidation pathways in both laboratory and biological systems [24–28], particularly those involving phosphatidylcholine and cholesterol. Measurement of hydroperoxides has also been employed to assess oxidative stress, antioxidant activity, and free radical-mediated DNA damage [29–33]. The post-column manifold for these applications is simple; the column eluate containing the separated analytes (oxidants) is merged with an aqueous reagent stream containing luminol or isoluminol, a catalyst (cytochrome C, microperoxidase, or other peroxidases), and a buffer (often at pH 10), in a T- or Y-piece prior to the detection flow cell. UV-absorbance detection can be incorporated in series (before the eluate is merged with the chemiluminescence reagent) [24–27], and mass spectroscopy can be used in parallel (by splitting the eluate) [28]. External standards are often used for calibration, but suitable internal standards have been reported [34,35]. Limits of detection for lipid hydroperoxides in blood plasma (including preliminary extraction procedures) have been stated as around 10^{-8} M [36,37]. Hui and coworkers reported a limit of detection of 1 pmol (i.e., 5×10^{-8} M in the 20 µL injection) using synthetic analyte standards [35].

A variety of biomolecules can be sensitively detected by coupling H_2O_2-generating enzymatic reactions with luminol chemiluminescence (providing detection limits generally between 10^{-8} and 10^{-6} M). The selectivity of these enzymatic reactions is very high, and this approach has often been applied without chromatographic separation [20]. However, in some cases, more than one concomitant species will form hydrogen peroxide and separation of sample components is required [38–41]. For example, Alam and coworkers determined histamine and N^{τ}-methylhistamine in the brain and peripheral tissues of rats, using reversed-phase LC with a post-column reactor containing immobilized diamine oxidase [39]. As with the organic hydroperoxides, the hydrogen peroxide generated in the reactor was detected by subsequent chemiluminescence reaction with luminol and a catalyst, which in this case was potassium hexacyanoferrate(III). The immobilized-enzyme column was stable for more than 1 month of operation (200 samples analyzed).

The highly sensitive detection afforded by luminol reagents has been applied to broad analyte classes through the development of derivatizing agents that contain the phthalazine-1,4-dione chemiluminophore [19,20,42]. In some cases, the

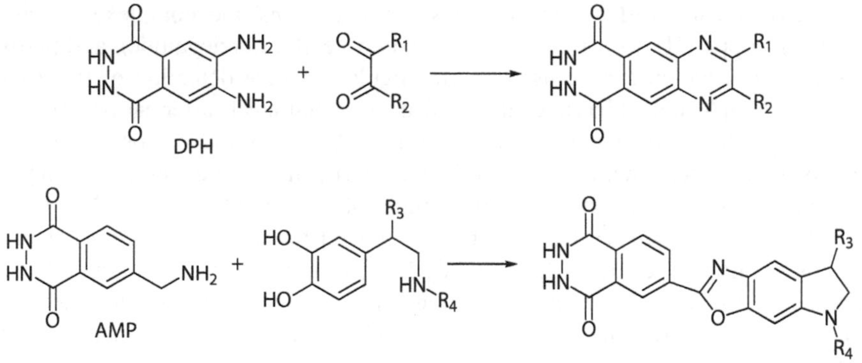

FIGURE 8.5 The reaction of chemiluminogenic reagents, DPH and AMP, with α-dicarbonyl compounds and catecholamines, respectively, to form highly chemiluminescent products.

product is far more chemiluminescent (>100 times) than the unreacted derivatizing agent (which is referred to as a *chemiluminogenic* reagent). Examples include 4,5-diaminophthalhydrazide (DPH) which forms highly chemiluminescent products with α-keto acids, α-dicarbonyl compounds and aromatic aldehydes, and 6-aminomethylphthalazine-1,4-dione (AMP), which has been used to detect 5-hydroxyindoles and catecholamines (Figure 8.5).

Alternatively, *chemiluminescence labeling* reagents can be used to bond an already effective chemiluminophore to a target class of analytes. Reagents of this type, such as the examples shown in Figure 8.6, have been developed for carboxylic acids, amines, amino acids, peptides, glycosides, and thiols.

Liquid chromatographic procedures incorporating initial derivatization with phthalazine-1,4-dione based reagents enable the highly sensitive detection of many analytes. Applications include the determination of amphetamines [43] and 3α,5β-tetrahydroaldosterone [44] in urine (using ABEI and DPH, respectively), and amantadine [45] and fatty acids [46] in plasma (using TPB-Suc and PROB). Typical chromatograms for the determination of amantadine in plasma are shown in Figure 8.7.

As with many fluorescence-labeling procedures, the preliminary off-line derivatizations often require considerable time (10–120 min) at elevated temperatures, but some proceed quite quickly at room temperature [47–49]. The post-column chemiluminescence reaction with the derivatized analytes has often been initiated by merging the column eluate with hydrogen peroxide and hexacyanoferrate(III), but electrochemical oxidation has also been reported. Limits of detection are typically between 0.2 and 200 fmol ($\sim$$10^{-11}$ to 10^{-8} M). Further details regarding luminol-type derivatization reagents can be found in the review by Yamaguchi and coworkers [19].

Many compounds enhance or inhibit the emission of light from luminol systems, which can also be exploited for quantitative analysis. An early approach was based on the suppression of chemiluminescence due to the interaction between analytes and the catalyst (e.g., Co(II) or Cu(II)). The column eluate is therefore first combined with a stream containing the catalyst, before merging with alkaline luminol and hydrogen peroxide solutions to initiate the chemiluminescence reaction. Limits of

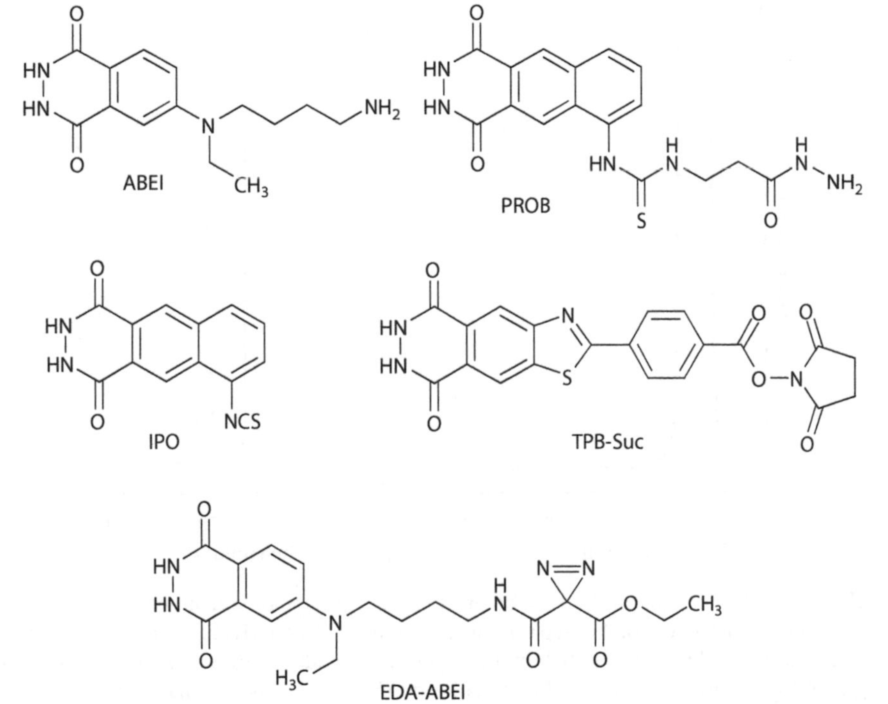

FIGURE 8.6 Selected luminol-type chemiluminescence labeling reagents: N-(4-aminobutyl)-N-ethylisoluminol (ABEI); 6-[N-(3-propionohydrazino)thioureido]benzo[g]-phthalazine-1,4(2H,3H)-dione (PROB); 6-isothiocyanatobenzo[g]-phthalazine-1,4(2H,3H)-dione (IPO); 4-(6,7-dihydro-5,8-dioxothiazolo[4,5-g]phthalazin-2-yl)benzoic acid N-hydroxysuccinimide ester (TPB-Suc); and 4-[(N-α-ethoxycarbonyldiazoacetyl)aminobutyl]-N-ethylisoluminol (EDA-ABEI).

detection for primary and secondary amino acids, amines, proteins, catecholamines, catechol, and aminoglycoside antibiotics after ion-exchange, ion-pair, or reversed-phase LC [50–53], were typically between 1 pmol and 2 nmol ($\sim 10^{-7}$ to 10^{-4} M).

It was later found that reducing agents, such as corticosteroids, phenols, and anilines, can react with radical intermediates of the light-producing pathway of various luminol systems and therefore influence the rate or efficiency of emission [54–57]. For post-column detection, the most common approach involves combining the column eluate stream with one solution containing luminol and base, and another containing hexacyanoferrate(III), but the order in which they are merged varies in different procedures [58–62]. A diverse range of other oxidants and catalysts (e.g., I_2, $Cu(II)/H_2O_2$, BrO^-, Co(II), and $[Cu(HIO_6)_2]^{5-}$) has also been used in procedures based on signal enhancement or inhibition [63–67].

An interesting variant of this approach is the assessment of antioxidant status, where measurement is based on the scavenging of reactive oxygen species, which inhibits the emission of light [68,69]. Appropriate selection of reagents and configuration of the instrument manifold can focus the assessment on specific oxygen species, such as hydrogen peroxide or the superoxide anion radical (produced from

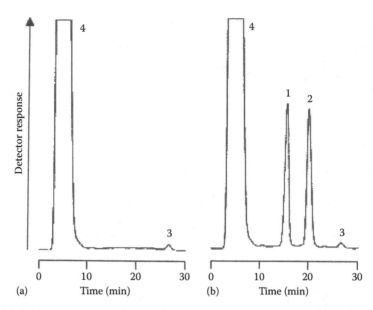

FIGURE 8.7 Chromatograms for the determination of amantadine in human plasma using reversed-phase chromatographic separation (C_{18} column) after derivatization with TPB-Suc (80°C, 20 min). (a) Drug-free sample without internal standard. (b) Sample taken 2 h after single oral administration of amantadine chloride (50 mg). Peaks: 1, amantadine; 2, internal standard [1-(1-adamantyl)ethylamine]; 3, an endogenous amino compounds; 4, reagent blank and other endogenous amino compounds. (Reprinted from *J. Chromatogr. A*, 907, Yoshida, H., Nakao, R., Matsuo, T., Nohta, H., and Yamaguchi, M., 4-(6,7-Dihydro-5,8-dioxothiazolo [4,5-g] phthalazin -2-yl) benzoic acid N-hydrosuccinimide ester as a highly sensitive chemiluminescence derivatization reagent for amines in liquid, 39, Copyright 2001, with permission from Elsevier.)

hypoxanthine and xanthine oxidase) [69]. This chemistry has been used to establish the overall antioxidant status of various samples [70], but when coupled with LC, each component of the sample can be individually assessed [68,69]. Figure 8.8 shows the chromatograms for a green tea extract using both UV-absorbance and chemiluminescence detection. In this case, the negative peaks in Figure 8.8b correspond to inhibition of the chemiluminescence signal from the reaction of luminol, hypoxanthine, and xanthine oxidase, therefore indicating which components are most effective scavengers of superoxide ion.

8.4.2 LUCIGENIN AND ACRIDINIUM ESTERS

The reaction of lucigenin (10,10′-dimethyl-9,9′-bisacridinium nitrate) with hydrogen peroxide in aqueous alkaline solution evokes a blue-green emission from *N*-methylacridone (Figure 8.9a) [71]. Certain transition metal ions are effective catalysts, but unlike the reaction of luminol with hydrogen peroxide, they are not essential for an intense emission of light with this reagent. Nevertheless, Chen and coworkers demonstrated that the quenching effect of amino acids on the cobalt(II)

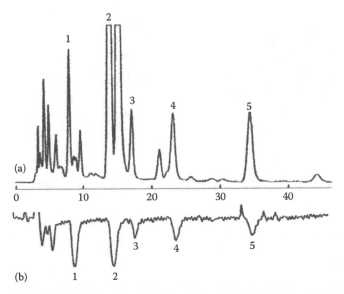

FIGURE 8.8 (a) UV-absorbance (265 nm) and (b) chemiluminescence inhibition (super-oxide scavenging) for a green tea extract after reversed-phase chromatographic separation (ODS column). Peaks: 1, epigallocatechin; 2, epigallocatechin gallate; 3, gallocatechin gallate; 4, epicatechin gallate; 5, catechin gallate. (Reprinted from *Talanta*, 60, Toyo'oka, T., Kashiwazaki, T., and Kato, M., On-line screening methods for antioxidants scavenging superoxide anion radical and hydrogen peroxide by liquid chromatography with indirect chemiluminescence detection, 467, Copyright 2003, with permission from Elsevier.)

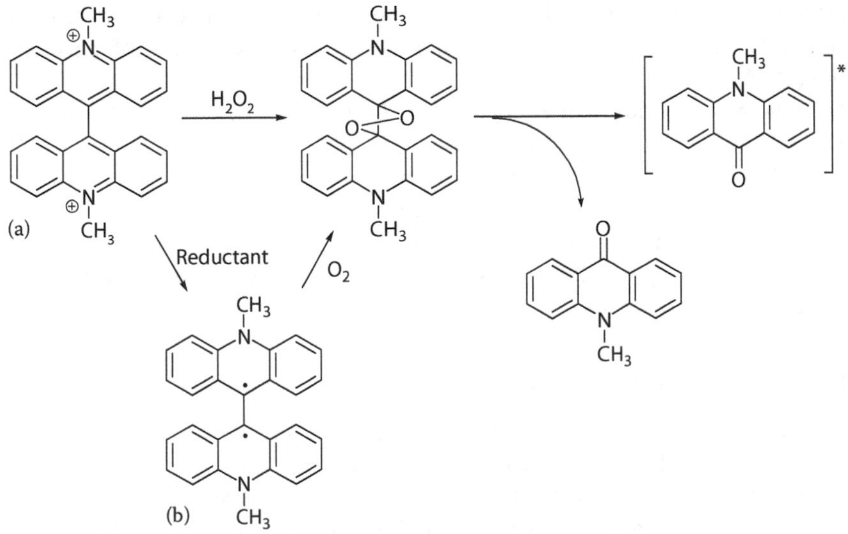

FIGURE 8.9 The chemiluminescent reaction of lucigenin in alkaline solution with (a) hydrogen peroxide and (b) a reducing agent and molecular oxygen.

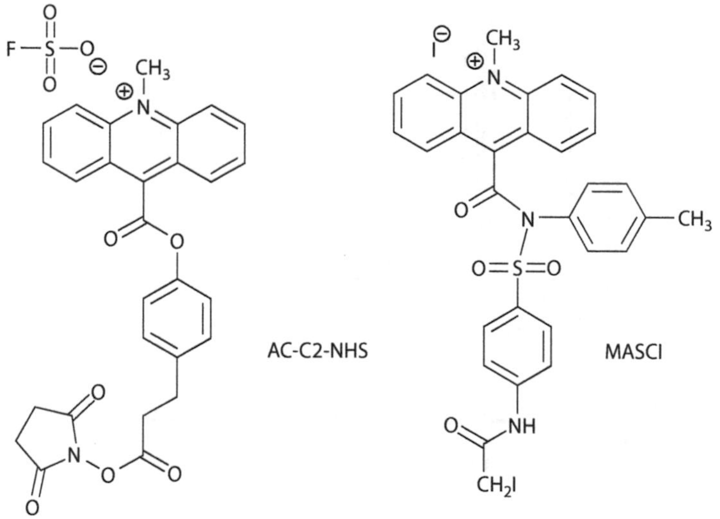

FIGURE 8.10 Acridinium esters: 4-(2-succinimidyloxycarbonylethyl) phenyl-10-methylacridinium-9-carboxylate fluorosulfonate (AC-C2-NHS) and 10-methyl-*N*-(*p*-tolyl)-*N*-(*p*-iodoacetamidobenzenesulfonyl)-9-acridinium carboxamide iodide (MASCI), for labeling amino acids [79], and carboxylic acids [80], respectively. (From Zhong, L. and Maloy, J.T., *Anal. Chem.*, 70, 1100, 1998; Steijger, O.M. et al., *J. Biolumin. Chemilumin.*, 13, 31, 1998.)

catalyzed oxidation of lucigenin could be used to detect these species at nanomolar levels without pre-column derivatization [72].

In the presence of dissolved oxygen, lucigenin will also produce light on reaction with a range of reducing agents (Figure 8.9b), which has been exploited for the post-column detection of ascorbic acid [73], glucose and other sugars [73], glucuronides (after enzyme-catalyzed hydrolysis to glucuronic acid) [74], and α-hydroxycarbonyl compounds (such as corticosteroids and *p*-nitrophenacyl esters of carboxylic acids) [75,76].

Like lucigenin, acridinium esters react with hydrogen peroxide in alkaline solution to form *N*-methylacridone in an electronically excited state. Several labeling reagents for HPLC and immunoassay incorporating acridinium esters have been developed (Figure 8.10) [77–80] and used to detect compounds such as chlorophenols [77], primary amines [79,81], and carboxylic acids [80]. In this approach, the column eluate is merged with base (NaOH or KOH) and hydrogen peroxide streams before entering the detector.

8.4.3 Peroxyoxalate Reagents

Certain diaryloxalates and diaryloxamides produce an intense emission of light when treated with hydrogen peroxide in the presence of a suitable fluorophore (Figure 8.11) [82–85]. Collectively known as *peroxyoxalate* chemiluminescence, these light-producing reactions have a complex multistep mechanism that has been extensively studied, but is yet to be fully elucidated [83,84]. Among the variety of postulated intermediates, perhaps the most contentious has been 1,2-dioxanedione

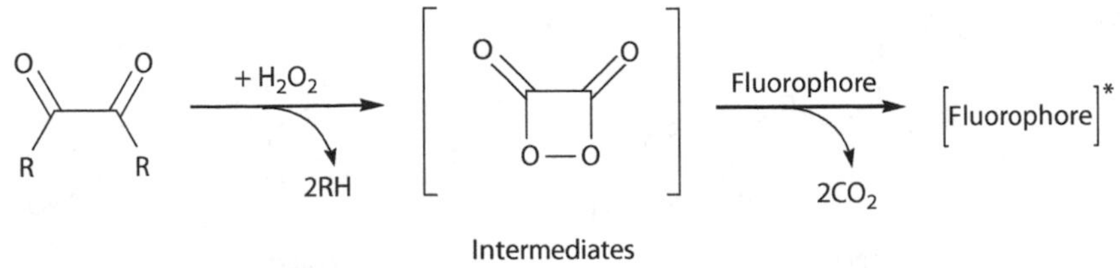

FIGURE 8.11 A simplified scheme for peroxyoxalate chemiluminescence.

(shown in Figure 8.11), first proposed over four decades ago [86], and only recently confirmed [87,88]. The reaction is subject to interrelated general-base and nucleophilic catalysis, for which imidazole is particularly effective.

Electronic excitation of the fluorophore is thought to involve: (1) the formation of a charge complex with the dioxanedione intermediate, which accepts an electron to form a radical anion; (2) decomposition of the radical anion into CO_2 and $[CO_2^{\bullet}]^{-}$; and (3) transfer of the electron back to the fluorophore at a higher energy level. For the greatest emission of light, the energy acceptor must therefore have both a low oxidation potential and efficient fluorescence. Polyaromatic hydrocarbons (particularly those with an amino substituent) are among the most effective.

Unlike the other light-producing reactions described in this chapter, peroxyoxalate chemiluminescence is not well suited to the predominantly aqueous conditions of reversed-phase HPLC, due to the limited solubility and stability of the reagents in water. There has been some progress in the development of peroxyoxalate reagents for aqueous solutions [84], but it has not yet reached the point of routine use in HPLC analysis. Coupling HPLC with contemporary peroxyoxalate detection therefore requires merging the column eluate with reagents prepared in organic solvents (such as acetonitrile or ethyl acetate). Efficient mixing is essential for high reproducibility and sensitivity. Acetonitrile is the most commonly used organic modifier for separation. Methanol should be avoided due to its strong quenching effect. The background emission is dependent on the proportion of water in the mobile phase and therefore solvent gradients can produce noisy and sloping baselines, but this can be minimized by a judicious selection of instrumental and reaction conditions [83].

The most commonly used peroxyoxalate reagents are shown in Figure 8.12. For any particular application, the best choice of reagent depends on the instrumental approach and the nature of the assay [83,89–92]. For example, DNPO reacts at faster rates than TCPO and produces greater chemiluminescence intensities [93], but this is only an advantage if the detection manifold is configured so that the most intense emission occurs within the flow cell [89,90]. TDPO is much more soluble in acetonitrile and more stable in the presence of hydrogen peroxide [94].

In spite of the complexity of utilizing peroxyoxalate chemiluminescence, it has been successfully employed for the sensitive post-column detection of many compounds. The target analytes normally act as energy acceptors and emitters, either due to their native fluorescence character, or after derivatization with fluorescence-labeling reagents. To a lesser extent, this chemistry has been used to detect compounds that produce hydrogen peroxide in chemical or photochemical reactions, and (in a few cases [95,96]) compounds that can enhance or inhibit the production of light.

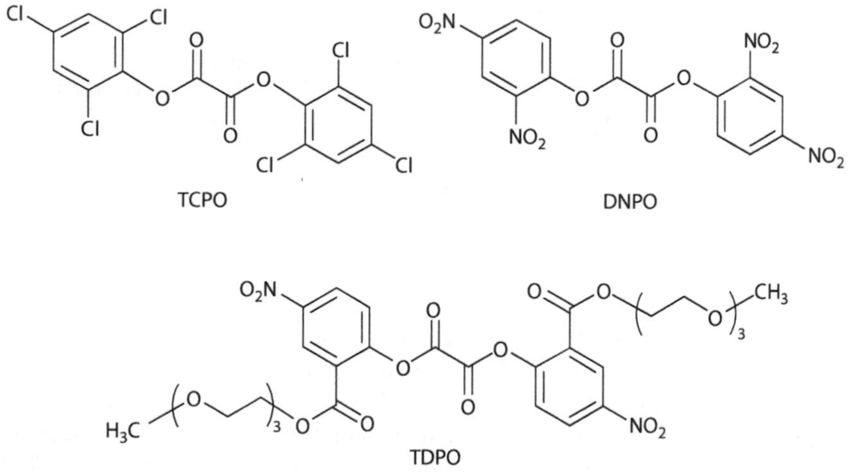

FIGURE 8.12   Selected peroxyoxalate reagents: bis(2,4,6-trichlorophenyl)oxalate (TCPO); bis(2,4-dinitrophenyl)oxalate (DNPO); and bis[2-(3,6,9-trioxadecyloxycarbonyl)-4-nitrophenyl] oxalate (TDPO).

For the detection of fluorescent compounds (native or derivatized), a reagent stream containing the diaryloxalate (in ethyl acetate or acetonitrile) is often combined and mixed with a solution of hydrogen peroxide (prepared by diluting the 30% v/v aqueous solution with acetonitrile, ethyl acetate or propan-2-ol) prior to merging with the column eluate. Alternatively, a single reagent stream, containing both the diaryloxalate and hydrogen peroxide can be used, but in this case the stability of the combined reagents should be established during the optimization of conditions [94,97]. Chromatographic separation is typically performed using reversed-phase C_{18} columns with isocratic mobile phases comprising of 20%–85% acetonitrile and an aqueous solution of imidazole adjusted to a pH between 6 and 8. However, a range of other conditions [98–100], including gradient elution [101–103], have been reported.

This mode of detection has been applied to polycyclic aromatic hydrocarbons (PAHs), particularly those with amino groups, for which the limits of detection are often 1–2 orders of magnitude better than those obtainable with fluorescence [104,105]. Perhaps more importantly, this approach has been adapted for the determination of nitro-substituted PAHs in airborne particulates [106–109]. In these procedures, the target analytes are reduced to the corresponding amino species, either off-line prior to analysis (e.g., by refluxing in ethanol in the presence of sodium hydrosulfide [107]) or online, in a reduction column packed with Pt/Rh coated alumina [108,109] or zinc immobilized on glass beads [106].

Many different derivatizing reagents have been used to extend peroxyoxalate chemiluminescence detection to new classes of analytes [84,85]. The most commonly used of these reagents feature the dansyl [i.e., 5-(dimethylamino)-naphthalene-1-sulfonyl] fluorophore, with chloro, hydrazino, or other functionality to allow binding with the analytes of interest (Figure 8.13). Applications include the determination of methamphetamine and its metabolites in a single hair or the urine of abusers

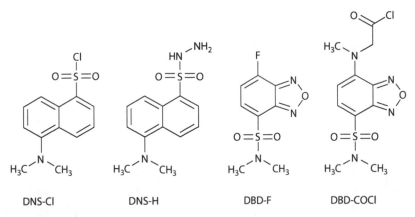

DNS-Cl   DNS-H   DBD-F   DBD-COCl

FIGURE 8.13 Some of the derivatizing reagents used with peroxyoxalate chemiluminescence detection: 5-(dimethylamino)-naphthalene-1-sulfonyl chloride (DNS-Cl); 5-(dimethylamino)-naphthalene-1-sulfonic hydrazide (DNS-H); 4-(N,N-dimethylaminosulfonyl)-7-fluoro-2,1,3-benzoxadiazole (DBD-F); and 4-(N,N-dimethylaminosulfonyl)-7-(N-chloroformylmethyl-N-methyl)amino-2,1,3-benzoxadiazole (DBD-COCl).

[101,110,111], ketosteroids in serum [112], and pesticide residues in vegetables and fruit juices [113,114]. Derivatizations generally require relatively long periods of time at elevated temperatures, and are therefore often performed off-line, prior to analysis. However, automated online derivatizations within heated mixing coils [115] or on solid-phase supports [111] have also been developed.

The hydrolysate of DNS-Cl also gives relatively intense chemiluminescence with peroxyoxalate reagents, which can interfere with the detection of analytes at low concentrations [98]. The hydrolysate of 4-(N,N-dimethylaminosulfonyl)-7-fluoro-2,1,3-benzoxadiazole (DBD-F), a related fluorescence-labeling agent for amines, is less readily formed and is less fluorescent, therefore reducing the potential for interference. DBD-F and its 7-hydrazino and 7-(N-chloroformylmethyl-N-methyl)amino analogues have been used in HPLC procedures with peroxyoxalate chemiluminescence detection, such as the determination of pharmaceuticals in biological fluids [98,116,117], and MDMA (ecstasy) and related compounds in hair (Figure 8.14) [118].

Catecholamines (epinephrine, norepinephrine, and dopamine) have been detected with peroxyoxalate chemiluminescence after derivatization with ethylene diamine or 1,2-bis(3-chlorophenyl)ethylenediamine [99,119]. An automated procedure incorporating (1) preliminary ion-exchange extraction, (2) chromatographic separation, (3) post-column fluorogenic derivatization, and (4) peroxyoxalate chemiluminescence detection has been developed (Figures 8.15 and 8.16) [120].

This approach has been adapted to quantify the 3-O-methyl metabolites, metanephrine, normetanephrine, and 3-methoxytyramine, by incorporating online electrochemical oxidation of the metabolites to the corresponding o-quinones [121,122], and to determine nitrocatecholamines, by introducing a reduction column into the flow manifold [123]. Isoproterenol, 4-methoxytyramine, and N-methyldopamine have been used as internal standards. Limits of detection of 0.3–2 fmol ($\sim 10^{-11}$ to 10^{-10} M) for catecholamines and 3-O-methyl metabolites have been reported [100,120,122].

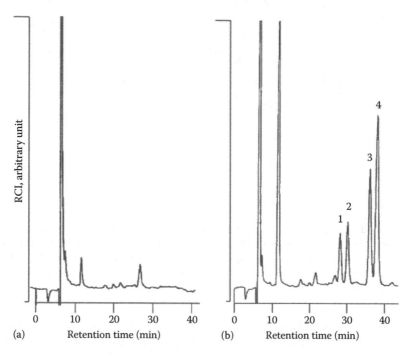

FIGURE 8.14 Chromatograms for the determination of amphetamines in (a) normal (drug-free) hair, and (b) drug abuser's hair. Sample preparation included digestion, extraction (*n*-heptane), and derivatization with DBD-F. Separation column: Capcellpak C18 UG120. Chemiluminescence reagent: bis(2,4,5-trichloro-6-carbopentoxyphenyl)oxalate (CPPO) and H_2O_2 in acetonitrile. Peaks: 1, amphetamine; 2, ecstasy; 3, methamphetamine; 4, internal standard (1-methyl-3-phenylpropylamine). (With kind permission from Springer Science+ Business Media: *Analytical and Bioanalytical Chemistry*, A sensitive semi-micro column HPLC method with peroxyoxalate chemiluminescence detection and column switching for determination of MDMA-related compounds in hair, 387, 2007, 1983, Nakamura, S., Wada, M., Crabtree, B.L., Reeves, P.M., Montgomery, J.H., Byrd, H.J., Harada, S., Kuroda, N., and Nakashima, K., Figures 6a and 7a, Copyright Springer-Verlag 2006.)

Like luminol, peroxyoxalate reagents can be used to detect the hydrogen peroxide generated in post-column immobilized-enzyme reactors, for the determination of substrates such as amino acids [124], choline and acetylcholine [125,126], polyamines [127,128], and glucose [126]. In general, a solution containing the peroxyoxalate reagent and suitable fluorophore is merged with the exit stream of the immobilized-enzyme reactor. A buffer stream is sometimes added between the separation column and reactor. Limits of detection are typically between 10^{-8} and 10^{-6} M.

Hydrogen peroxide can also be generated in post-column photochemical reactions, from species such as aliphatic alcohols and molecular oxygen, sensitized by quinones [129]. Because of the catalytic nature of the reaction, each quinone molecule can produce many molecules of hydrogen peroxide as it passes through the reactor, which (when coupled with peroxyoxalate chemiluminescence detection) provides sub-picomole detection limits for species such as 9,10-anthraquinone and menadione (vitamin K_3). Originally, the fluorophore (rubrene) was added to the peroxyoxalate

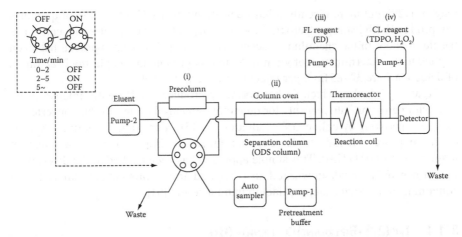

FIGURE 8.15 Flow manifold for the determination of catecholamines in rat plasma. (From Takezawa, K., Tsunoda, M., Murayama, K., Santa, T., and Imai, K., Automatic semi-microcolumn liquid chromatographic determination of catecholamines in rat plasma utilizing peroxyoxalate chemiluminescence reaction, *Analyst*, 125, 293, 2000. Reproduced by permission of The Royal Society of Chemistry.)

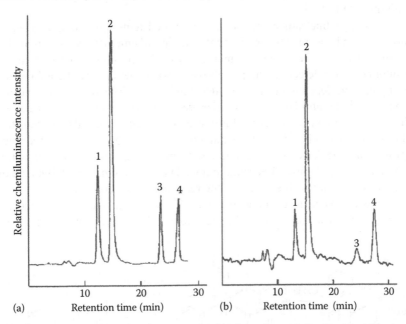

FIGURE 8.16 Chromatograms for the determination of catecholamines in (a) a standard solution and (b) rat plasma. Procedure includes online extraction, separation (Capcell Pak C18 UG120), derivatization (ethylenediamine) and chemiluminescence detection (TDPO and H_2O_2). Peaks: 1, norepinephrine; 2, epinephrine; 3, dopamine; 4, internal standard (*N*-methyldopamine). (From Takezawa, K., Tsunoda, M., Murayama, K., Santa, T., and Imai, K., Automatic semi-microcolumn liquid chromatographic determination of catecholamines in rat plasma utilizing peroxyoxalate chemiluminescence reaction, *Analyst*, 125, 293, 2000. Reproduced by permission of The Royal Society of Chemistry.)

reagent [129], but more recently it has been shown that UV irradiation of quinones can produce both hydrogen peroxide and a highly fluorescent species that can act as the fluorophore in the chemiluminescence detection [130]. This approach has been applied to the determination of vitamin K homologues in human plasma [92]; limits of detection were 32–85 fmol per injection (i.e., 3.2–8.5 nM).

Organic peroxides, which do not normally produce a sufficiently intense emission with peroxyoxalate reagents for quantitative analysis, can also be converted to hydrogen peroxide by online UV irradiation [91,131]. For example, online UV irradiation improved the limit of detection for cumene hydroperoxide by five orders of magnitude (to 1 μM) [131]. This simple approach has been applied to the determination of benzoyl peroxide in wheat flour [131], and artemisinin (an antimalarial drug containing an internal endoperoxide) in human serum [91].

8.4.4 TRIS(2,2′-BIPYRIDINE)RUTHENIUM(III)

The reduction of tris(2,2′-bipyridine)ruthenium(III) with certain amines (particularly aliphatic tertiary amines), organic acids, conjugated dienes, and various other organic and inorganic compounds evokes an orange luminescence (λ_{max} = 610–620 nm) that matches the characteristic photoluminescence of tris(2,2′-bipyridine) ruthenium(II) [132,133].

Tris(2,2′-bipyridine)ruthenium(III) has a limited temporal stability in aqueous solution, but can be formed by oxidizing the stable ruthenium(II) form (Figure 8.17), either shortly before the reagent is merged with the column eluate or within the chemiluminescence detector. There are several ways that this can be achieved; the most appropriate depends on the intended application. The stability of the tris(2,2′-bipyridine)ruthenium(III) is greater in more acidic solution [134], but the analyte-dependent optimum pH of the chemiluminescence reaction must also be considered. Solid lead dioxide can be added off-line and then removed by filtration as the solution is aspirated into the detector. Some of the best limits of detection have been obtained by generating tris(2,2′-bipyridine)ruthenium(III) in this manner [133], but lead dioxide is quite toxic and also breaks down the reagent over several hours.

Online oxidation, using chemical (cerium(IV) or permanganate), electrochemical (at an electrode surface), or photochemical (using persulfate) approaches, ensures

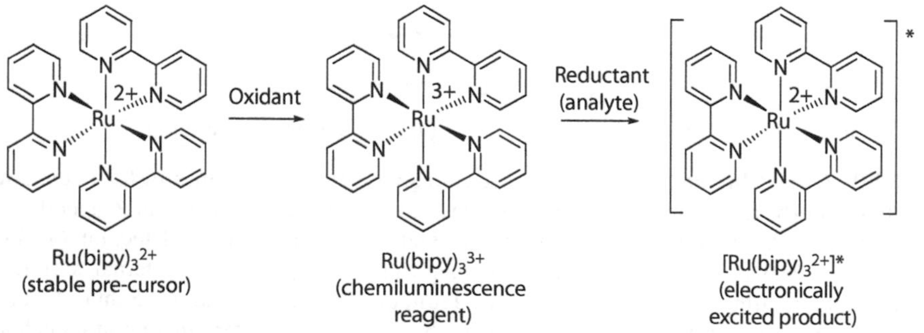

$Ru(bipy)_3^{2+}$
(stable pre-cursor)

$Ru(bipy)_3^{3+}$
(chemiluminescence reagent)

$[Ru(bipy)_3^{2+}]^*$
(electronically excited product)

FIGURE 8.17 Preparation and chemiluminescent reduction of tris(2,2′-bipyridine) ruthenium(III).

reproducible preparation, but increases the complexity of the flow manifold. When permanganate is used to prepare the reagent, the excess oxidant may also undergo chemiluminescence reactions (see the following section) [135]. Most commonly used organic modifiers, such as acetonitrile and methanol, are compatible with this reagent, although both positive and negative changes in chemiluminescence intensity have been reported.

Reversed-phase HPLC coupled with tris(2,2′-bipyridine)ruthenium(III) chemiluminescence or electrochemiluminescence detection has been used to determine oxalate [136], hydroxyproline [137], and various pharmaceuticals [138–142] in biological fluids, aminopolycarboxylic acids in natural waters [143], and various other compounds of interest (e.g., quinolizidine alkaloids, glycoalkaloids, quinolones, and pipecolic acid) in biological fluids, plants, and/or foods (Figure 8.18) [144–147]. In several cases, separation has been performed using commercially available *monolithic* columns, consisting of a single rod of porous silica with C_{18} (endcapped) surface modification [145,147,148]. Compared to conventional 5 μm particulate columns, the monolithic columns allow higher flow rates, while maintaining separation efficiency [149]. While the main advantage of the higher flow rates is a significant

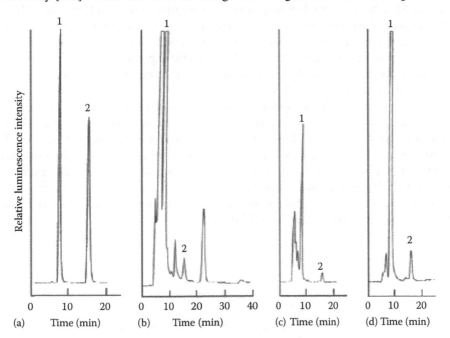

FIGURE 8.18 Chromatograms for the determination of pipecolic acid in (a) a standard solution, (b) human serum, (c) cow milk, and (d) beer. Samples were filtered to remove protein, diluted, and acidified before injection. Column: 2 × Chromolith Performance RP-18e. The tris(2,2′-bipyridine)ruthenium(III) reagent was generated by online electrochemical oxidation of the ruthenium(II) form. Peaks: 1, proline; 2, pipecolic acid. (Reprinted from *J. Chromatogr. A*, 1140, Kodamatani, H., Komatsu, Y., Yamazaki, S., Saito, K., Highly sensitive and simple method for measurement of pipecolic acid using reverse-phase ion-pair high performance liquid chromatography with tris(2,2′-bipyridine) ruthenium(III) chemiluminescence detection, 88, Copyright 2007, with permission from Elsevier.)

reduction in overall run time, they can also improve the sensitivity of fast chemiluminescence reactions, such as those with tris(2,2'-bipyridine)ruthenium(III), due to
the more rapid propulsion of the column eluate and reagent mixture into the flow-
through detector [148].

Chemiluminescence detection with this reagent has been extended through the
chemical or photochemical modification of compounds that would not otherwise
elicit an intense emission. For example, treatment of amino acids with dansyl chloride (DNS-Cl; Figure 8.13) increased the electrogenerated chemiluminescence
response with tris(2,2'-bipyridine)ruthenium(II) by three orders of magnitude, due to
the greater substitution on the amino nitrogen and the presence of a tertiary amine on
the derivatizing agent [150]. Other pre-column derivatizing agents have been specifically designed to add tertiary amines to compounds with alcohol, carboxylic acid, or
primary amine functionality (Figure 8.19) [151]. The most sensitive approach (using
IDHPIA) provided an on-column detection limit of 0.5 fmol (5×10^{-11} M in the 10 μL
injection) for myristic acid [151]. Typical chromatograms for the determination of
free fatty acids in human plasma are shown in Figure 8.20.

Divinylsulfone (Figure 8.21), which converts primary amines into 1,4-thiazane-
1,1-dioxides, has been used in procedures to determine amino acids in serum [152].
The derivatization of carbonyl compounds with methylmalonic acid, and amines with
diketene, to form species that produce light with tris(2,2'-bipyridine)ruthenium(III)
has also been demonstrated [153,154]. Although these procedures provide sensitive
detection, the preliminary off-line derivatization of the target analytes increases the
overall analysis time by 20–120 min.

Automated, post-column derivatization has also been demonstrated. Primary or
secondary amines have been converted (online) to tertiary amines using acryloni-
trile or epichlorohydrin [155,156]. Limits of detection for mono-, di- and tri-ethanol-
amine using this approach were 30, 25, and 40 pmol, respectively [156]. Post-column
chemical or photochemical degradation of diols, amines, amino acids, aromatic
compounds, N-nitrosamines, and N-methylcarbamates has been utilized to form species such as oxalate and certain amines that can be more sensitively detected with
tris(2,2'-bipyridine)ruthenium(III) [157–159].

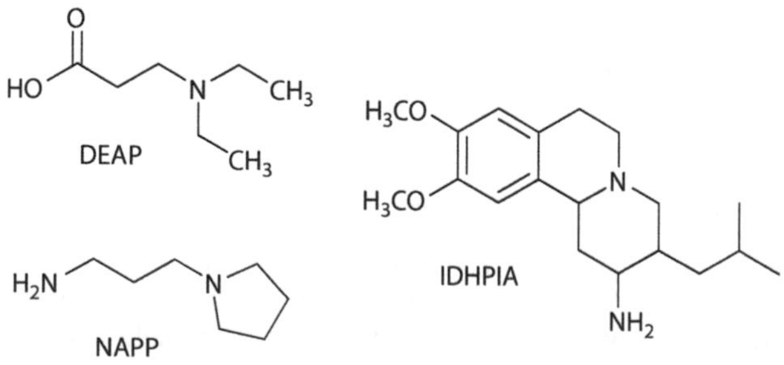

FIGURE 8.19   Pre-column derivatizing agents that add a tertiary amine group for detection with tris(2,2'-bipyridine)ruthenium(III): 3-(diethylamino)propionic acid (DEAP);
N-(3-aminopropyl)pyrrolidine (NAPP); and 3-isobutyl-9,10-dimethoxy-1,3,4,6,7,11b-hexahydro-
2H-pyrido[2,1-a]isoquinolin-2-ylamine (IDHPIA).

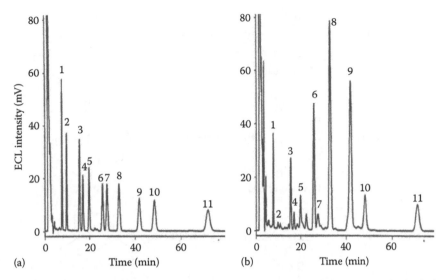

FIGURE 8.20 Chromatograms for the determination of fatty acids in (a) a standard solution (2 pmol) and (b) human serum extract (chloroform/*n*-hexane, 1:1 v/v). Analytes were derivatized off-line with IDHPIA. Column: Inertsil C8. The tris(2,2'-bipyridine)ruthenium(III) reagent was generated by on-line electrochemical oxidation of the ruthenium(II) form. Peaks: 1, lauric acid; 2, myristoleic acid; 3, myristic acid; 4, R-linolenic acid; 5, palmitoleic acid; 6, R-linoleic acid; 7, arachidonic acid; 8, palmitic acid; 9, oleic acid; 10, internal standard (margaric acid); 11, stearic acid. (Reprinted with permission from Morita, H. and Konishi, M., Electrogenerated chemiluminescence derivatization reagent, 3-isobutyl-9,10-dimethoxy-1,3,4,6,7,11b-hexahydro-2H-pyrido[2,1-a]isoquinolin-2-ylamine, for carboxylic acid in high-performance liquid chromatography using tris(2,2'-bipyridine)ruthenium(II). *Anal. Chem.*, 75, 940, 2003. Copyright 2003 American Chemical Society.)

The tris(2,2'-bipyridine)ruthenium(III) reagent can be regenerated by oxidation of the reaction product, and there has been considerable effort made to develop stable immobilized reagents, particularly on the surface of electrodes [133]. An electrochemiluminescence sensor with reagent immobilized in titania-Nafion nanocomposite films coated on an electrode has been used for the determination of phenothiazine derivatives and erythromycin in human urine [160].

8.4.5 POTASSIUM PERMANGANATE

The reaction of potassium permanganate with a wide variety of organic compounds (particularly phenols, polyphenols, catechols, and indoles) in acidic aqueous solution evokes a broad red emission (λ_{max} = 734 nm) from an excited manganese(II) species [11,161]. Formaldehyde/formic acid or polyphosphates are commonly used to improve the sensitivity of this reagent (the presence of polyphosphates also shifts the maximum intensity to 689 nm) [162].

Acidic potassium permanganate chemiluminescence is well suited to reversed-phase HPLC, including procedures with mobile-phase gradients. A reagent solution

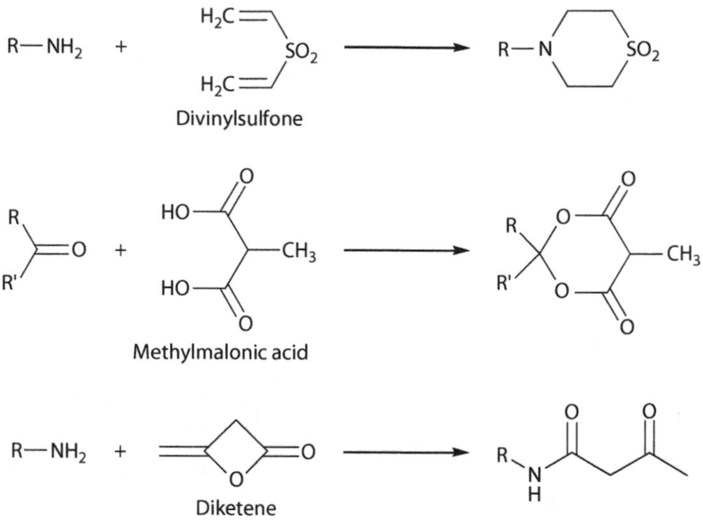

FIGURE 8.21   Pre-column derivatizing agents that convert amines or carbonyl compounds to other functionality that is suitable for detection with tris(2,2′-bipyridine)ruthenium(III).

containing both permanganate and polyphosphate enhancer (adjusted to pH 2–3) is reasonably stable, and therefore only a single reagent line is required [163]. Enhancement with formaldehyde or formic acid is more problematic, because these species slowly react with permanganate and must be merged with the reagent or column eluate shortly prior to the final confluence point [164,165]. The chemiluminescent reactions with the target analytes are often very rapid, and therefore the time taken to present the reacting mixture to the detection flow cell must be kept as short as possible. Methanol is the most commonly used organic modifier for chromatographic separations coupled with this reagent; acetonitrile should be avoided due its strong quenching effect [11].

Applications include the determination of opiate alkaloids in body fluids and industrial process streams [166], neurotransmitter metabolites in urine [163], polyhydroxybenzenes in river water [165], adrenergic amines in weight-loss products [167], arbutin and L-ascorbic acid in whitening cosmetics [164], and alkylthiouracils in human serum [168].

Figure 8.22 shows previously unpublished data from the development of a rapid HPLC procedure for the determination of opiate alkaloids in process streams, using a monolithic column and gradient elution [148]. The superior selectivity of the chemiluminescence detection enabled baseline resolution of both morphine and oripavine within 1.5 min, which could not be achieved with UV-absorbance detection.

8.4.6   OTHER LIQUID-PHASE CHEMILUMINESCENCE REAGENTS

Various phenol [169–171], thiol [172–175], and related compounds [175] have been detected using chemiluminescence reactions with cerium(IV) in acidic aqueous solution. In each system, an enhancer (rhodamine B, rhodamine 6G, quinine) is merged

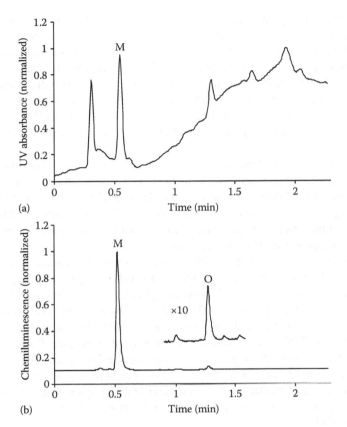

FIGURE 8.22 Chromatograms for the determination of morphine (M) and oripavine (O) in an industrial process sample using HPLC with a monolithic column (Chromolith SpeedROD RP-18e) and (a) UV-absorbance or (b) acidic potassium permanganate chemiluminescence detection.

with the column eluate and the mixture is combined with the oxidant stream shortly prior to entering the detection coil. The dominant light-producing pathway involves the formation of high-energy intermediates that transfer energy to the enhancers, which emit light at their characteristic wavelengths. However, the exact nature of the intermediates is yet to be confirmed. Limits of detection using these systems are typically between 0.01 and 1.0 μM.

Several other oxidant/enhancer combinations have been explored. Tetracyclines can be detected in acidic solution with permanganate, sulfite and β-cyclodextrin [176], or with cerium(IV) and rhodamine B (incorporating photoirradiation of the column eluate) [177]. Polyphenols can be detected in alkaline solution with hydrogen peroxide and 1-ethyl-3-(3-dimethylaminopropyl)carbodiimide [178]. Pyrethroid and benzoylurea insecticides have been determined in vegetables by combining chromatographic separation, post-column irradiation with UV light, and chemiluminescence detection with potassium hexacyanoferrate(III) in alkaline solution. The chemiluminescence signal increased with the percentage of acetonitrile in the reaction medium [179,180].

Like potassium permanganate (described previously), manganese(III) and manganese(IV) oxidants react with many molecules in acidic solution to produce an excited manganese(II) species that emits red light, and some preliminary demonstrations of post-column detection with these alternative manganese oxidants have emerged [181]. "Soluble" manganese(IV) can be formed by dissolving freshly precipitated manganese dioxide in 3.0 M orthophosphoric acid. Manganese(III) can be generated by online electrochemical oxidation of manganese(II). Interestingly, the selectivities of these reagents differ markedly from acidic potassium permanganate [181].

Alkaline hypobromite has been used for the detection of arginine and arginine-containing peptides separated with monolithic column chromatography [182]. Although the limit of detection was only 0.1 μM for arginine and 4 μM for bradykinin, the reagent has been found to be selective toward these monosubstituted guanidine species (and urea) in the presence of many common amino acids, peptides, and other low-molecular-weight biological compounds.

The photoirradiation of compounds such as anthracenes, naphthalenes, halogenated biphenyls, and Rose Bengal produces molecular oxygen in its singlet excited state. Although this product itself can emit light (λ_{max} = 634, 703, and 1268 nm) to return to the triplet ground state [183], it can also be detected through the formation of a dioxetane with 1,2-diethoxyethylene or ethyl vinyl ether [184]. Thermal degradation (70°C) of the dioxetane in the presence of an efficient fluorophore, such as 9,10-dibromoanthracene or 9,10-dibromoanthracene-2-sulfonate, elicits relatively intense chemiluminescence. This chemistry has been applied to the determination of toxic polychlorinated biphenyls in spiked herring oil samples using reversed-phase and normal-phase LC. Limits of detection were between 2 and 9 nM (for normal-phase LC) [184].

8.5   CONCLUDING REMARKS

HPLC with chemiluminescence detection has been utilized for the sensitive determination of a wide range of compounds, including many that do not possess a strong chromophore or native fluorophore. Furthermore, the high selectivity of some chemiluminescence reagents can simplify the analysis of complex samples. This provides the opportunity for more rapid separation, which can enhance the improvements in separation speed obtained through recent innovations in column technology, such as monolithic stationary phases. When a suitable reagent for direct chemiluminescence detection cannot be found, derivatization with chemiluminophores (or fluorophores that can be excited by high-energy reaction intermediates) can provide greater sensitivity than traditional fluorescence-labeling approaches. A large majority of the chemiluminescence detection systems for HPLC published thus far have been based on a few well-established reagent classes. The greatest potential for new developments may therefore lie with a more complete understanding of their respective light-producing pathways and the exploration of new chemiluminescence reactions.

REFERENCES

1. García-Campaña, A.M.; Baeyens, W.R.G.; Eds.; *Chemiluminescence in Analytical Chemistry*; Marcel Dekker: New York, 2001.
2. Barnett, N.W.; Francis, P.S.; In: *Encyclopedia of Analytical Science*, Vol. 1, 2nd edn.; Worsfold, P.J.; Townshend, A.; Poole, C.F.; Eds.; Elsevier: Oxford, U.K., 2005, pp. 506–510.
3. Su, Y.; Chen, H.; Wang, Z.; Lv, Y.; *Applied Spectroscopy Reviews*, 2007, 42, 139.
4. Francis, P.S.; Hogan, C.F.; In: *Advances in Flow Injection Analysis and Related Techniques*, Comprehensive Analytical Chemistry Series, Vol. 54; McKelvie, I.D.; Kolev, S.D.; Eds.; Elsevier: Oxford, U.K., 2008, pp. 343–373.
5. Nieman, T.A.; In: *Chemiluminescence and Photochemical Reaction Detection in Chromatography*; Birks, J.W.; Ed.; VCH: New York, 1989, pp. 99–123.
6. Nieman, T.A.; In: *Luminescence Techniques in Chemical and Biochemical Analysis*, Vol. 12; Baeyens, W.R.G.; De Keukeleine, D.; Korkidis, K.; Eds.; Marcel Dekker: New York, 1991, pp. 523–565.
7. Hage, D.S.; In: *HPLC Detection: Newer Methods*; Patonay, G.; Ed.; VCH: New York, 1992, pp. 57–75.
8. Kuroda, N.; Kai, M.; Nakashima, K.; In: *Chemiluminescence in Analytical Chemistry*; García-Campaña, A.M.; Baeyens, W.R.G.; Eds.; Marcel Dekker: New York, 2001, pp. 393–425.
9. Li, F.; Zhang, C.; Guo, X.; Feng, W.; *Biomedical Chromatography*, 2003, 17, 96.
10. Tsukagoshi, K.; *Science and Engineering Review of Doshisha University*, 2005, 45, 168.
11. Adcock, J.L.; Francis, P.S.; Barnett, N.W.; *Analytica Chimica Acta*, 2007, 601, 36.
12. Howard, A.L.; Thomas, C.L.B.; Taylor, L.T.; *Analytical Chemistry*, 1994, 66, 1432.
13. Shi, H.; Taylor, L.T.; Fujinari, E.M.; Yan, X.; *Journal of Chromatography A*, 1997, 779, 307.
14. Yurek, D.A.; Branch, D.L.; Kuo, M.-S.; *Journal of Combinatorial Chemistry*, 2002, 4, 138.
15. Lane, S.; Boughtflower, B.; Mutton, I.; Paterson, C.; Farrant, D.; Taylor, N.; Blaxill, Z.; Carmody, C.; Borman, P.; *Analytical Chemistry*, 2005, 77, 4354.
16. Styslo-Zalasik, M.; Li, W.; *Journal of Pharmaceutical and Biomedical Analysis*, 2005, 37, 529.
17. Idowu, A.D.; Dasgupta, P.K.; *Analytical Chemistry*, 2007, 79, 9197.
18. Ojanperä, S.; Tuominen, S.; Ojanperä, I.; *Journal of Chromatography B*, 2007, 856, 239.
19. Yamaguchi, M.; Yoshida, H.; Nohta, H.; *Journal of Chromatography A*, 2002, 950, 1.
20. Marquette, C.A.; Blum, L.; *Journal of Analytical and Bioanalytical Chemistry*, 2006, 385, 546.
21. Barni, F.; Lewis, S.W.; Berti, A.; Miskelly, G.M.; Lago, G.; *Talanta*, 2007, 72, 896.
22. Jones, P.; Williams, T.; Ebdon, L.; *Analytica Chimica Acta*, 1989, 217, 157.
23. Badocco, D.; Pastore, P.; Favaro, G.; Maccà, C.; *Talanta*, 2007, 72, 249.
24. Kumakura, K.; Kitada, M.; Horie, T.; Awazu, S.; *Analytical Letters*, 1997, 30, 1483.
25. Adachi, J.; Asano, M.; Naito, T.; Ueno, Y.; Tatsuno, Y.; *Lipids*, 1998, 33, 1235.
26. Nakamura, A.; Ohori, Y.; Watanabe, K.; Sato, Y.; Boger, P.; Wakabayashi, K.; *Pesticide Biochemistry and Physiology*, 2000, 66, 206.
27. Adachi, J.; Yoshioka, N.; Funae, R.; Nagasaki, Y.; Naito, T.; Ueno, Y.; *Lipids*, 2004, 39, 891.
28. Tagiri-Endo, M.; Nakagawa, K.; Sugawara, T.; Ono, K.; Miyazawa, T.; *Lipids*, 2004, 39, 259.
29. Saeki, R.; Inaba, H.; Suzuki, T.; Miyazawa, T.; *Journal of Chromatography*, 1992, 606, 187.
30. Miyazawa, T.; Suzuki, T.; Fujimoto, K.; Kinoshita, M.; *Journal of Biochemistry [Tokyo]*, 1993, 114, 588.
31. Park, D.K.; Song, J.H.; *Korean Biochemistry Journal*, 1994, 27, 473.
32. Henderson, D.E.; Slickman, A.M.; Henderson, S.K.; *Journal of Agriculture and Food Chemistry*, 1999, 47, 2563.

33. Adachi, J.; Tomita, M.; Yamakawa, S.; Asano, M.; Naito, T.; Ueno, Y.; *Free Radical Research*, 2000, 33, 321.
34. Adachi, J.; Kudo, R.; Ueno, Y.; Hunter, R.; Rajendram, R.; Want, E.; Preedy, V.R.; *Journal of Nutrition*, 2001, 131, 2916.
35. Hui, S.-P.; Chiba, H.; Sakurai, T.; Asakawa, C.; Nagasaka, H.; Murai, T.; Ide, H.; Kurosawa, T.; *Journal of Chromatography B*, 2007, 857, 158.
36. Miyazawa, T.; *Free Radical Biology and Medicine*, 1989, 7, 209.
37. Frei, B.; Yamamoto, Y.; Niclas, D.; Ames, B.N.; *Analytical Biochemistry*, 1988, 175, 120.
38. Kiba, N.; Goto, Y.; Furusawa, M.; *Journal of Chromatography B*, 1993, 620, 9.
39. Alam, M.K.; Sasaki, M.; Watanabe, T.; Maeyama, K.; *Analytical Biochemistry*, 1995, 229, 26.
40. Kiba, N.; Oyama, Y.; Kato, A.; Furusawa, M.; *Journal of Chromatography A*, 1996, 724, 354.
41. Kiba, N.; Saegusa, K.; Furusawa, M.; *Journal of Chromatography B*, 1997, 689, 393.
42. Ohba, Y.; Kuroda, N.; Nakashima, K.; *Analytica Chimica Acta*, 2002, 465, 101.
43. Nakashima, K.; Suetsugu, K.; Yoshida, K.; Imai, K.; Akiyama, S.; *Analytical Science*, 1991, 7, 815.
44. Ishida, J.; Sonezaki, S.; Yamaguchi, M.; *Analyst*, 1992, 117, 1719.
45. Yoshida, H.; Nakao, R.; Matsuo, T.; Nohta, H.; Yamaguchi, M.; *Journal of Chromatography A*, 2001, 907, 39.
46. Yoshida, H.; Ureshino, K.; Ishida, J.; Nohta, H.; Yamaguchi, M.; *Analytical Science*, 1999, 15, 937.
47. Sano, A.; Nakamura, H.; *Analytical Science*, 1998, 14, 731.
48. Ishida, J.; Yakabe, T.; Nohta, H.; Yamaguchi, M.; *Analytica Chimica Acta*, 1997, 346, 175.
49. Yakabe, T.; Ishida, J.; Yoshida, H.; Nohta, H.; Yamaguchi, M.; *Analytical Science*, 2000, 16, 545.
50. Hara, T.; Toriyama, M.; Ebuchi, T.; *Bulletin of the Chemical Society of Japan*, 1985, 58, 109.
51. MacDonald, A.; Nieman, T.A.; *Analytical Chemistry*, 1985, 57, 936.
52. Koerner, P.J., Jr.; Nieman, T.A.; *Mikrochimica Acta*, 1987, 2, 79.
53. Ci, Y.; Tie, J.; Wang, Q.; Chang, W.; *Analytica Chimica Acta*, 1992, 269, 109.
54. Kubo, H.; Toriba, A.; *Analytica Chimica Acta*, 1997, 353, 345.
55. Navas Díaz, A.; García Sánchez, F.; González García, J.A.; *Journal of Bioluminescence and Chemiluminescence*, 1995, 10, 175.
56. Navas Díaz, A.; García Sánchez, F.; González García, J.A.; *Journal of Photochemistry and Photobiology A*, 1998, 113, 27.
57. Navas Díaz, A.; García Sánchez, F.; González García, J.A.; *Journal of Bioluminescence and Chemiluminescence*, 1998, 13, 75.
58. Toriba, A.; Kubo, H.; *Journal of Liquid Chromatography and Related Technologies*, 1997, 20, 2965.
59. Zhou, J.; Cui, H.; Wan, G.; Xu, H.; Pang, Y.; Duan, C.; *Food Chemistry*, 2004, 88, 613.
60. Vázquez, B.I.; Feas, X.; Lolo, M.; Fente, C.A.; Franco, C.M.; Cepeda, A.; *Luminescence*, 2005, 20, 197.
61. Huang, C.; Zhou, G.; Peng, H.; Gao, Z.; *Analytical Science*, 2005, 21, 565.
62. Chen, F.-N.; Zhang, Y.-X.; Zhang, Z.-J.; *Chinese Journal of Chemistry*, 2007, 25, 942.
63. Serrano, J.M.; Silva, M.; *Journal of Chromatography A*, 2006, 1117, 176.
64. Zhang, Y.; Zhang, Z.; Sun, Y.; *Journal of Chromatography A*, 2006, 1129, 34.
65. Zhang, Y.; Zhang, Z.; Sun, Y.; Wei, Y.; *Journal of Agriculture and Food Chemistry*, 2007, 55, 4949.
66. Nalewajko, E.; Wiszowata, A.; Kojlo, A.; *Journal of Pharmaceutical and Biomedical Analysis*, 2007, 43, 1673.

67. Nakashima, K.; Kawaguchi, S.; Akiyama, S.; Schulman, S.G.; *Biomedical Chromatography*, 1993, 7, 217.
68. Dapkevicius, A.; van Beek, T.A.; Niederländer, H.A.G.; de Groot, A.; *Analytical Chemistry*, 1999, 71, 736.
69. Toyo'oka, T.; Kashiwazaki, T.; Kato, M.; *Talanta*, 2003, 60, 467.
70. Magalhães, L.M.; Segundo, M.A.; Reis, S.; Lima, J.L.F.C.; *Analytica Chimica Acta*, 2008, 613, 1.
71. Maskiewicz, R.; Sogah, D.; Bruice, T.C.; *Journal of the American Chemical Society*, 1979, 101, 5347.
72. Chen, G.N.; Xu, X.Q.; Duan, J.P.; Lin, R.E.; Zhang, F.; *Analytical Communications*, 1996, 33, 99.
73. Veazey, R.L.; Nieman, T.A.; *Journal of Chromatography*, 1980, 200, 153.
74. Klopf, L.L.; Nieman, T.A.; *Analytical Chemistry*, 1985, 57, 46.
75. Takeda, M.; Maeda, M.; Tsuji, A.; *Biomedical Chromatography*, 1990, 4, 119.
76. Maeda, M.; Tsuji, A.; *Journal of Chromatography*, 1986, 352, 213.
77. Novak, T.J.; Grayeski, M.L.; *Microchemistry Journal*, 1994, 50, 151.
78. Rollag, J.G.; Liu, T.; Hage, D.S.; *Journal of Chromatography A*, 1997, 765, 145.
79. Zhong, L.; Maloy, J.T.; *Analytical Chemistry*, 1998, 70, 1100.
80. Steijger, O.M.; Kamminga, D.A.; Lingeman, H.; Brinkman, U.A.T.; *Journal of Bioluminescence and Chemiluminescence*, 1998, 13, 31.
81. Nelson, N.C.; Reynolds, M.A.; Arnold, L.J., Jr.; *Nonisotopic DNA Probe Techniques*; Academic Press: San Diego, CA, 1992, pp. 275–310.
82. Givens, R.S.; In: *Chemiluminescence and Photochemical Reaction Detection in Chromatography*; Birks, J.W.; Ed.; VCH: New York, 1989, pp. 125–147.
83. Hadd, A.G.; Birks, J.W.; In: *Selective Detectors*; Sievers, R.E.; Ed.; John Wiley & Sons: New York, 1995, pp. 209–240.
84. Stigbrand, M.; Jonsson, T.; Pontén, E.; Irgum, K.; Bos, R.; In: *Chemiluminescence in Analytical Chemistry*; García-Campaña, A.M.; Baeyens, W.R.G.; Eds.; Marcel Dekker: New York, 2001, pp. 141–173.
85. Tsunoda, M.; Imai, K.; *Analytica Chimica Acta*, 2005, 541, 13.
86. Rauhut, M.M.; Bollyky, L.J.; Roberts, B.G.; Loy, M.; Whitman, R.H.; Iannotta, A.V.; Semsel, A.M.; Clarke, R.A.; *Journal of the American Chemical Society*, 1967, 89, 6515.
87. Bos, R.; Barnett, N.W.; Dyson, G.A.; Lim, K.F.; Russell, R.A.; Watson, S.P.; *Analytica Chimica Acta*, 2004, 502, 141.
88. Tonkin, S.A.; Bos, R.; Dyson, G.A.; Lim, K.F.; Russell, R.A.; Watson, S.P.; Hindson, C.M.; Barnett, N.W.; *Analytica Chimica Acta*, 2008, 614, 173.
89. De Jong, G.J.; Lammers, N.; Spruit, F.J.; Dewaele, C.; Verzele, M.; *Analytical Chemistry*, 1987, 59, 1458.
90. Weber, A.J.; Grayeski, M.L.; *Analytical Chemistry*, 1987, 59, 1452.
91. Amponsaa-Karikari, A.; Kishikawa, N.; Ohba, Y.; Nakashima, K.; Kuroda, N.; *Biomedical Chromatography*, 2006, 20, 1157.
92. Ahmed, S.; Kishikawa, N.; Nakashima, K.; Kuroda, N.; *Analytica Chimica Acta*, 2007, 591, 148.
93. Mohan, A.G.; Turro, N.J.; *Journal of Chemical Education*, 1974, 51, 528.
94. Imai, K.; Matsunaga, Y.; Tsukamoto, Y.; Nishitani, A.; *Journal of Chromatography*, 1987, 400, 169.
95. Van Zoonen, P.; Bock, H.; Gooijer, C.; Velthorst, N.H.; Frei, R.W.; *Analytica Chimica Acta*, 1987, 200, 131.
96. Katayama, M.; Taniguchi, H.; Matsuda, Y.; Akihama, S.; Hara, I.; Sato, H.; Kaneko, S.; Kuroda, Y.; Nozawa, S.; *Analytica Chimica Acta*, 1995, 303, 333.
97. Imaizumi, N.; Hayakawa, K.; Miyazaki, M.; Imai, K.; *Analyst*, 1989, 114, 161.
98. Uzu, S.; Imai, K.; Nakashima, K.; Akiyama, S.; *Analyst*, 1991, 116, 1353.

99. Higashidate, S.; Imai, K.; *Analyst*, 1992, 117, 1863.
100. Tsunoda, M.; Takezawa, K.; Yanagisawa, T.; Kato, M.; Imai, K.; *Biomedical Chromatography*, 2001, 15, 41.
101. Hayakawa, K.; Miyoshi, Y.; Kurimoto, H.; Matsushima, Y.; Takayama, N.; Tanaka, S.; Miyazaki, M.; *Biological and Pharmaceutical Bulletin*, 1993, 16, 817.
102. Lin, M.; Huie, C.W.; *Journal of Liquid Chromatography and Related Technologies*, 1997, 20, 681.
103. Meseguer Lloret, S.; Molins Legua, C.; Verdú Andrés, J.; Campíns-Falcó, P.; *Journal of Chromatography A*, 2004, 1035, 75.
104. Sigvardson, K.W.; Kennish, J.M.; Birks, J.W.; *Analytical Chemistry*, 1984, 56, 1096.
105. Hayakawa, K.; Kitamura, R.; Butoh, M.; Imaizumi, N.; Miyazaki, M.; *Analytical Science*, 1991, 7, 573.
106. Murahashi, T.; Hayakawa, K.; *Analytica Chimica Acta*, 1997, 343, 251.
107. Hayakawa, K.; Nakamura, A.; Terai, N.; Kizu, R.; Ando, K.; *Chemistry and Pharmaceutical Bulletin*, 1997, 45, 1820.
108. Hayakawa, K.; Noji, K.; Tang, N.; Toriba, A.; Kizu, R.; Sakai, S.; Matsumoto, Y.; *Analytica Chimica Acta*, 2001, 445, 205.
109. Hayakawa, K.; Lu, C.; Mizukami, S.; Toriba, A.; Tang, N.; *Journal of Chromatography A*, 2006, 1107, 286.
110. Takayama, N.; Tanaka, S.; Hayakawa, K.; *Biomedical Chromatography*, 1997, 11, 25.
111. Molins-Legua, C.; Campíns-Falcó, P.; Sevillano-Cabeza, A.; *Analyst*, 1998, 123, 2871.
112. Appelblad, P.; Jonsson, T.; Bäckström, T.; Irgum, K.; *Analytical Chemistry*, 1998, 70, 5002.
113. Orejuela, E.; Silva, M.; *Journal of Chromatography A*, 2003, 1007, 197.
114. Orejuela, E.; Silva, M.; *Analytical Letters*, 2004, 37, 2531.
115. Cobo, M.; Silva, M.; *Journal of Chromatography A*, 1999, 848, 105.
116. Hamachi, Y.; Nakashima, M.N.; Nakashima, K.; *Journal of Chromatography B*, 1999, 724, 189.
117. Funato, K.; Imai, T.; Nakashima, K.; Otagiri, M.; *Journal of Chromatography B*, 2001, 757, 229.
118. Nakamura, S.; Wada, M.; Crabtree, B.L.; Reeves, P.M.; Montgomery, J.H.; Byrd, H.J.; Harada, S.; Kuroda, N.; Nakashima, K.; *Analytical and Bioanalytical Chemistry*, 2007, 387, 1983.
119. Ragab, G.H.; Nohta, H.; Kai, M.; Ohkura, Y.; Zaitsu, K.; *Journal of Pharmaceutical and Biomedical Analysis*, 1995, 13, 645.
120. Takezawa, K.; Tsunoda, M.; Murayama, K.; Santa, T.; Imai, K.; *Analyst*, 2000, 125, 293.
121. Tsunoda, M.; Takezawa, K.; Santa, T.; Imai, K.; *Analytical Biochemistry*, 1999, 269, 386.
122. Tsunoda, M.; Nagayama, M.; Funatsu, T.; Hosoda, S.; Imai, K.; *Clinica Chimica Acta*, 2006, 366, 168.
123. Tsunoda, M.; Uchino, E.; Imai, K.; Hayakawa, K.; Funatsu, T.; *Journal of Chromatography A*, 2007, 1164, 162.
124. Jansen, H.; Brinkman, U.A.T.; Frei, R.W.; *Journal of Chromatography*, 1988, 440, 217.
125. Honda, K.; Miyaguchi, K.; Nishino, H.; Tanaka, H.; Yao, T.; Imai, K.; *Analytical Biochemistry*, 1986, 153, 50.
126. Emteborg, M.; Irgum, K.; Gooijer, C.; Brinkman, U.A.T.; *Analytica Chimica Acta*, 1997, 357, 111.
127. Wada, M.; Kuroda, N.; Ikenaga, T.; Akiyama, S.; Nakashima, K.; *Analytical Science*, 1996, 12, 807.
128. Kamei, S.; Ohkubo, A.; Saito, S.; Takagi, S.; *Analytical Chemistry*, 1989, 61, 1921.
129. Poulsen, J.R.; Birks, J.W.; *Analytical Chemistry*, 1990, 62, 1242.
130. Ahmed, S.; Fujii, S.; Kishikawa, N.; Ohba, Y.; Nakashima, K.; Kuroda, N.; *Journal of Chromatography A*, 2006, 1133, 76.

131. Wada, M.; Inoue, K.; Ihara, A.; Kishikawa, N.; Nakashima, K.; Kuroda, N.; *Journal of Chromatography A*, 2003, 987, 189.
132. Gerardi, R.D.; Barnett, N.W.; Lewis, S.W.; *Analytica Chimica Acta*, 1999, 378, 1.
133. Gorman, B.A.; Francis, P.S.; Barnett, N.W.; *Analyst*, 2006, 131, 616.
134. Gerardi, R.D.; Barnett, N.W.; Jones, P.; *Analytica Chimica Acta*, 1999, 388, 1.
135. Lenehan, C.E.; Barnett, N.W.; Lewis, S.W.; Essery, K.M.; *Australian Journal of Chemistry*, 2004, 57, 1001.
136. Skotty, D.R.; Lee, W.-Y.; Nieman, T.A.; *Analytical Chemistry*, 1996, 68, 1530.
137. Ikehara, T.; Habu, N.; Nishino, I.; Kamimori, H.; *Analytica Chimica Acta*, 2005, 536, 129.
138. Holeman, J.A.; Danielson, N.D.; *Journal of Chromatographic Science*, 1995, 33, 297.
139. Hori, T.; Hashimoto, H.; Konishi, M.; *Biomedical Chromatography*, 2006, 20, 917.
140. Yoshida, H.; Hidaka, K.; Ishida, J.; Yoshikuni, K.; Nohta, H.; Yamaguchi, M.; *Analytica Chimica Acta*, 2000, 413, 137.
141. Chiba, R.; Fukushi, M.; Tanaka, A.; *Analytical Science*, 1998, 14, 979.
142. Ridlen, J.S.; Skotty, D.R.; Kissinger, P.T.; Nieman, T.A.; *Journal of Chromatography B*, 1997, 694, 393.
143. Pérez-Ruiz, T.; Martínez-Lozano, C.; García, M.D.; *Journal of Chromatography A*, 2007, 1169, 151.
144. Yi, C.; Li, P.; Tao, Y.; Chen, X.; *Microchimica Acta*, 2004, 147, 237.
145. Kodamatani, H.; Saito, K.; Niina, N.; Yamazaki, S.; Tanaka, Y.; *Journal of Chromatography A*, 2005, 1100, 26.
146. Wan, G.-H.; Cui, H.; Pan, Y.-L.; Zheng, P.; Liu, L.-J.; *Journal of Chromatography B*, 2006, 843, 1.
147. Kodamatani, H.; Komatsu, Y.; Yamazaki, S.; Saito, K.; *Journal of Chromatography A*, 2007, 1140, 88.
148. Costin, J.W.; Lewis, S.W.; Purcell, S.D.; Waddell, L.R.; Francis, P.S.; Barnett, N.W.; *Analytica Chimica Acta*, 2007, 597, 19.
149. Mistry, K.; Grinberg, N.; *Journal of Liquid Chromatography and Related Technologies*, 2005, 28, 1055.
150. Lee, W.-Y.; Nieman, T.A.; *Journal of Chromatography*, 1994, 659, 111.
151. Morita, H.; Konishi, M.; *Analytical Chemistry*, 2003, 75, 940.
152. Uchikura, K.; *Chemical and Pharmaceutical Bulletin*, 2003, 51, 1092.
153. Uchikura, K.; *Chemical Letters*, 2003, 32, 98.
154. Uchikura, K.; *Analytical Science*, 2000, 16, 453.
155. Yamazaki, S.; Ban'i, K.; Tanimura, T.; *Journal of High Resolution Chromatography*, 1999, 22, 487.
156. Niina, N.; Kodamatani, H.; Uozumi, K.; Kokufu, Y.; Saito, K.; Yamazaki, S.; *Analytical Science*, 2005, 21, 497.
157. Yokota, K.; Saito, K.; Yamazaki, S.; Muromatsu, A.; *Analytical Letters*, 2002, 35, 185.
158. Kodamatani, H.; Shimizu, H.; Saito, K.; Yamazaki, S.; Tanaka, Y.; *Journal of Chromatography A*, 2006, 1102, 200.
159. Pérez-Ruiz, T.; Martínez-Lozano, C.; García, M.D.; *Journal of Chromatography A*, 2007, 1164, 174.
160. Choi, H.N.; Cho, S.-H.; Park, Y.-J.; Lee, D.W.; Lee, W.-Y.; *Analytica Chimica Acta*, 2005, 541, 49.
161. Adcock, J.L.; Francis, P.S.; Smith, T.A.; Barnett, N.W.; *Analyst*, 2008, 133, 49.
162. Barnett, N.W.; Hindson, B.J.; Jones, P.; Smith, T.A.; *Analytica Chimica Acta*, 2002, 451, 181.
163. Adcock, J.L.; Barnett, N.W.; Costin, J.W.; Francis, P.S.; Lewis, S.W.; *Talanta*, 2005, 67, 585.
164. Wei, Y.; Zhang, Z.; Zhang, Y.; Sun, Y.; *Chromatographia*, 2007, 65, 443.
165. Fan, S.-L.; Zhang, L.-K.; Lin, J.-M.; *Talanta*, 2006, 68, 646.

166. Francis, P.S.; Adcock, J.L.; Costin, J.W.; Purcell, S.D.; Pfeffer, F.M.; Barnett, N.W.; *Journal of Pharmaceutical and Biomedical Analysis*, 2008, 48, 508.
167. Slezak, T.; Francis, P.S.; Anastos, N.; Barnett, N.W.; *Analytica Chimica Acta*, 2007, 593, 98.
168. Wei, Y.; Zhang, Z.-J.; Zhang, Y.-T.; Sun, Y.-H.; *Journal of Chromatography B*, 2007, 854, 239.
169. Zhang, Q.; Cui, H.; *Journal of Separation Science*, 2005, 28, 1171.
170. Zhang, Q.; Cui, H.; Myint, A.; Lian, M.; Liu, L.; *Journal of Chromatography A*, 2005, 1095, 94.
171. Zhang, Q.; Lian, M.; Liu, L.; Cui, H.; *Analytica Chimica Acta*, 2005, 537, 31.
172. Li, H.-N.; Ci, Y.-X.; Huang, L.; *Analytical Science*, 1997, 13, 821.
173. Zhao, Y.N.; Baeyens, W.R.G.; Zhang, X.R.; Calokerinos, A.C.; Nakashima, K.; Van der Weken, G.; Van Overbeke, A.; *Chromatographia*, 1997, 44, 31.
174. Zhang, Z.; Baeyens, W.R.G.; Zhang, X.; Zhao, Y.; Van Der Weken, G.; *Analytica Chimica Acta*, 1997, 347, 325.
175. Ouyang, J.; Baeyens, W.R.G.; Delanghe, J.; Van Der Weken, G.; Van Daele, W.; De Keukeleire, D.; Garcia Campana, A.M.; *Analytica Chimica Acta*, 1999, 386, 257.
176. Wan, G.-H.; Cui, H.; Zheng, H.-S.; Zhou, J.; Liu, L.-J.; Yu, X.-F.; *Journal of Chromatography B*, 2005, 824, 57.
177. Santiago Valverde, R.; Sánchez Pérez, I.; Franceschelli, F.; Martínez Galera, M.; Gil García, M.D.; *Journal of Chromatography A*, 2007, 1167, 85.
178. Ma, L.; Nakazono, M.; Ohba, Y.; Zaitsu, K.; *Analytical Science*, 2002, 18, 1163.
179. Martínez Galera, M.; Gil García, M.D.; Santiago Valverde, R.; *Journal of Chromatography A*, 2006, 1113, 191.
180. Gil García, M.D.; Martínez Galera, M.; Santiago Valverde, R.; *Analytical and Bioanalytical Chemistry*, 2007, 387, 1973.
181. Brown, A.J.; Francis, P.S.; Adcock, J.L.; Lim, K.F.; Barnett, N.W.; *Analytica Chimica Acta*, 2008, 624, 175.
182. Francis, P.S.; Adcock, J.L.; Costin, J.W.; Agg, K.M.; *Analytical Biochemistry*, 2005, 336, 141.
183. Shakhashiri, B.Z.; Williams, L.G.; *Journal of Chemical Education*, 1976, 53, 358.
184. Niederländer, H.A.G.; Nuijens, M.J.; Dozy, E.M.; Gooijer, C.; Velthorst, N.H.; *Analytica Chimica Acta*, 1994, 297, 349.

9 Multidimensional High-Performance Liquid Chromatography

Coleen S. Milroy, Paul G. Stevenson,
Mariam Mnatasakyan, and R. Andrew Shalliker

CONTENTS

9.1 Introduction 252
9.2 Multidimensional HPLC 253
    9.2.1 Sample Dimensionality 254
    9.2.2 Selective and Nonselective Displacements 257
    9.2.3 Measures of Separation Displacement—Orthogonality 260
        9.2.3.1 Determining Orthogonality 260
        9.2.3.2 Mathematical Tools for Determining Orthogonality 262
        9.2.3.3 Applications of Mathematical Approach for
                  Determining Orthogonality Low-Molecular-Weight
                  Polystyrenes 266
        9.2.3.4 Peptides 267
        9.2.3.5 Pharmaceuticals 269
9.3 Two-Dimensional Chromatographic Systems 269
    9.3.1 Comprehensive and Heart-Cutting Separations 269
    9.3.2 Two-Dimensional System Designs 270
    9.3.3 Data Collection 271
        9.3.3.1 Peak Retention Time Determination 272
        9.3.3.2 Peak Retention Time Determination
                  in a Two-Dimensional HPLC Chromatogram 273
        9.3.3.3 Two-Dimension Peak Matching 274
9.4 Applications in Selectivity Screening and Fingerprint
    Analysis: Café Expresso 275
9.5 Conclusion 282
References 282

9.1 INTRODUCTION

Chemical fingerprinting is a means of providing a chemical profile or signature that represents the components that are present in a sample and describes a range of methods where the primary aim is to provide a unique graphical representation of the sample by identifying the chemical elements within a matrix in comparison to similar matrices. Chemical fingerprinting may provide the characterization, quantification, differentiation, and the identification of complex mixtures based on their chemical composition and is particularly important in such fields as pharmaceuticals, natural products, food, forensic, and environmental sciences and will become the ideal requirement for fields of study, such as, metabolomics: Accordingly, techniques that purport to acquiring chemical fingerprinting information require high levels of reproducibility and accuracy.

Many methods exist in which a chromatographic approach is used for the resolution of the sample matrix and detection methods such as mass spectrometry (MS), infrared (IR), and nuclear magnetic resonance spectroscopy (NMR) are used to assist in the overall analysis and identification. Although MS offers high selectivity, the overall sensitivity of MS in the analysis of complex mixtures is restricted as the ionization of weaker compounds may be suppressed by those of stronger compounds. Isomers are also difficult to differentiate by MS as the molecular ions are identical. The expense of MS, both the initial purchase price and subsequent running costs also detracts from its routine application base, although the power of the technique has rightly placed it as an essential detection method in any laboratory undertaking serious chemical analysis. IR detection suffers from solvent interference effects, limiting its application base. NMR is expensive, relatively insensitive, but essentially absolute in its ability to provide information that relates directly to the identity of a substance. In combination with LC, these three hyphenated methods of detection yield a combined process of analysis that is unsurpassed in its ability to provide qualitative and quantitative sample information, yet as a whole, only a few laboratories worldwide can afford to accommodate all three methods of analysis: The cost and upkeep of such instruments being the limitation. This chapter explores alternative methods for obtaining chemical fingerprints, that being, multidimensional high-performance liquid chromatography (MDHPLC).

HPLC is the most commonly used analytical separation technique for the determination of components in complex mixtures as it offers high sensitivity and can be highly selective. Unidimensional HPLC is undertaken such that a single separation process is responsible for the retention and consequently the separation of the sample constituents. However, separation displacement in a single dimension is not a unique characteristic to any specific compound. Therefore, many compounds could potentially co-elute and in complex samples they do, even despite the fact that modern HPLC column technology has lead to vastly improved separations, which under specialized conditions can lead to nearly 1 million plates per separation [1]. Indeed as chromatographers are continually attempting to resolve increasingly complex samples, the problem of co-elution becomes even more important and this to some extent explains the recent drive toward ultra-performance liquid chromatography.

Irrespective of whether conventional HPLC or the ultra-performance mode is employed, the separation space in a unidimensional system is limited by the peak width and determined by the efficiency of the column. This is highly dependent upon the number of theoretical plates (N) available for the separation and therefore the peak capacity. The peak capacity is the measure of the number of components that can be theoretically resolved side by side over the entire separation space without peak overlap [2] and is generally proportional to the square root of the theoretical plates available. However, complex samples contain multifaceted components, the chromatographic behavior of which depends highly on their chemical nature. As such components within complex samples tend to be randomly distributed, resulting in a substantial decrease in the theoretical peak capacity due to statistical component overlap [3,4]. Furthermore, compounds with similar chemical structures elute within similar retention windows, further crowding the separation space and placing greater demands on the separation power required for complete resolution. Therefore, unique displacement in a single dimension can never be guaranteed, even at very high plate numbers. Hence the ability of a unidimensional separation to serve as a fingerprinting tool is very limited.

When dealing with complex samples, a reduction in the complexity of the sample is usually warranted in order to overcome the limitation in peak capacity. This is usually achieved using a variety of sample pretreatments (solvent partitioning, selective precipitation) or by employing multiple chromatographic selectivity steps where fractions are collected after elution from one phase, followed by subsequent solvent elimination and reinjection onto another phase. Each successive fraction can then be further analyzed by HPLC to yield detailed sample information. This process can be very time consuming, labor intensive, and may have additional problems such as sample loss/recovery and component stability. This process may also be compromised if some constituents are labile, in either heat or incompatible solvents.

Unidimensional HPLC with gradient elution is another means by which complex samples can be analyzed. Here the mobile phase composition is changed in a predetermined and continuous manner. The advantage of this method for complex sample mixtures is that the resolution between components is greatly improved as components that are usually weakly retained or strongly retained are able to be separated in a single run. As a result of the continually increasing solvent strength, bands eluting under the influence of a gradient will decrease in peak width and as a consequence the separation space (peak capacity) is increased and theoretically more components may be resolved. While gradient elution does allow for an increase in the separation power and there is less probability of component overlap, retention times are still not unique to any given compound.

The limitations of unidimensional HPLC have been extensively reviewed by Guiochon [5].

9.2 MULTIDIMENSIONAL HPLC

The need for increased resolving power, driven by the demands of chemical profiling has been in part the driving force behind the development of the technique referred to as multidimensional HPLC. Multidimensional HPLC refers to separation methods that employ more than one type of separation step and for HPLC generally implies two dimensions. The additional separation step(s) offer retention mechanisms that may be very different to that of the first dimension. Thereby, judicious selection of the various separation steps can be made by considering the nature of the sample and subsequently the separation can be essentially tuned to the various sample attributes. Ultimately, this type of separation process can lead to very high levels of selectivity and hence the probability of component overlap in the two-dimensional (2D) domain decreases. Thus a chemical profiling environment can be established. Discussion on the concepts and practice of multidimensional HPLC fits within the context of this book largely because the technique can in fact be considered as a hyphenated method of analysis. The first dimension serves as the separation step, much like in LC-MS/MS and the second separation dimension acts as a selective detection system, tuned to resolve the components transported to the second dimension according to their selective interactions with the chromatographic environment (further discussion illustrating "selective detection" using multidimensional HPLC will be covered in Section 9.4). Multidimensional HPLC has been comprehensively reviewed by numerous workers [6–9].

9.2.1 SAMPLE DIMENSIONALITY

Sample dimensionality can be described as the number of features of the sample that can be utilized for separation purposes. That is, which sample attributes dominate solvent and/or stationary phase behavior? Then to separate a n-dimensional sample, a n-dimensional chromatographic separation system should be employed [4], one dimension for each sample attribute, ultimately yielding the greatest degree of separation power and maintaining greatest control on separation order. Characteristics that could be considered for sample dimensionality could for instance be; molecular weight, pK_a, and the presence of isomers; including structural, diastereomers, and enantiomers. For example, the length of an alkyl chain may be described by the molecular weight or define the hydrophobicity of the molecule. If this chain is branched, then the degree and location of the branching may be a second and third dimension. Double and triple bonds, a fourth or fifth dimension. The position and/or number of certain functional groups, perhaps on an aromatic ring or rings, even the bonding of these rings within the structure are possibilities for describing the sample dimensionality and hence exploiting these sample attributes from a separation sense.

Two simple examples of sample dimensionality are found for the samples of: (1) polynuclear aromatic hydrocarbons (PAHs) and (2) low-molecular-weight polymers (see Figure 9.1). In the case of the PAHs, if we consider the members of the homologous series, naphthalene, anthracene, 2,3-benzanathracene, and pentacene, we see that each increases in size through the addition of a single aromatic ring—the first sample dimension. If we then consider the structural isomers of say the four-ring homologue (chrysene, pyrene, 2,3-benzanthracene and benz[a]anthracene), we then have the second dimension.

In the same way, we can describe the multidimensional sample characteristics of low-molecular-weight polymers [10]. First, the number of monomers that make up the chain determines the polymer's molecular weight and hence the first sample

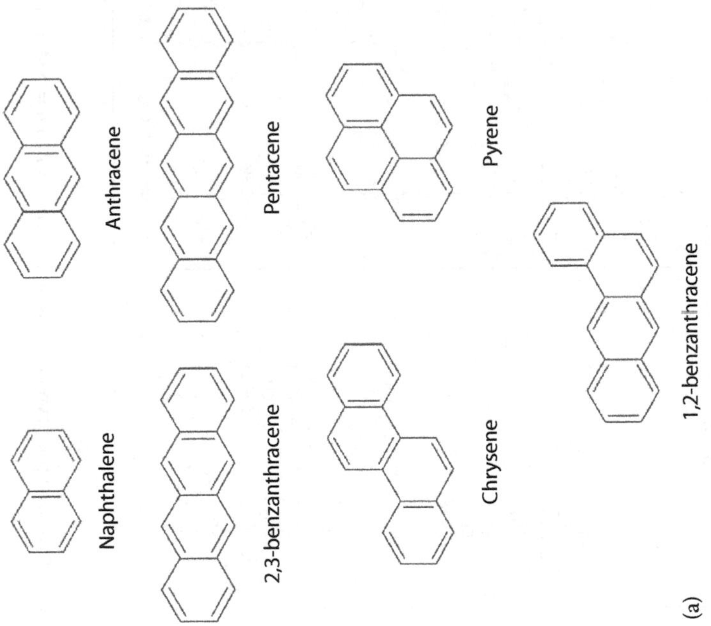

(continued)

FIGURE 9.1 Illustration of sample dimensionality (a) polycyclic aromatic hydrocarbons (size and shape) and

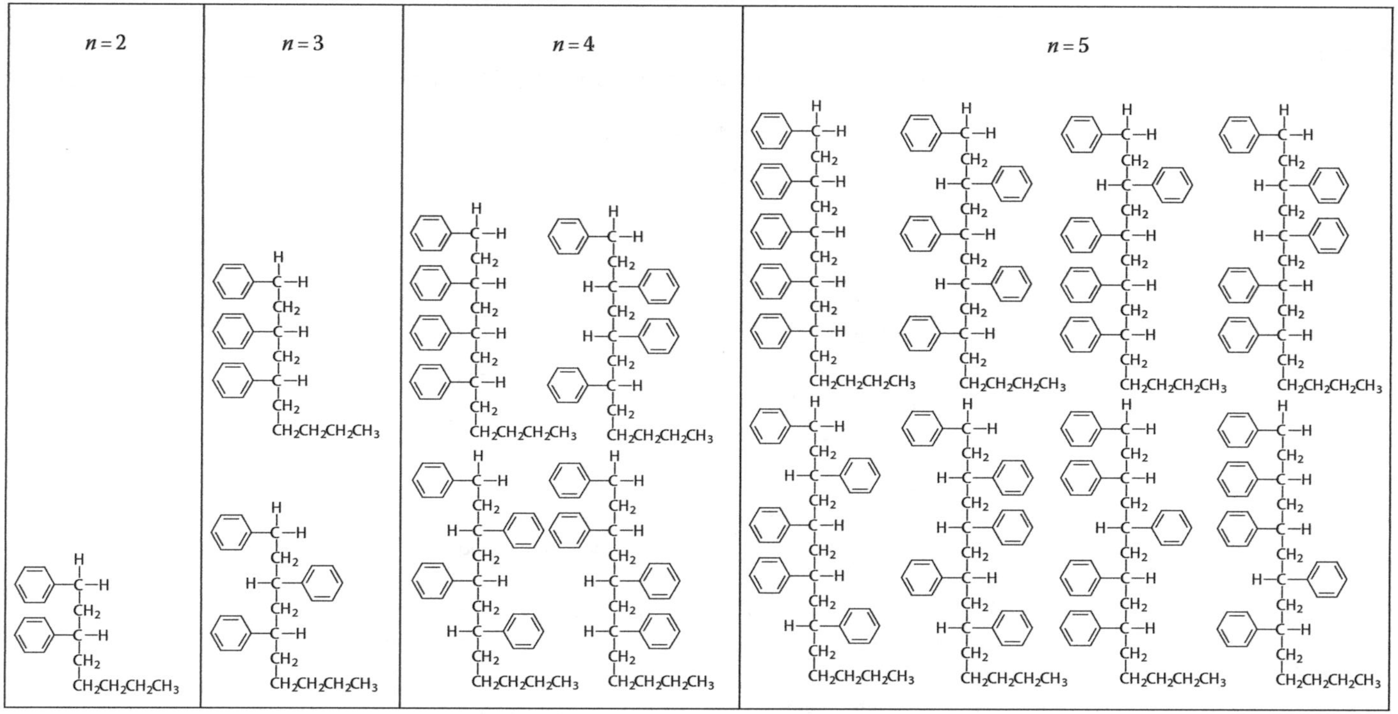

FIGURE 9.1 (continued)   (b) diastereoisomers of *n*-butyl polystyrene oligomers with *n* = 2 to *n* = 5 styrene repeating units.

dimension. Then if the monomer contains a site of stereochemistry, such as is the case for polystyrene, then the tacticity of the polymer can be used to describe the second dimension. The third dimension could be described by the enantiomers of each of the diastereomers.

Irrespective of how the sample is described, the key to separation is then expressing each sample attribute chromatographically. If the sample attribute cannot be chromatographically expressed, then essentially that sample dimension does not exist. For example, enantiomers co-elute in all chromatographic environments, except those that are chiral, and even then, the chiral environment must be sensitive to the sample. Therefore, in an achiral environment, this sample dimension does not exist, and will not exist unless a chiral environment is included in the separation process.

9.2.2 SELECTIVE AND NONSELECTIVE DISPLACEMENTS

This brings us to the next phase of discussion regarding sample dimensionality and its relationship to retention, that is, selective and nonselective displacements. Displacements in 2D column HPLC are sequential: Separation must occur in the first dimension or column followed by the analysis of discrete sections of the first dimension in the second dimension. An important class of sequential displacements are discrete displacements. Discrete separations are those in which a small discrete sample is applied to a corner of the 2D separation plane, akin to an injection onto the first dimension. Separation occurs along each subsequent axis, producing discrete elliptical zones [11].

Discrete one-dimensional (1D) displacements underlying the 2D displacements fall into two categories [11]: Selective (S) and Nonselective (N). Selective displacements occur when separation of the sample components occur in each subsequent phase of the multidimensional system. That is the selectivity factors (α) observed for the sample components are greater than 1 (Figure 9.2a and b). Selective displacements are separative displacements. Nonselective displacements result in no separation, $\alpha = 1$. These S and N displacements can be combined in a number of ways (Table 9.1) corresponding to the displacement along each axis of the 2D plane [11].

Maximum separation in the 2D plane occurs when each dimension offers selective displacement, particularly when the separation mechanisms are totally independent. Figure 9.2a shows an $S \times S_I$ (I = independent) separation in which complete separation of components is achieved. A benefit of 2D systems is that although sample components co-elute in any single dimension the component zone can be well resolved in two dimensions. When the retention mechanisms are identical or correlated ($S \times S_c$), most of the 2D space becomes unavailable for separation and the separation will converge back to a 1D separation, as shown in Figure 9.2b by the alignment of data along the main diagonal [11]. When the separation dimensions are partially correlated, the 2D space available for the separation of the components is reduced. The separated components begin to cluster closer to the main diagonal, similar, but not as severe as the example illustrated in Figure 9.2b. In this instance, resolution and peak capacity between sample components would be expected to decrease. Discrete $N \times S$ can in some instances be considered analogous since no gain in the selectivity factor is observed in the first dimension as no components are separated in the first

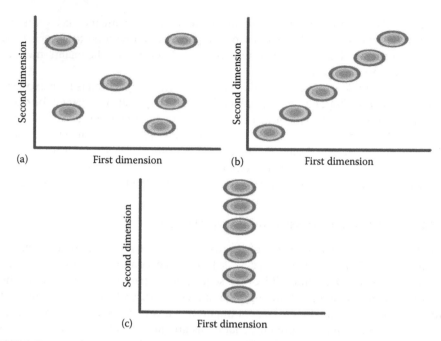

(a)   First dimension   (b)   First dimension

(c)   First dimension

FIGURE 9.2 Illustration of the combinations of discrete selective (S) and nonselective (N) displacements: (a) 2D $S \times S_I$ displacement; (b) 2D $S \times S_c$ displacement; (c) 2D $N \times S$ displacement. (From Shalliker, R.A. and Gray, M.J., *Adv. Chromatogr.*, 44, 177, 2006; Giddings, J.C.R.E., *Anal. Chem.*, 56, 1258A, 1984.)

TABLE 9.1
Types of Discrete Displacement Combinations and Their Effect on 2D Peak Capacity

Displacement Pair	Displacement Separation Peak Capacity
$S \times S_I$	$n_2 \sim n_1^2$
$S \times S_C$	$n_2 \sim n_1$
$S \times N$	$n_2 \sim n_1$
$N \times N$	$n_2 = n_1$

---

dimension. For discrete $N \times S$ combinations, the column of separated components in the second dimension appear at a uniform separation time in the first dimension (Figure 9.2c). The effectiveness of these processes is shown in Table 9.1.

In many instances, both selective and nonselective displacements can be observed in one liquid chromatographic dimension, because the retention behavior of the sample within that dimension is dominated by a particular sample attribute, in which case the sample dimension within that dimension of the system is essentially equal to 1. While Table 9.1 predicts that an $N \times S$ displacement will not lead to an increase in

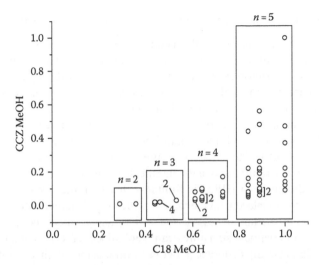

FIGURE 9.3 Normalized 2D plot of the C18(methanol)/CCZ (methanol) system in the separation of the 58 oligostyrene isomer mix. Each boxed section represents isomeric components containing the same number of configurational repeat units. The numbers adjoining the data points indicate the number of components co-eluting. (From Gray, M.J. et al., *J. Chromatogr. A*, 1015, 89, 2003.)

separation power, it does not predict that a nonselective displacement with respect to say one of three sample attributes can in fact lead to an improved separation process because the order of component elution can be more predictable, and this is very important in the separation of complex samples.

For example, consider the 2D separation of a mixture of 58 low-molecular-weight polystyrenes [10]. Selective displacement was observed for the diastereomers belonging to the oligostyrenes of varying molecular weights, and also within these groups selectivity differences were observed between oligostyrenes with different end groups, either *tert-, sec-,* or *n*-butyl. From plotting the normalized retention data for both dimensions (Figure 9.3), distinct columns were evident for the diastereomers indicating a nonselective displacement in the first dimension for the diastereomers resolved in the second dimension. Essentially, the first dimension was capable of separating according to two sample attributes, that of molecular weight and that of end groups and the second dimension able to resolve according to stereoselectivity only [10]. Hence the concept of the 2D system operating as HPLC × HPLC, whereby the second dimension was the selective diastereomer analyzer [10].

Alcohol ethoxylates have both distribution in ethylene oxide units and also a distribution in the length of hydrophobic (alkyl) endgroups [12]. In the separation of Neodol 25-12, Murphy et al. [12] demonstrated that selective displacement on a normal phase LC system occurred based upon the distribution of ethylene oxide while a nonselective displacement occurred in the RP first dimension based upon the length of hydrophobic alkyl chains in the alcohol ethoxylates, which are resolved in the second dimension. It was therefore necessary in both instances to combine selective and nonselective displacements so as to deter chaotic 2D component separation.

9.2.3 MEASURES OF SEPARATION DISPLACEMENT—ORTHOGONALITY

To this point, we have discussed the need to separate components in a multidimensional sense according to the various sample attributes that can be utilized for separative purposes. However, some measure of retention divergence across each separation dimension is useful in order to describe the power of the multidimensional system. By knowing the power of the separation system, we gain confidence in the degree of uniqueness in component displacement and hence in the chemical profile. A 2D system in which the separation displacements of each dimension are very similar will yield a system that has limited separation power because of the high correlation and hence limited scope for chemical signature work for exactly the same reason that unidimensional HPLC is limited—that is limited peak capacity. Yet a system in which the displacement mechanisms are very different will afford a higher degree of surety in the uniqueness of separation displacement, hence a high peak capacity and better chemical signatures can be obtained. A term that has commonly been used to describe the differences between dimensions is that of orthogonality [13,14], which is an important measure of the separation capability of a 2D HPLC system. Strictly speaking, orthogonality is a binary property, but over the years has become a measure of divergence. The higher the degree of divergence, approaching that of an orthogonal separation, the lower the correlation between each dimension. A low degree of correlation translates to a maximized peak capacity, as in an entirely orthogonal 2D system the peak capacity of the system is equal to the product of the peak capacity in each dimension. Thus in such a system, a maximum number of components may be resolved as a result of low correlation between dimensions as the peak capacity of the system is increased and as a consequence, the probability of a component occupying a space that is unique to that component is improved.

As a consequence of the expanded separation space, there is a lowered probability of component overlap occurring at elution, hence, the two sets of retention data (one for each dimension) for a given sample can be plotted against each other and displayed as a contour plot or scatter graph where the individual retention times provide distinctive identifying markers useful for chemical fingerprinting. If there are existing standards, these can also be used to create a fingerprint by plotting the retention data for both dimensions and then the sample's fingerprint can be overlayed to establish degrees of similarity or in some cases a more simplistic approach, degrees of difference.

9.2.3.1 Determining Orthogonality

In order to determine the orthogonality and maximize the separation space available for the separation of complex samples, numerous approaches have been developed for 2D HPLC systems. Selectivity studies, where the comparison of different stationary phase and mobile phase combinations are examined, are generally the most simplistic means of determining the most orthogonal and least correlated systems and thus most effective system for separation. Selectivity can be controlled through changes in not only stationary phases but also through the use of combinations of mobile phases references [15–18], mobile phase additives [19,20], pH modifications [19,20], and through temperature adjustments [20,21].

The change of elution order of solutes in the individual dimensions is one means of assessing orthogonality [22]. Other workers have used the comparison of retention times of each dimension to determine the most orthogonal systems for their particular criteria [23–25]. Gilar et al. [19] used the comparison of retention times and correlation coefficient r^2 to determine the orthogonality of an off-line RP 2D HPLC system for the analysis of peptides. The retention times of the first dimension were plotted in the x axis and the retention times of the second dimension in the y axis. Comparison of the 2D plots yielded information to the selectivity differences and correlation of differing 2D systems, which were varied by adjusting the pH of the mobile phases in both dimensions.

Cacciola and coworkers [26] compared the degree of similarity between the 2D RP systems by measuring the correlation between the retention factors (k). The correlation between the retention factors of phenolic antioxidants for each column was compared to determine the degree of similarity of the columns used in this study. Eighteen of twenty compounds could be separated using this system, which was also successfully applied for the analysis of hop, beer, and tea samples.

Murahashi et al. [27] evaluated the retention factors and also the selectivity factor of 1-nitropyrenes on several columns. Reverse elution order was observed between the alkyl stationary phases and the stationary phases that had π electrons in the ligands. The selectivity factor (α) indicated either PAH or nitro-PAH selectivity. The differing retention behavior and also the large variance in selectivity factor (α) ratios assisted in the selection of the columns for the 2D system for the analysis of 1-nitropyrene in extracts from automobile exhaust particulate matter.

Murahashi [28] plotted the relationship between carbon numbers and the logarithm of the retention factors of different analytes on a range of RP columns. For all columns, linear correlations were observed between the carbon number and the log k for all alkanes, alkylbenzenes, PAHs, and nitro-PAHs (NPAHs). On all columns, the slopes of the regression equations in the alkyl benzenes were similar to those of the alkanes, in contrast the slopes of the PAHs were similar to those of the NPAHs, but were not similar to those of the alkanes for most columns. The slope of alkanes/slope of PAHs (S_A/S_P) ratios were used to determine the retentive behavior and the degree of similarity between the different stationary phases. The first-dimensional column was selected for a 2D system for the separation of PAHs in extracts from gasoline and gasoline exhaust according to the lowest ratio to that of the alkyl columns that were employed as the second-dimensional column.

The determination of orthogonal HPLC systems is sometimes used not for coupling of dimensions in a 2D system but purely used to exhibit columns that can be useful in providing divergent separation mechanisms. This is particularly relevant for the pharmaceutical industry that requires at least two very different analysis methods for the qualification of pharmaceuticals and their impurities. In general, analytes are not separated on 2D systems but analyzed on two separate columns either through separate injections or by flow splitting. However, the two following studies may prove beneficial for future studies in establishing the orthogonality of 2D systems that could fulfill the obligatory requirements of the pharmaceutical industry.

Van Gysegham et al. [13] determined the orthogonality of 11 chromatographic systems for the evaluation of set of 68 drugs by calculating the Pearson correlation coefficients (r) between the retention factors for all combinations of each system. The retention results of the different systems were plotted against each other for visual inspection of the selectivity differences. The values obtained in this study close to |1| indicated total correlation where the solutes were eluting in a similar order and similar retention times; whereas values close to zero indicated large selectivity differences possibly due to differing retention mechanisms. Although the correlation assessments were valid for most systems, there were some systems in which this method was not successful. It was therefore also necessary to visually ascertain the correlation between systems as outliers in some plots lowered the correlation although true correlation would actually exist. Principal component analysis (PCA) and OPTICS color maps were useful for visualization, as were colored contour plots that were used to determine the orthogonality between two systems. The differences between the systems could be achieved by interpreting the differences in the color patterns, although the eye tends to focus on specific colors, which can lead to a false impression on the degree of difference between the systems.

Pellett et al. [29] developed a procedure for an orthogonal method for the separation of pharmaceutical samples. Their approach saw an existing method being employed complemented by an "orthogonal" separation method. The basis for determining orthogonality was that of selectivity changes, where components that overlapped in the first dimension were separated such that when separated on the second dimension, they changed elution order. Another criterion was that all the peaks in the second dimension should be resolved preferably with baseline resolution ($R_s > 1.5$). The required change in selectivity was determined to correspond to an increase in R_s of at least one unit. A change in the selectivity factor can move an overlapped band either toward or away from the center of the overlapping band, so a starting point for ascertaining orthogonality was that of a change in the absolute value of $|\delta \log \alpha|_{avg} \geq 0.10$ by varying experimental factors that affect selectivity.

9.2.3.2   Mathematical Tools for Determining Orthogonality

There are several mathematical processes that can be utilized to evaluate the orthogonality of 2D systems; these include informational entropy, percentage of synentropy, peak spreading angle, and factor analysis. The normalization of retention data gained from individual dimensions is an important starting point for the determination of a chromatographic system's orthogonality as selectivity studies generally use different columns under differing conditions. Using this approach, the retention data of each chromatographic dimension is first normalized according to Equation 9.1 [30,31]:

$$X_a = \frac{Rt_i - Rt_0}{Rt_f - Rt_0} \tag{9.1}$$

where
  X_a is the normalized retention time
  Rt_i is the retention time of the component
  Rt_0 is the retention time of the least retained solute
  Rt_f is the retention time of the final solute in the sample

Normalization allows the scaled retention factors (X_a) to be compared for different systems and as such compensates for differences between the columns: such as manufacturer, base silica, and particle size. Comparison between normalized 2D scatter graphs gives an indication of the differences in selectivity, but these differences can be quantified using information theory (IT).

Informational entropy, I, which is mathematically described later, is the measurable information content of a signal or band. In IT, "information" is defined as a measure of the uncertainty of the incidence of an event [32]. In this instance, the "information" or informational entropy, I, is a measure of the reduction in the uncertainty about the nature of the substance and is a quantity that is measured in units of bits [31]. IT allows a mathematical evaluation of qualitative methods by the calculation of the expected or average amount of information obtained from an analysis [33]. Steuer et al. [30] compared HPLC, SFC, and CZE for the analysis of drugs using IT to describe the Informational Orthogonality between the chromatographic systems. Huber et al. [34] applied IT to retention data to determine the optimal selection of gas chromatographic columns for the analysis of chemical warfare agents. IT has also been used to describe the "Informational Orthogonality" of 2D chromatographic separations of complex mixtures [31,35].

The informational entropy of a measurement whose unit of measure is the "bit," is a probabilistic quantity described by Equation 9.2 [31,32]:

$$I = \sum_{d=1}^{n} -p(x)\log_2 p(x) \tag{9.2}$$

where $p(x)$ is the probability of the incidence of a single possible result, x, out of n possible results [31,32]. Using the method of Slonecker et al. [31], the informational entropy (I) is calculated for each chromatographic step in a 2D separation individually and then for two dimensions. Since retention correlation invariably exists in the majority of coupled chromatographic systems, the 2D information gained is reduced due to *mutual information* [31]. Then by calculating the fractional information content, h, the informational similarity (H) can be calculated according to Equations 9.3 and 9.4:

$$h(1,2) = 1 - \frac{I(1;2)}{I(1,2)} \tag{9.3}$$

$$H(1,2) = [1 - h^2(1,2)]^{1/2} \tag{9.4}$$

where
$I(1;2)$ is the mutual information between the two chromatographic dimensions (1,2)
$I(1,2)$ is the total 2D informational entropy

The informational similarity, as described by Slonecker et al. [31], is used as a measure of solute crowding on a normalized [30] 2D retention plot as shown in Figure 9.4.

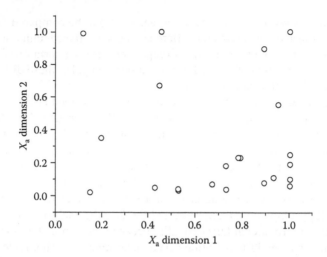

FIGURE 9.4 Normalized 2D plot.

The retention times are scaled according to the method of Steuer et al. [30] demonstrated by Equation 9.1, to allow comparison of a diverse range of chromatographic systems. Values of informational similarity range between zero and one. A value of one represents the highest level of solute crowding and occurs in conjunction with total correlation. Also determined using the method of Slonecker et al. [31] is the percentage Synentropy, which is a measure of the retention mechanism equivalency between two chromatographic systems. This is calculated according to Equation 9.5:

$$\%\text{synentropy} = \left(\frac{I_a}{I(1,2)}\right) \times 100 \tag{9.5}$$

where I_a is the informational entropy of the data aligned (to within some arbitrary boundary) along the unit diagonal of a normalized 2D retention plot. Values of the percentage Synentropy range between 0 and 100, with a value of 100 indicating that the chromatographic systems in a 2D combination are 100% equivalent.

Factor analysis is useful for examining large data sets and for determining the orthogonality and practical peak capacity of 2D chromatographic systems [36]. Correlation matrices can be constructed from the scaled retention times of solutes from each of the dimensions and in this way the practical peak capacity is able to be visualized. The correlation matrix (C) is calculated according to Equation 9.6 [36]:

$$C = \left(\frac{1}{N-1}\right)M'^{\text{T}}M' \tag{9.6}$$

where
  N is the number of scaled retention times
  M' is a matrix of scaled retention times
  M'^{T} is the transposed matrix of the matrix of scaled retention times

This yields a square correlation matrix in the form of Equation 10.7 [36]:

$$C = \begin{vmatrix} 1 & C_{12} \\ C_{21} & 1 \end{vmatrix}$$ (9.7)

where $C_{12} = C_{21}$ and is a measure of the correlation between two sets of retention time data and the orthogonality of a 2D system. Complete correlation exists in a chromatographic system when C_{21} = unity. When C_{21} = zero, a totally orthogonal chromatographic system is evident.

The product of the peak capacities of the individual dimensions theoretically predicts the peak capacity of a 2D system in truly orthogonal systems. However, the practical peak capacity is much smaller than the theoretical value when some degree of correlation is present. For most 2D separations, some correlation does exist and there is a reduction in the separation space available. Another way in which to approximate the practical peak capacity is by using peak spreading angles (β) [36] in which the region of the correlation is calculated and is then subtracted from the product of the theoretical peak capacity of each dimension. The creation of a geometric plot (Figure 9.5) using these calculations demonstrates the unavailability of the 2D retention space due to correlation. The practical peak capacity is given by Equation 9.8 [36]:

$$N_P = N_T - (A + C)$$ (9.8)

where

N_P is the practical 2D peak capacity

N_T the theoretical 2D peak capacity

A and C are the unavailable area in Figure 9.5 due to correlation

The gridded region in Figure 9.5 is the area available for separation to occur in a partially correlated 2D system. Values of the spreading angle range between 0° and 90°. A spreading angle of 90° indicates a maximum peak capacity in which true orthogonality exists for the 2D system. A spreading angle of zero indicates a highly correlated 2D system equivalent to that of a 1D system.

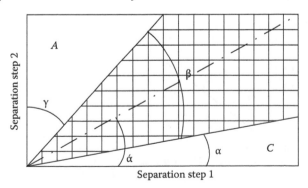

FIGURE 9.5 Effective non-orthogonal, two-dimension retention space when the peak spreading angle is β. (From Liu, Z. et al., *Anal. Chem.*, 67, 3840, 1995.)

9.2.3.3   Applications of Mathematical Approach for Determining Orthogonality Low-Molecular-Weight Polystyrenes

Demonstrating the 2D separation of low-molecular-weight polystyrenes is a useful means to illustrate the concept of sample dimensionality and its relationship to multidimensional separations. Oligostyrene and diastereomer separations in 2D separations have been developed where the resolution afforded far exceeds any separation that has previously been undertaken in a unidimensional system [37–40]. In order to optimize the 2D separations of low-molecular-weight oligostyrenes Gray et al. [35] applied both information theory and factor analysis to determine the correlation between systems and determine the solute crowding, useable separation region and system peak capacities. These studies employed complex mixtures of either 32 oligostyrenes consisting of 5 configurational repeat units with end groups of either *n*-, *sec*-, or *tert*-butyl, or a 58-component mixture that also included the addition of oligomers containing 2, 3, and 4 configurational repeating units. For both sample sets, a C18 column was used in the first dimension, with a carbon-clad zirconia column used in the second dimension. In the best performing system, 100% methanol was used as the mobile phase in the first dimension, while 100% acetonitrile was used in the second dimension. The normalized 2D retention plot shown in Figure 9.6 illustrates that the elution of the bands from the 32-component sample occurs predominantly in the lower right quadrant of the 2D separation plane. This resulted in a moderate degree of solute crowding (similarity value equal to 0.56). Despite this, moderate degree of solute crowding the %synentropy was 3.0, indicating that each dimension offered vastly different retention processes. This resulted to a high peak spreading angle (75°) and a resulting high degree of space utilization (90%). The theoretical peak capacity was 60, with a practical peak capacity of 56, allowing for the resolution of 26 of the 32 components. In the case of the 58-component sample,

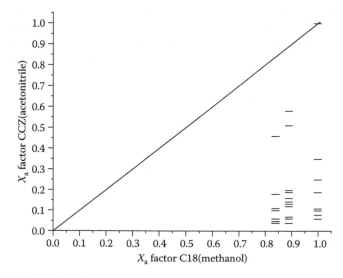

FIGURE 9.6   Normalized 2D plot of C18(methanol)/CCZ(acetonitrile) system in the separation of the 32 oligostyrene isomer mix. (From Gray, M. et al., *J. Chromatogr. A*, 975, 285, 2002.)

the separation was still largely confined to the lower right quadrant of the separation plane, with the addition of the sample molecular weight attribute increasing the degree of solute crowding (informational similarity 0.74). However, the %synentropy remained essentially the same (1.9). The increase in solute crowding did, however, affect the space utilization, with a consequent reduction in the peak spreading angle to 64° and a 78% space utilization. The practical peak capacity was 206 and the total number of separated bands was 47.

The high degree of divergence in the retention processes for two reversed-phase systems is somewhat unusual. However, the basic principle by which separation takes place in each dimension is very different. On the CCZ phase, retention is very spacially dependent, hence the separation is sensitive to diastereomers, whilst on the C18 phase, retention is governed by the hydrophobic nature of the compounds, hence retention increases systematically with molecular weight. Therefore, these two phases intrinsically are sensitive to very different aspects of the sample characteristics—molecular weight and molecular shape.

Using multiple methods in which to evaluate the resolving power of a 2D system can give a clear indication of the orthogonality and theoretical separation space that is available in comparison to other 2D systems. This is particularly important when presented with complex mixtures that contain closely related structures that may cluster in a 2D space as demonstrated by Gray et al. [35]. For example, when the mobile phase in the first dimension separation of the 32-component mixture was 100% acetonitrile, the practical peak capacity was expanded enormously to 156 (compared to 53 C18/MeOH); however, because the C18/acetonitrile system was also selective toward diastereomers (the methanol mobile phase suppressed this dimension), the solute crowding increased significantly. Consequently, the useable separation space decreased to 56% (compared to 90% C18/MeOH) and the spreading angle also decreased to 41° (compared to 75° C18/MeOH). Despite this, the geometric approach to factor analysis predicted that 26 out of the 32 components would be resolved. However, because of the high degree of solute crowding, the practical application of this separation was significantly more difficult than when the C18/methanol system was employed.

9.2.3.4 Peptides

Gilar et al. [41] proposed a different geometric approach to describe orthogonality and employed a sample matrix of 196 peptides to demonstrate the technique. Their technique was based upon a peak surface coverage across the 2D separation plane. In this method, the normalized 2D retention data was plotted on a square separation space divided into sections called bins, the number of bins were equivalent to the number of data points. Each data point represented a normalized peak area (peak width was measured at 4σ, 13.4% of peak height) and each bin corresponded to a peak area. The degree of the area coverage described the orthogonality of a 2D separation, with greater coverage referring to the greatest divergence between dimensions. The hypothetical 2D plots illustrated in Figure 9.7 shows this geometric approach to orthogonality. These plots represent the 2D separation space between two different chromatographic environments, with peak capacities set to 100. When the 2D systems are totally correlated, the data points align along the main diagonal and the surface coverage is 10% (Figure 9.7a). The data points in a truly orthogonal

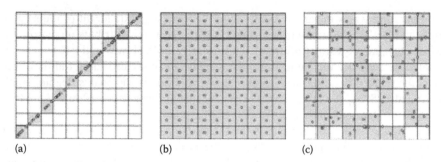

(a)   (b)   (c)

FIGURE 9.7   Geometric orthogonality concept. Hypothetical separation of 100 analytes in 10 × 10 normalized separation space. (a) Non-orthogonal system, 10% area coverage represents 0% orthogonality. (b) Hypothetical ordered system, full area coverage. (c) Random, ideally orthogonal, system; area coverage is 63% representing 100% orthogonality. (From Gilar, M. et al., *Anal. Chem.*, 77, 6426, 2005.)

system, where bands are uniformly spaced allows 100% coverage of the surface (Figure 9.7b). When solutes are, however, randomly spaced across the separation domain, co-elution of some components may occur and hence these components occupy the same bin, leaving some bins unused (Figure 9.7c).

Following from the hypothetical example illustrated in Figure 9.7, this geometric approach to orthogonality was applied to the 2D separation of peptides. The normalized retention data of 196 peptides were plotted in a space of 14 × 14 bins (total peak capacity of 196). The orthogonality of such 2D systems were calculated according to Equation 9.9:

$$O = \frac{\Sigma^{\text{bins}} - \sqrt{P_{\text{max}}}}{0.63\,P_{\text{max}}} \tag{9.9}$$

where

O is orthogonality
Σ^{bins} is the number of bins in the 2D plot containing data points
P_{max} is the total peak capacity as a sum of all bins

The orthogonality was expressed as a percentage with 0% representing a perfectly correlated system and 100% an orthogonal 2D system.

The practical peak capacity was calculated by using the knowledge of the 2D surface coverage as shown in Equation 9.10:

$$N_P = P_{2D}\frac{\sum^{\text{bins}}}{P_{\text{max}}} \tag{9.10}$$

where N_p is lower than the theoretical peak capacity P_{2D} defined by Equation 9.11 as not all of the surface is used for a separation.

$$P_{2D} = P_1 P_2 \tag{9.11}$$

9.2.3.5   Pharmaceuticals

Xue et al. [42] developed a fully automated comprehensive orthogonal method evaluation technology (COMET) system employing orthogonal HPLC separations and hyphenated UV-MS detection for impurities in pharmaceutical drugs. The orthogonality and practical peak capacity was determined by applying the geometric factor analysis approach to the chromatographic data. The nine methods with the highest practical peak capacity and maximum orthogonality were chosen for the COMET screening. The retention time correlation coefficients r^2 between any two of the methods chosen were used to indicate orthogonality, with a correlation coefficient r^2 closer to zero indicating orthogonality.

9.3   TWO-DIMENSIONAL CHROMATOGRAPHIC SYSTEMS

9.3.1   COMPREHENSIVE AND HEART-CUTTING SEPARATIONS

Depending upon the goal of the analysis, 2D separations can be carried out using either a heart-cutting process or comprehensively. Either of these techniques could feasibly be employed in a screening process, where the goal of the separation is perhaps to search for the appearance of certain chemicals and less interest is paid to compounds of no relevance to the analyst. In such a situation, the 2D separation could be fine-tuned to target those particular compounds, sacrificing total peak capacity, but maximizing resolution in the region that is most important.

In the case of fingerprinting, however, comprehensive separations are considered more suitable as the entire sample is generally subjected to 2D analysis, unless the analytical result can be substantiated with less information, in which case a heart-cutting approach could be feasible.

The process of heart cutting involves the transport of a discrete area of interest from the first dimension to the second dimension for further separation. This may even involve several heart-cut fractions from the first dimension being transported to the second dimension. Comprehensive chromatography involves the transfer of the entire first dimension to the second dimension for further separation. However, the transportation of the entire sample from the first dimension to the second has many disadvantages, most significant of which is the physical limitations associated with undertaking the second-dimension separation within a time frame appropriate for the first dimension. The second dimension must therefore be fast in order to avoid wraparound affects where solute from the latter cut from the first dimension becomes mixed with the previous cut in the second dimension. This results in chaotic band displacement, and potential co-elution of compounds that were previously separated in the first dimension—negating the power of the 2D system. In order to gain speed in the second dimension, peak capacity is often sacrificed, with the flow down affect being less discrimination in the fingerprinting result. As such, there is a delicate balance that may be played between how many peaks can be separated and how many peaks need to be separated in order to show a chemical signature match. Is a higher resolution 2D separation of less components better than a lower resolution 2D separation containing more components, or visa versa. A way of compromise is to employ a modified version of the comprehensive approach—that being off-line comprehensive 2DHPLC. In this technique, sample fractions from the first dimension are collected, stored, and then when convenient, run in the second dimension. In this mode of operation, the second dimension can have a high peak capacity as there is no time limitation associated with the analysis. However, the drawback is that sample fractions collected from the first dimension may need to be reconcentrated prior to injection into the second dimension. There is also the risk of labile compounds degrading while waiting for analysis in the second dimension. Furthermore, the off-line approach is slow, but here speed is sacrificed for peak capacity.

A variation in the off-line approach mentioned earlier, is the comprehensive heart-cutting approach. In this technique, sample is injected in the first dimension, a small aliquot from the first dimension is transported to the second dimension via appropriate switching valves, and then analysis takes place in the second dimension. Once separation in both dimensions is complete, another aliquot of sample is loaded into the first dimension, and this time a different fraction is heart cut to the second dimension, and the process repeated. Depending on how many samples are injected into the first dimension, how small the aliquot sampled to the second dimension is, and how much of the first dimension is actually sampled—will determine the quality of the chemical signature. An advantage of this process, however, is that very high theoretical peak capacities can be obtained since there is no speed limitation in the second dimension, other than that dictated by the patience of the operator.

A disadvantage of the comprehensive approach, either online or off-line is the creation of the enormous amount of information that is collected, and hence must be analyzed because many components, not just those of interest in the sample, would be resolved. This creates more complicated chromatograms as large amounts of data need to be converted from the recorded data acquisitions, which may prove time consuming and problematic and in our modern world of information collection, it is sometimes the analysis of the data that proves to be the limiting factor. Here is in fact the advantage of a targeted heart-cutting approach: Only the components of interest are specifically analyzed, be they predetermined contaminants whose analysis is dictated by regulatory authorities, or components within the sample whose presence largely describes the quality of the sample. Applied in a targeted approach, heart cutting is useful for improving the resolution of components by simplifying the matrix as only the bands of interest are cut from the first dimension and transported to the second dimension, but since not all the sample is analyzed, the speed in analysis is somewhat faster. A limitation of this technique is, however, that some previous knowledge of the components of interest may be required to ascertain the area(s) to be heart cut and may require some additional prework. Hence this form of 2D HPLC is very useful for continuous screening of samples.

9.3.2   Two-Dimensional System Designs

The use of an automated online 2D system eliminates the need for manual sample handling such as fraction collection and reinjection such as required for an off-line method; however, some consideration to the system design and experimental objectives are required. The most common way of interfacing columns for 2D HPLC systems is that of either 4-, 6-, or 10- port, two-position automated switching valves.

The switching valves essentially allow the dimensions to operate independently from one another without loss of the resolution achieved in the first dimension. 2D configurations generally incorporate either two sample loops and switching valves; two sample traps and switching valves; or switching valves with a dual or quad column configuration in the second dimension. The use of sample loops allows eluent to be collected from the first dimension while eluent held on an additional loop is loaded on the second-dimensional column. This process is controlled by the precise timing of the switching valves and is generally computer controlled online. Almost any HPLC system can be converted to a 2D system through the addition of switching valves and further expanded upon by and the use of multiple HPLC pumps, an injector either manual or automatic, and suitable detection.

One of the earliest 2D HPLC systems was developed by Erni and Frei [43] where two loops were connected to an eight-port switching valve. While one loop was being filled from the first dimension, the second loop was being loaded onto the second-dimensional column, alternating with each valve switch. Many systems have been developed based on this pioneering concept, with adaptations of this design by Bushey and Jorgenson [44] being used for comprehensive 2D separations [45]. This design does have a problem, which is related to differences in the retention times of certain components, particularly when loop sizes become large as the sample is forward flushed on one loop and reversed flush on the other loop [46]. Ten-port switching valves with two loops [46,47] and six-port switching valves are also used for comprehensive 2D systems [48]. This allows for continuous collection and reinjection of the first-dimensional eluent. A 2D system comprised of four six-port switching valves that connected the two dimensions was developed for the separation of oligostyrenes [49]. By reducing the six-port switching valves to two, this system is capable of operating under heart-cutting conditions.

9.3.3 DATA COLLECTION

Another important factor relates to how data is collected and then subsequently represented as 2D chromatographic information. If a heart cutting or "semi-comprehensive" approach is employed, then this is achieved through detection of sample in each dimension and the data is collected and either represented as conventional unidimensional chromatograms or transferred to a spreadsheet and converted to a contour plot or a 2D plot for visual interpretation, depending on how many "cuts" are analyzed in the second dimension. When a comprehensive approach is employed, it is necessary only to detect the output from the second dimension. A typical output is intensity as a function of frequency data set. This information is collected in a continuous output over the entire duration of the 2D separation. At the end of the separation, the unidimensional data stream is converted to a matrix according to the frequency of sample modulating from the first to the second dimension. This information is then presented as a contour plot.

Contour plots have been used in many applications to display the acquired information derived from 2D systems. Given the expanded separation space afforded by 2D HPLC, retention times within this 2D space have a higher probability of being unique to any particular component, and thus, contour plots are a convenient way in which to view compositional changes within sample sets. Examples of the application of 2D contour plots are found for the separations of proteins and peptides [44,50,51], alkyl benzenes [16] mixtures of amines, acids and other substances [15], PAHs [28], and hydrocarbon and benzene derivatives [17].

A variety of softwares have been employed for data analysis in 2DHPLC, but essentially these are limited to self-derived programs, with limited commercial programs yet been made available. Murphy et al. [12] used a FORTRAN program to process the 2D data for the analysis of alcohol ethoxylates, producing exceptional graphical representation. Chen et al. [45] used an in-house program to interpret the data collected for the analysis of *Rhizoma chuanxiong* and the 2D data was graphically presented with the aid of Fortner Transform.

In the results presented here, a Mathematica program was employed to plot the 2D data, following which a peak picking function was utilized to determine the location and number of components separated. Ultimately, this process can be coupled to statistical methods that allow measurement in the degree of divergence between dimensions in various 2D systems; information theory, geometrical approach to factor analysis, and/or the bin theory. While it may be a challenging task for quantitative information to be extracted from a 2D chromatogram, such a process is simplified greatly through the utilization of peak picking functions, such as that described here. For statistical analysis methods to be applied, the retention time for each peak must be determined in all separation dimensions (in the case of 2D HPLC, the retention times in both first and second dimensions). This can be achieved manually only if the number of compounds being analyzed is small and there are few cuts from the first dimension to the second. This, however, is not the case when performing 2D HPLC where potentially there are over a 100 chromatograms to analyze and hundreds of component peaks. For timely analysis of 2D HPLC data, and to ensure accuracy in measurement, methods to automate the data analysis need to be employed.

An overview of the steps taken to determine the 2D retention time of peaks in a 2D chromatogram are as follows:

1. Determine the threshold of peak detection. This can be achieved using a representative 1D chromatogram. This parameter is user defined and can be adjusted to select major components only and hence minimize noise, or the threshold could be lowered so as to not exclude potentially important information that is found within the lower concentrated components. It is important to realize though, that setting the threshold can and most likely will influence the type of information that is obtained from the 2D data: Impacting chemical signatures and statistical analysis.
2. Find retention times of all peaks in each unidimensional chromatogram (i.e., find 2D retention times).
3. Determine if peaks that are detected in adjacent chromatograms are from the same source (i.e., two-dimension peak matching).

9.3.3.1   Peak Retention Time Determination

The first step in determining the retention times of a 2D HPLC separation is to define the thresholds of a single chromatogram that distinguish chromatogram noise from peaks. Peak retention times are found in a 1D chromatogram by the magnitude of detector response, and by examining the first and second derivatives of the chromatogram [52].

The chromatogram is filtered to disregard peaks that have a detector response less than a given threshold, which may be caused by noise. This threshold also serves to define preliminary peak regions that will be refined by the first derivative. The peak region begins when the chromatogram signal increases above the threshold and ends when the chromatogram returns to below the threshold. Adjacent peaks that have a valley (i.e., minimum point between two adjacent peaks) that passes below the threshold will have two distinct peak regions. When the valley does not pass below the threshold, the peaks have the same peak region. An alternative threshold may be used at the analyst's discretion that is based on the chromatogram noise. The noise is determined for each data point by calculating the signal response subtracted by the mean of the responses of the two neighboring detector samples. The chromatogram noise is then determined to be the median of these values [52].

The Savitzky–Golay smoothing filter [53,54] is capable of calculating the derivatives of experimental data and also compensates for noise amplification when the first and second derivatives are calculated [52]. The first derivative of the chromatogram is used to refine the peak regions with an adjustable first derivative threshold [52]. The beginning of the peak region is defined when the first derivative curve crosses the first derivative threshold while possessing a positive gradient (i.e., as the chromatogram is tracked from left to right, the y-coordinate is increasing). The end of the peak region is when the curve crosses the negative of the first derivative threshold, again with a positive gradient. By defining the peak regions with the first derivative, it is possible to distinguish between peaks that elute near each other and have peak valleys that do not pass below the peak threshold.

The point where the first derivative crosses the y-axes between the peak region, maximum and minimum with a negative gradient is the retention time of the peak. However, the second derivative of the chromatogram should also be examined, as the first derivative is unable to distinguish between peaks that strongly overlap and shoulder [52]. When observing the second derivative of a chromatogram, the region of this curve that is below the second derivative threshold (note that the second derivative threshold is a negative number) represents a peak. The minimum point of this negative region is the retention time of the peak. By determining the retention time of peaks with the second derivative, subtle irregularities in the curve are noticed and the retention times of strongly co-eluting peaks can be identified. The second derivative threshold can be adjusted by the analyst to remove any false positives cause by irregularities from other sources, such as detector noise and sampling frequency.

9.3.3.2   Peak Retention Time Determination in a Two-Dimensional HPLC Chromatogram

To find the 2D retention time, the process of examining the chromatogram and its first and second derivatives is applied first to a single one-dimension chromatogram. A chromatogram is selected by the analyst that is a good representation of all the 2D chromatograms to define thresholds that are appropriate for the separation, or for the type of analysis being undertaken. With these defined thresholds, the 2D separation is then analyzed. This is easily completed with a computational mathematics software package, where the algorithm used to find the peak coordinates in one dimension can be placed in a loop to process every unidimensional chromatogram in the separation. The 2D peak coordinates are then output by the algorithm described with three coordinates, i.e., first dimension retention time (or cut time), second dimension retention time, and detector response. More information about the peaks can also be supplied in the form of peak region boundaries.

9.3.3.3   Two-Dimension Peak Matching

In comprehensive 2D HPLC, compounds have the potential to be detected in multiple neighboring first-dimension slices. For a more accurate representation of the 2D separation, peaks that are detected need to be compared to neighboring peaks to determine if they are caused by the same compound, and are thus a single 2D peak. Peters et al. [55] developed an algorithm that compares the two-dimension retention times of peaks in adjacent cuts in GC × GC separations to determine if the peak is caused by the same compound. This algorithm can also be applied in LC × LC scenarios. Adjacent peaks are matched based on their degree of peak region overlap and by then examining the peak maximum profile.

The overlap of adjacent peak regions is compared as the initial test to determine if peaks in adjacent 1D chromatograms are from the same source [55]. Assuming that the peak in question is peak A, and peak A is being compared to peak B, Peters et al. [55] described five ways that the peak region of A will overlap with that of B. These are stated as [55]

1. Both the peak regions of A and B start and stop at the same time
2. Peak A starts later than peak B and also ends later
3. Peak A starts earlier than peak B and also ends earlier
4. Peak A starts earlier than peak B, but also ends later
5. Peak B starts earlier than peak A, but also ends later

Overlap is determined by measuring the amount of time the two peaks overlap and dividing this value by the total peak region time of peak A. If the overlap is greater than a user-defined overlap threshold (e.g., 90%), peaks A and B are assumed to both be part of the same 2D peak. Peaks that overlap according to scenarios (4) and (5) are always selected as candidate peaks. If more than one peak adjacent to peak A is a candidate peak, the peak that has the closest second-dimension retention time to peak A is chosen to be in the same 2D peak.

After adjacent peaks have been grouped together on the basis of overlap threshold, the peak maximum profile of these peaks is examined (visually, this is a plot of 1D retention times of each peak vs. detector response) [55]. It is required that each 2D peak have only one maximum and a minimum is indicative of multiple 2D peaks. The peaks either side of the minimum are separated into separate 2D peaks. The 1D peak that represents the minimum in detector response is assigned to the 2D peak that has the closest second-dimension retention time.

The final process to determine the 2D retention time for the multidimensional peaks is an extension of the peak maximum profile, where the data between points on the peak maximum profile is interpolated [55]. This is done as, potentially, the low number of cuts in comprehensive analysis does not provide a detailed peak maximum profile. More information about the peak maximum profile is obtained by supplementing the first-dimension data with data found in the second dimension according to Equation 9.12 [55]:

$$h_{int} = \frac{h_x d_y + h_y d_x}{d_x + d_y} \tag{9.12}$$

where
  x and y are two consecutive peak maxima from the unimodality criteria
  h_{int} is the interpolated detector response
  h_x and h_y are the detector responses of the two peaks at their respective first-dimension and second-dimension retention times
  d_x and d_y are the time differences between the interpolated point to points x and y, respectively

The larger the second dimension time difference between the two one-dimension peaks, the more points will be interpolated for the peak maximum profile, with each point representing one detector sample. Two-dimensional peaks are then determined based on the position of minimum values in the interpolated peak maximum profile.

By following the processes outlined here, the multitude of 2D data collected from potentially hundreds of 1D HPLC injections can be turned into meaningful information. This information (i.e., first-dimension retention time, second-dimension retention time, and detector response) is a qualitative description of the retention of all the peaks that have been resolved in the multidimensional separation and can be applied to the statistical methods described earlier to describe the quality of the separation. As an example, Figure 9.8 illustrates a contour plot of a 2D separation of café expresso. Each line that adjoins dots indicates components that are derived from the same source in the first dimensions and represent the distribution of a single compound. Further details of this separation are presented in Section 9.4.

9.4 APPLICATIONS IN SELECTIVITY SCREENING AND FINGERPRINT ANALYSIS: CAFÉ EXPRESSO

The unique taste, fragrance, and stimulating properties of Coffee brews make them one of the most popular beverages throughout the world. Coffee-based drinks contribute to 64% of the total antioxidant intake, followed by fruits, berries, tea, wines, cereals, and vegetables [56]. In recent years, there has been an increasing interest in possible health beneficial properties of coffee consumption [57], although the findings are contradictory [58,59]. Both, coffee brews and green coffee beans represent extremely complex chemical compositions containing large numbers of components of various composition and size, some of which may be strong antioxidants with beneficial physiological, physicochemical, and anticarcinogenic properties for human health.

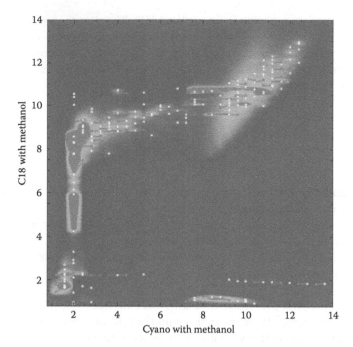

FIGURE 9.8 A 2D contour plot of a café expresso separation employing a cyano column in the first dimension and a C18 column in the second dimension. Each line that adjoins dots indicates components that are derived from the same source in the first dimensions and represent the distribution of a single compound. White dots indicate the retention time of the peak maxima.

As for the beneficial effects of coffee, in both green and roasted coffees, compounds possessing antioxidant and radical scavenging activity play the main key responsibility [60–62]. Unprocessed green coffee beans are one of the richest dietary sources of certain natural antioxidants, mainly hydroxycinnamic acid derivatives; chlorogenic acid (or 5-caffeoylquinic acid, CGA) and its two major positional isomers, 3-CQA and 4-CQA [63], accounting for up to 10% of the dry weight of green coffee [57,64] and others like caffeic acid, ferulic acid, *p*-coumaric acid [65,66]. The content of these beneficial compounds varies between the coffee tree [67], geographical origin [68], coffee preparation [69], and degree of the roasting process [70]. The roasting process in coffee production is necessary to develop the typical sensory characteristics of coffee, markedly affecting its final composition. A considerable number of phenolic compounds have been identified in roasted coffee, either derived from chlorogenic acid [71], such as chlorogenic acid isomers and their *di*-esters, or related to other hydroxycinnamic acid conjugates like feruloyl-quinic acids and caffeoyl-tyrosine [72]. Nevertheless, depending on the roasting conditions, compounds with antioxidant properties decompose to some extent [66]. A decrease in protein, amino acids, and other compounds is also described following roasting [57]. However, development of new compounds during thermal treatment, including Maillard reaction products, like water-soluble polymer melanoidin antioxidants [65], balance the thermal degradation of naturally occurring phenolics and maintain or even enhance the overall antioxidant properties of coffee brew [57,73,74]. This means that the overall physiological properties of roasted coffee are expected to be dependent on the extent of the Maillard reaction, the degree of which determines either formation of pro-oxidant compounds, like acrylamide, in the early stage of the reaction, [75] or on contrary in the advanced stages of roasting, antioxidant products, like melanoidins seem to prevail [70]. These compositional changes complicate the chemical matrix of coffee, and consequently, the coffee profile becomes even more complex. An analytical technique, to provide reliable analysis of the coffee composition, therefore could constitute a useful tool to understand the complexity of coffee composition from both sensory and health point of view.

Some of the classical techniques used for the analytical determination of compositional matrix of real-life samples, sometimes are limited in providing sufficient information about the real sample composition, emphasizing the requirement of techniques with higher resolving power. Detailed discussion in the previous sections of this chapter emphasized the separation power of MDHPLC, underlining the importance of divergent, ideally non-correlated retention processes in each of the dimensions to yield high peak capacity separations, suited to the separation of complex samples.

Numerous 2D-HPLC applications on natural products have been published, e.g., references [76–80], including using the combination of CN × C18 as an attractive approach for separation of natural antioxidants [79–81]. The combination of the cyano and C18 phases offers a likely difference in their retention mechanisms, due to differences in hydrophobic, dipolar, and electronic interactions. Despite these differences, the complexity of the sample negates the ability of either separation media, when operated unidimensionally, to resolve adequately the components in the sample. Furthermore, when dealing with such complex samples, how best to determine the orthogonality of coupled systems? One solution would be to prepare an extensive set of standard compounds and test their behavior in different unidimensional systems. From this, the most orthogonal systems could be coupled. Limitations to this approach include, how many compounds and of which specific compound classes best reflect the actual 2D behavior of the complex natural sample. This may be a difficult question to resolve. Another solution to the problem may be to use the sample itself as the test "standard." The limitation to this approach is how do you describe changes in separation selectivity if the complexity of the sample and the resulting separation is such that changes in retention order cannot be deduced from basic unidimensional separation information? An advantage, however, is that the resulting separations exactly reflect that of the sample. To overcome the limitations of peak capacity in measuring the change in selectivity of the system, multidimensional HPLC can, in itself, be used as a pseudo-hyphenated method of analysis. By setting the second dimension to constant retention, changes in the selectivity of the first dimension (either through different stationary phases or mobile phases) are reflected by the retention order in the second dimension, visually depicted by contour plots of the 2D couple. Differences, however, are a reflection of the change made in the first dimension. Using this approach, selectivity changes can be observed, even for very complex samples and application of the peak picking process described in Section 9.2.3 determines the number of components isolated and then statistical methods can be applied to provide a metric for the best system.

Here, 2D HPLC was used to scout for selective dimensions in the analysis of café expresso, and subsequently illustrate differences between a café expresso and a decaffeinated "café expresso." The two expresso coffees, "Ristretto" and "Decaffeinato" are known for their organoleptic differences as "subtle fruity full bodied" and "aroma of red fruit," respectively. Each expresso coffee was prepared using a Nespresso café machine using the respective cartridges. The expresso was diluted in water 1:4 prior to analysis, both unidimensionally and two-dimensionally.

The results presented here represent separations obtained on a Luna CN phase, a Luna hexyl-phenyl phase, and a SphereClone C18 phase (all columns 150 × 4.60 mm, 5 µm pd). While only the results for these three columns, running just one mobile phase, are presented here, a detailed selectivity study was undertaken on five columns and three mobile phases. The chromatograms shown in Figure 9.9 are 1D

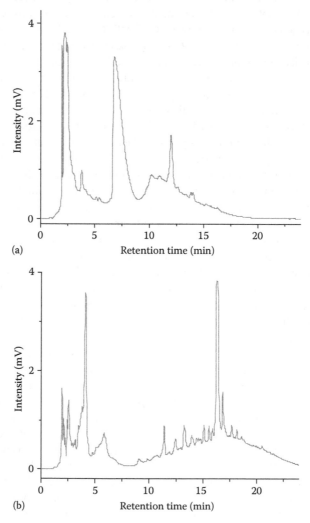

FIGURE 9.9   One-dimensional chromatograms café expresso on (a) cyano, (b) hexyl-phenyl, and

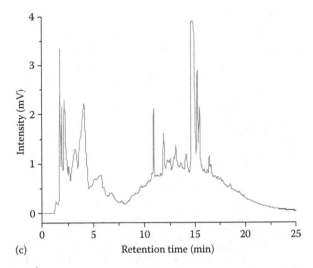

FIGURE 9.9 (continued) (c) C18 stationary phases. All columns 150 × 4.6 mm, 5 µd, mobile phase ϕ_i (100/0) water/MeOH to ϕ_f (0/100) water/MeOH at 5% per minute. Flow rate = 1 mL/min detection at 280 nm.

gradient separations employing aqueous/methanol mobile phases. Each separation is run under identical conditions, starting with 100% water and running to 100% methanol at a rate of 5% per minute. The flow rate in each case was 1 mL/min. Clearly, there are differences between the separations on each phase, but these differences are difficult to quantify since the separation space is saturated.

Selectivity changes between the cyano phase and the hexyl-phenyl phase were measured by incorporating these columns into a 2D system, whereby, in each case, the second dimension was the C18 column. Differences between the cyano phase and the hexyl-phenyl phase were therefore detected by observing the change in the 2D retention plots. As an example, the 2D separation of the Ristretto sample on the cyano-C18 couple is illustrated in Figure 9.10a, and that of the hexyl-phenyl phase is illustrated in Figure 9.10b. The distribution of bands across the 2D plane, and the absence of data aligned along the main diagonal for the cyano-C18 couple (Figure 9.10a) verifies the selective differences between the cyano and C18 dimensions for the coffee separation. The separation shown in Figure 9.10b shows how this separation has changed when the hexyl-phenyl column replaced the cyano column. Data at longer retention times in both dimensions is now aligned along the main diagonal, showing correlation between the hexyl-phenyl phase and the C18 phase (see highlighted region (A). However, the region highlighted as (B) is distinctly different to that observed on the cyano-C18 couple, illustrating the selective differences between the cyano and hexyl-phenyl phases, and also, the distinct difference between the hexyl-phenyl phase and the C18 phase.

The 2D application was then applied to the analysis of two different café expressos. That is, the analysis of the Ristretto and the Decaffeinato. The analysis of both these coffees is shown in Figure 9.11a and b. Both analyses were undertaken using exactly the same conditions, employing the cyano column in

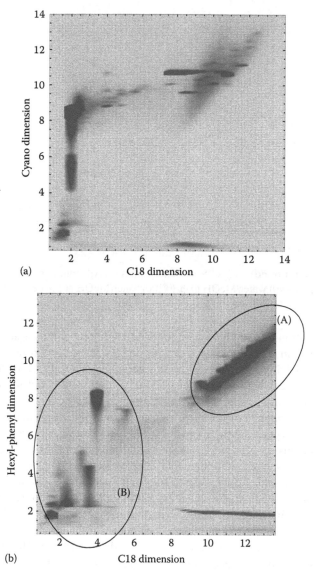

FIGURE 9.10 Two-dimensional separation of Ristretto café expresso. Comparison between the first dimensions of (a) cyano and (b) hexyl-phenyl phases. Both mobile phases where aqueous/methanol gradient separations in both dimension 1 and 2. All conditions identical for each phase system. Details stated in the text.

the first dimension and the C18 column in the second dimension with aqueous/methanol gradient elution as per Figure 9.10. Clearly, these coffee brews can be distinguished by their differences in these 2D separations.

The separations depicted in Figures 9.9 through 9.11 were undertaken on Waters 600E Multi Solvent Delivery LC System equipped with Waters 717 plus auto injector, Waters 600E pumps, Waters 2487 series UV/vis detectors, Waters 600E system

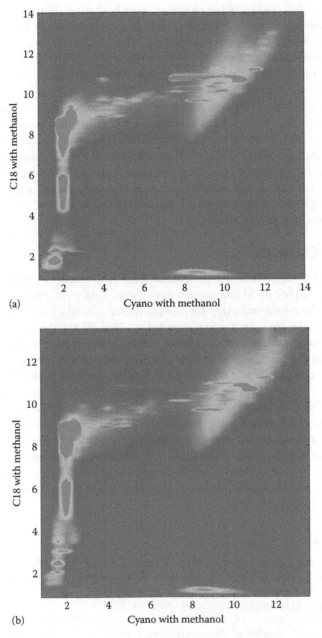

FIGURE 9.11 Two-dimensional separation of (a) Ristretto café expresso and (b) Decaffenito café expresso. Separations undertaken on a cyano-C18 system, both dimensions running aqueous/methanol mobile phases, in gradient mode. Conditions stated in the text.

controller. Column switching was achieved by electronically controlled six-port, dual positioned valves. All injection volumes were 100 μL and UV detection was set at 280 nm. Gradient elution was undertaken, with both dimensions running aqueous/methanol mobile phases with initial mobile phases 100% water running to 100% methanol in 10 min. Flow rates were 1 mL/min. A comprehensive heart-cutting approach was used to express the 2D peak displacement, by which a 200 μL heart-cut section was transferred to the second dimension, with subsequent second-dimension separation being undertaken. The first-dimensional separation was repeated, following which another 200 μL first-dimension fraction was transferred to the second dimension. This was repeated at every 0.4 mL across the entire first-dimension separation, i.e., the first-dimension separation was repeated a total of 34 times over a 20 h period. Application of this type of comprehensive heart-cutting analysis yields potentially very high peak capacity separations that may yield chemical signature information that reflects the sample identity; particularly useful in systems that are then to be employed for the preparative targeted isolation of "key" components from within the complex sample matrix.

9.5   CONCLUSION

While LC × LC separations are not conventional with respect to hyphenated methods of analysis, they do in fact serve that purpose. Selectivity changes for complex samples can be monitored using the second dimension as a means to "detect" the changes that take place in the first dimension. This allows the analyst to develop separation protocols for complex samples that truly reflect the characteristics of the sample, rather than being based on a set of model compounds, which may or may not adequately represent the real sample. Furthermore, the high peak capacity of a 2D system allows for chemical signatures to be obtained. These signatures are readily achieved without the need for more complex hyphenated methods of analysis, and could be used as a first step in the classification of the sample, perhaps then limiting the number of samples that require further, more detailed hyphenated methods of analysis. This would greatly reduce the cost and speed in the analysis of complex samples. This type of analysis is ideally suited to the needs of the analyst who also requires that key sample components be extracted and collected from the more complex native sample.

REFERENCES

1. Miyamoto, K.; Hara, T.; Kobayashi, H.; Morisaka, H.; Tokuda, D.; Horie, K.; Koduki, K.; Makino, S.; Nez, O.; Yang, C.; Kawabe, T.; Ikegami, T.; Takubo, H.; Ishihama, Y.; Tanaka, N.; *Analytical Chemistry*, 2008, 80, 8741.
2. Giddings, J.C.; *Analytical Chemistry*, 1967, 39, 1027.
3. Davis, J.M.; Giddings, J.C.; *Analytical Chemistry*, 1983, 55, 418.
4. Giddings, J.C.; *Journal of Chromatography A*, 1995, 703, 3.
5. Guiochon, G.; *Journal of Chromatography A*, 2006, 1126, 6.
6. Guiochon, G.; Marchetti, N.; Mriziq, K.; Shalliker, R.A.; *Journal of Chromatography A*, 2008, 1189, 109.
7. Shalliker, R.A.; Gray, M.J.; *Advances in Chromatography*, 2006, 44, 177.

8. Stolla, D.R.; Lia, X.; Wanga, X.; Carr, P.W.; Sarah, E.G.; Porter, S.E.G.; Rutan, S.C.; *Journal of Chromatography A*, 2007, 1168, 3.
9. Schure, M.R.; In: Cohen, S.A.; Schure, M.R.; Eds.; *Multidimensional Liquid Chromatography: Theory, Instrumentation and Applications*, Wiley, New York, 2008.
10. Gray, M.J.; Dennis, G.R.; Slonecker, P.J.; Shalliker, R.A.; *Journal of Chromatography A*, 2003, 1015, 89.
11. Giddings, J.C.R.E.; *Analytical Chemistry*, 1984, 56, 1258A.
12. Murphy, M.R.; Schure, M.; Foley, J.P.; *Analytical Chemistry*, 1998, 70, 4353.
13. Van Gyseghem, E.; Van Hemelryck, S.; Daszykowski, M.; Questier, F.; Massart, D.L.; Vander Heyden, Y.; *Journal of Chromatography A*, 2002, 988, 77.
14. Guttman, A.; Varoglu, M.; Khandurina, J.; *Drug Discovery Today*, 2004, 9, 136.
15. Venkatramani, C.J.; Zelechonok, Y.; *Analytical Chemistry*, 2003, 75, 3484.
16. Ikegami, T.; Hara, T.; Kimura, H.; Kobayashi, H.; Hosoya, K.; Cabrera, K.; Tanaka, N.; *Journal of Chromatography A*, 2006, 1106, 112.
17. Tanaka, N.; Kimura, H.; Tokuda, D.; Hosoya, K.; Ikegami, T.; Ishizuka, N.; Minakuchi, H.; Nakanishi, K.; Shintani, Y.; Furuno, M.; Cabrera, K.; *Analytical Chemistry*, 2004, 76, 1273.
18. Gray, M.J.; Sweeney, A.P.; Dennis, G.R.; Wormell, P.; Shalliker, R.A.; *Journal of Liquid Chromatography and Related Technologies*, 2004, 27, 2905.
19. Gilar, M.; Olivova, P.; Daly, A.E.; Gebler, J.C.; *Journal of Separation Science*, 2005, 28, 1694.
20. Bashir, W.; Tyrrell, E.; Feeney, O.; Paull, B.; *Journal of Chromatography A*, 2002, 964, 113.
21. Bolliet, D.; Poole, C.F.; *Analyst*, 1998, 123, 295.
22. Turowski, M.; Morimoto, T.; Kimata, K.; Monde, H.; Ikegami, T.; Hosoya, K.; Tanaka, N.; *Journal of Chromatography A*, 2001, 911, 177.
23. Valkó, K.; Espinosa, S.; Du, C.M.; Bosch, E.; Rosés, M.; Bevan, C.; Abraham, M.H.; Unique selectivity of perfluorinated stationary phases with 2,2,2-trifluoroethanol as organic mobile phase modifier. *Journal of Chromatography A*, 2001, 933, 73.
24. Fields, S.M.; Ye, C.Q.; Zhang, D.D.; Branch, B.R.; Zhang, X.J.; Okafo, N.; *Journal of Chromatography A*, 2001, 913, 197.
25. Neue, U.D.; Alden, B.A.; Walter, T.H.; *Journal of Chromatography A*, 1999, 849, 101.
26. Cacciola, F.; Jandera, P.; Blahova, E.; Mondello, L.; *Journal of Separation Science*, 2006, 29, 2500.
27. Murahashi, T.; Tsuruga, F.; Sasaki, S.; *Analyst*, 2003, 128, 1346.
28. Murahashi, T.; *Analyst*, 2003, 128, 611.
29. Pellett, J.; Lukulay, P.; Mao, Y.; Bowen, W.; Reed, R.; Ma, M.; Munger, R.C.; Dolan, J.W.; Wrisley, L.; Medwid, K.; Toltl, N.P.; Chan, C.C.; Skibic, M.; Biswas, K.; Wells, K.A.; Snyder, L.R.; *Journal of Chromatography A*, 2006, 1101, 122.
30. Steuer, W.; Grant, I.; Erni, F.; *Journal of Chromatography*, 1990, 507, 125.
31. Slonecker, P.J.; Li, X.; Ridgway, T.H.; Dorsey, J.G.; *Analytical Chemistry*, 1996, 68, 682.
32. Huber, J.F.K.; Kenndler, E.; Reich, G.; *Journal of Chromatography*, 1979, 172, 15.
33. Massart, D.L.; Vandeginste, B.G.M.; Buydens, L.M.C.; Jong, S.D.; Lewi, P.J.; Smeyers-Verbeke, J.; *Handbook of Chemometrics and Qualimetrics: Part A. Data Handling in Science and Technology*, In: Vandeginste, B.G.M.; Ed., Elsevier Science, Amsterdam, the Netherlands, 1997.
34. Huber, J.F.K.; Kenndler, E.; Reich, G.; Hack, W.; Wolf, J.; *Analytical Chemistry*, 1993, 65, 2903.
35. Gray, M.; Dennis, G.R.; Wormell, P.; Shalliker, R.A.; Slonecker, P.; *Journal of Chromatography A*, 2002, 975, 285.
36. Liu, Z.; Patterson, D.G.; Lee, M.L.; *Analytical Chemistry*, 1995, 67, 3840.

37. Sweeney, A.P.; Wyllie, S.G.; Shalliker, R.A.; *Journal of Liquid Chromatography and Related Technologies*, 2001, 24, 2559.
38. Sweeney, A.P.; Wong, V.; Shalliker, R.A.; *Chromatographia*, 2001, 54, 24.
39. Shalliker, R.A.; *Journal of Separation Science*, 32, 17, 2903.
40. Gray, M.J.; Sweeney, A.P.; Dennis, G.R.; Slonecker, P.J.; Shalliker, R.A.; *Analyst*, 2003, 128, 598.
41. Gilar, M.; Olivova, P.; Daly, A.E.; Gebler, J.C.; *Analytical Chemistry*, 2005, 77, 6426.
42. Xue, G.; Bendick, A.D.; Chen, R.; Sekulic, S.S.; *Journal of Chromatography A*, 2004, 1050, 159.
43. Erni, F.; Frei, R.W.; *Journal of Chromatography*, 1978, 149, 561.
44. Bushey, M.M.; Jorgenson, J.W.; *Analytical Chemistry*, 1990, 62, 161.
45. Chen, X.; Kong, L.; Su, X.; Fu, H.; Ni, J.; Zhao, R.; Zou, H.; *Journal of Chromatography A*, 2004, 1040, 169.
46. van der Horst, A.; Schoenmakers, P.J.; *Journal of Chromatography A*, 2003, 1000, 693.
47. Dugo, P.; Favoino, O.; Luppino, R.; Dugo, G.; Mondello, L.; *Analytical Chemistry*, 2004, 76, 2525.
48. Murahashi, T.; Kawabata, M.; Sugiyama, H.; Hasei, T.; Watanabe, T.; Hirayama, T.; *Journal of Health Science*, 2004, 50, 635.
49. Gray, M.J.; Dennis, G.R.; Slonecker, P.J.; Shalliker, R.A.; *Journal of Chromatography A*, 2004, 1041, 101.
50. Opiteck, G.J.; Lewis, K.C.; Jorgenson, J.W.; *Analytical Chemistry*, 1997, 69, 1518.
51. Opiteck, G.J.; Ramirez, S.M.; Jorgenson, J.W.; Moseley, M.A.; *Analytical Biochemistry*, 1998, 295, 349.
52. Vivó-Truyols, G.; Torres-Lapasió, J.R.; Vander Heyden, A.M.; Massart, D.L.; *Journal of Chromatography A*, 2005, 1096, 133.
53. Savitzky, A.; Golay, M.J.E.; *Analytical Chemistry*, 1964, 36, 1627.
54. Steinier, J.; Termonia, Y.; Deltour, J.; *Analytical Chemistry*, 1972, 44, 1906.
55. Peters, S.; Vivó-Truyols, G.; Marriott, P.J.; Schoenmakers, P.J.; *Journal of Chromatography A*, 2007, 1156, 14.
56. Svilaas, A.; Sakhi, A.K.; Andersen, L.F.; Svilaas, T.E.C.; Ström, E.-C.; Jacobs, D.R.; *Journal of Nutrition*, 2004, 134, 562.
57. Gómez-Ruiz, J.A.; Ames, J.M.; Leake, D.S.; *European Food Research Technology*, 2008, 227, 1017.
58. Verhoef, P.; Pasman, W.J.; van Vliet, T.; Urgert, R.; Katan, M.B.; *American Journal of Clinical Nutrition*, 2002, 76, 1244.
59. Taylor, S.R.; Demming-Adams, B.; *Nutrition and Food Science*, 2007, 37, 406.
60. Ito, H.; Gonthier, M.-P.; Manach, C.; Morand, C.; Mennen, L.; Remesy, C. et al.; *British Journal of Nutrition*, 2005, 94, 500.
61. Lajolo, F.; Saura-Calixto, F.; Penna, E.; Wenzel, E.; *Fibra Dietética En Iberoamérica: Tecnología Y Salud Editorial*, Sao Paulo, Brazil, Varela, 2001.
62. Rawel, H.M.; Kulling, S.E.; *Journal of Consumer Protection and Food Safety*, 2007, 2, 399.
63. Risso, M.E.; Peres, R.G.; Amaya-Farfan, J.; *Journal of Food Chemistry*, 2007, 105, 1578.
64. Clifford, M.N.; *Journal of the Science of Food and Agriculture*, 2000, 80, 1033.
65. Delgado-Andrade, C.; Morales, F.; *Journal of Agriculture and Food Chemistry*, 2005, 53, 1403.
66. Votavová, L.; Voldřich, M.; Ševčík, R.; Čížková, H.; Mlejnecká, J.; Stolař, M.; Fleišman, T.; *Czech Journal of Food Science*, 2009, 27, S49.
67. Richelle, M.; Tavazzi, I.; Offord, E.; *Journal of Agricultural and Food Chemistry*, 2001, 49, 3438.
68. Belay, A.; Ture, K.; Redi, M.; Asfaw, A.; *Food Chemistry*, 2008, 108, 310.
69. Sanchez-Gonzalez, I.A.; Jimenez-Escrig, A.; Saura-Calixto, F.; *Food Chemistry*, 2005, 90, 133.

70. Manzocco, L.; Calligaris, S.; Mastrocola, D.; Nicoli, M.C.; Lerici, C.R.; *Trends in Food Science & Technology*, 2001, 11, 340.
71. Sarrazin, C.; Lequére, J.L.; Gretsch, C.; Liardon, R.; *Food Chemistry*, 2000, 70, 99.
72. Clifford, M.N.; *Journal of. Science Food and Agriculture*, 1999, 79, 362.
73. Del Castillo, M.D.; Ames, J.M.; Gordon, M.H.; *Journal of Agricultural and Food Chemistry*, 2002, 50, 3698.
74. Andueza, S.; Cid, C.; Nicoli, M.C.; *Lebensmittel-Wissenschaft und-Technologie*, 2004, 37, 893.
75. Murkovic, M.; Derler, K.; *Journal of Biochemical and Biophysical Methods*, 2006, 69, 25032.
76. Blahová, E.; Jandera, P.; Cacciola, F.; Mondello, L.; *Journal of Separation Science*, 2006, 29, 555.
77. Liu, Y.; Xue, X.; Guo, Z.; Xu, Q.; Zhang, F.; Liang, X.; *Journal of Chromatography A*, 2008, 1208, 133.
78. de Souza, L.M.; Cipriani, T.R.; Sant'Ana, C.F.; Iacomiini, M.; Gorin, Ph.A.J.; Sassaki, G.L.; *Journal of Chromatography A*, 2009, 1216, 99.
79. Wong, V.; Shalliker, R.A.; *Journal of Chromatography A*, 2004, 1036, 15.
80. Wong, V.; Sweeney, A.P.; Shalliker, R.A.; *Journal of Separation Science*, 2004, 27, 47.
81. Kivilompolo, M.; Hyötyläinen, T.; *Journal of Chromatography A*, 2007, 1145, 155.

Index

A

Acridinium esters, 230–232
Amperometric detectors, 188
Anti-Stokes and Stokes lines, of CCl$_4$ spectrum, 103–104
Atmospheric pressure chemical ionization (APCI), 6–7
Atmospheric pressure ionization, 5
Attenuated total reflectance (ATR) effect, 111
Automated solid-phase extraction system, 113, 115

C

Chemical ionization (CI), 5
  mass spectra, 40–41
  positive ion, formation of, 39
Chemiluminescence; *see also* Liquid-phase chemiluminescence detection, HPLC
  definition, 221
  detection, liquid chromatography, 223–224
  indirect/sensitized, 222
  process, 221
  quantum yield, 222
Chemometrics, 25–27
Cold plasma, 168
Comprehensive heart-cutting approach, 270
Comprehensive orthogonal method evaluation technology (COMET) system, 269
Concentric flow nebulizer, 121–123
Conductometric detectors, 188
Continuous-flow LC-NMR mode
  hardware setup, 69
  uses, 70
Corona-CAD detector
  applications, 158
  ceramide IIB peak, 157–158
  vs. ELSD, calibration curves, 157
  operating principle, 156
Coulometric detectors, 188
Cyclic voltammetry, HPLC-ED
  description, 204, 206
  electrochemically reversible redox couple, 206–207
  pH effect, 206–207, 209
Cylindrical internal reflectance cells (CIRCLE), 111

D

2D chromatographic systems
  comprehensive heart-cutting approach, 270
  comprehensive separation, 269–270
  data collection, 271–272
  design, 270–271
  heart cutting process, 269–270
  peak matching, 274–275
  peak retention time determination, 272–274
Diffuse reflectance infrared Fourier transform (DRIFT), 115–116

E

Electron ionization (EI), 4–5
  electron impact, 37–39
  full-scan and selected-ion monitoring modes, 36–37
Electrospray ionization, 5–6
Electrospray nebulization, 126–127
ELSD, *see* Evaporative light scattering detector (ELSD)
Energy level diagram, 101–102
Evaporative light scattering detector (ELSD)
  applications
    micro-chromatography, 155–156
    pharmaceutical industry, 155
    plant extract profiling, 155
  characteristics, 150–151
  Corona-CAD detector, 156–158
  glucose, calibration curves, 150–151
  log-log transformation, 152
  operation
    light scattering, 149–150
    nebulization process, 146–148
    response *vs.* solute concentration, 150–152
    solvent evaporation, 148–149
  principle, 145
  response parameters
    drift tube temperature, 153
    gas pressure and mobile phase flow rate, 152
    mobile phase composition, 153–154
  sensitivity, 151

F

Fast atom bombardment (FAB), 7–8
Ferulic acid, biosynthetic formation of, 81–82
Flavanoid determination, HPLC-ED
   catechins determination, pretreatment
      plasma sample, 216–217
      tea sample, 216
   catechin structures, 208, 211
   chromatographic separation, 194, 196
   CLC-ED system
      chromatographic baseline noise,
         211–212
      description, 208–209, 212
      optimization conditions, 212–214
   column connection and temperature, 211
   gasket thickness, electrochemical flow cell,
      212–213
   half-wave potentials, 207, 210
   methanol effects, 197
   mobile phase preparation, 211
   pathway effects, 211
   structure, 194–195
   validation method
      accuracy, 214–215
      detection limit, 214–216
      linearity, 214–215
Flow cells
   infrared spectroscopy
      advantages, 109
      attenuated total reflectance (ATR)
         effect, 111
      chemigrams, 113
      detection, solvent exchange, 113–114
      evanescent wave, 111
      optical materials, as windows, 110
      ultramicro-cylindrical internal
         reflectancecells (CIRCLE),
         111–112
   Raman spectroscopy, 129–130
   windowless flow cell, 134–135
Flow-throug chemiluminescence detector, 225

G

Gas chromatography-mass spectrometry
   (GC-MS)
   chemical ionization
      mass spectra, 40–41
      positive ion, formation of, 39
   data evaluation
      class identification, in kerosene sample,
         51–52
      isotope dilution mass spectrometry,
         of naphthalene, 56–68
      lavender essential oil, characterization
         of, 53–56

   electron ionization
      electron impact, 37–39
      full-scan and selected-ion monitoring
         modes, 36–37
   mass analyzer
      ion trap, 46–47
      magnetic sector, 45
      quadrupole, 43–44
      role of, 42
      tandem mass spectrometry, 47–48
      time-of-flight (TOF), 48–50
   procedure, 33–35
Glassycarbon (GC) electrode polishing
      procedure, 199–200

H

Hard ionization technique, 37–39
High-performance liquid chromatography-
      electrochemical detection (HPLC-ED)
   baseline noise analysis
      composition and dimensions,
         electrochemical flow cells, 203
      power spectral analysis, 200–201
      pump selection, 200–203
      working electrode material, 203–204
   cyclic voltammetry
      description, 204, 206
      electrochemically reversible redox couple,
         206–207
      pH effect, 206–207, 209
   description, schematic illustration, 189
   detector features, 188
   electroactive natural products, 217–218
   flavanoid determination
      catechin structures, 208, 211
      CLC-ED system, 208–209, 212–214
      column connection and temperature, 211
      gasket thickness, electrochemical flow
         cell, 212–213
      mobile phase preparation, 211
      pathway effects, 211
      sample pretreatment, 216–217
      validation method, 214–216
   hydrodynamic voltammetry, 208
   instrumentation
      baseline noise analysis, 200–204
      catechins chromatogram, 191, 193
      columns characteristics, 191–192
      cross and radial flow electrochemical
         cells, 199
      electrochemical oxidation reaction,
         197–198
      epigallocatechin gallate, detection limits,
         191, 193
      Faradaic and non-Faradaic noise
         current, 197

GC electrode polishing, 199–200
injectors, 198
mobile phase, 192, 194–197
modes of, 191
piston pump characteristics, 198
method tuning and optimization scheme,
204–205
signal current, 190–191
HPLC-ED, *see* High-performance liquid
chromatography-electrochemical
detection (HPLC-ED)
HPLC–ICP MS coupling
HILIC–ICP MS, 173
HPLC separation mechanisms, 171
ion-exchange HPLC–ICP MS, 171
post-column splitting and dilution, 170
post-column volatilization, 174
reversed-phase HP LC–ICP MS, 172–173
size-exclusion LC–ICP MS, 171–172
standard configuration, 170
Hydrodynamic voltammetry, HPLC-ED, 208
Hydrophilic interaction chromatography
(HILIC)–ICP MS, 173

I

ICP MS, *see* Inductively coupled plasma mass
spectrometry (ICP MS)
Inductively coupled plasma mass spectrometry
(ICP MS)
advantages, 162
application
chemical warfare agents, detection of, 181
metal complex speciation, 179–180
metallodrugs, 180–181
microorganisms, plants and food, metal
speciation in, 179
organoarsenic compounds, speciation
of, 178
organoselenium compounds, speciation
of, 178–179
redox species analysis, 177–178
biomolecules quantification
non-speciated isotope dilution analysis
(IDA), 176
speciated isotope dilution analysis (IDA),
176–177
capillary and nanoflow HPLC coupling,
174–175
HPLC–ICP MS coupling
HILIC–ICP MS, 173
HPLC separation mechanisms, 171
ion-exchange HP LC–ICP MS, 171
post-column splitting and dilution, 170
post-column volatilization, 174
reversed-phase HP LC–ICP MS,
172–173

size-exclusion LC–ICP MS, 171–172
standard configuration, 170
interface and mass analyzers
quadrupole analyzers, 165–166
sector field mass spectrometry, 166–167
time-of-flight mass spectrometry, 166
interferences
collision/reaction cells, 168–169
high-resolution mass analyzers, 169
low-temperature plasma, 168
mathematical corrections, 167
ionization, 164
liquid samples, 163–164
Infrared spectroscopy
concentric flow interfaces, 121–124
electrospray interfaces, 126
evaporation interfaces, 116–117
flow cells
advantages, 109
attenuated total reflectance (ATR)
effect, 111
chemigrams, 113
detection, solvent exchange, 113–114
evanescent wave, 111
optical materials, as windows, 110
ultramicro-cylindrical internal
reflectancecells (CIRCLE), 111–112
instrumental factors, 108
for nitroethane, 102–103
particle beam interfaces
MAGIC-FTIR interface, 129
monodisperse aerosol generator, 128
pneumatic interfaces, 120–121
solvent elimination, 114–116
thermospray interfaces
moving belt LC-FT-IR interface,
118–119
spray patterns, 118
ultrasonic interfaces, 124–126
Ion-exchange HPLC–ICP MS, 171
Ion trap mass analyzer
gas chromatography-mass spectrometry,
46–47
mass spectrometry, 17–18

L

Light scattering, ELSD, 149–150
Liquid chromatography (LC)
infrared spectroscopy
concentric flow interfaces, 121–124
electrospray interfaces, 126
evaporation interfaces, 116–117
flow cells, 109–114
particle beam interfaces, 126, 128–129
pneumatic interfaces, 120–121
solvent elimination, 114–116

thermospray interfaces, 118–120
ultrasonic interfaces, 124–126
Raman spectroscopy
concentric flow interface, 137–138
evaporative methods, 136
flow cells, 129–130
solvent elimination methods, 135–136
UV-resonance Raman detection, 133–134
waveguides, 131–132
Liquid chromatography-nuclear magnetic
resonance (LC-NMR) spectroscopy
applications
in biosynthetic studies, 81–82
drug metabolism, degradation
and impurities, 86–88
environmental samples, 83–85
polymer research, 91–92
reaction monitoring and chemical product
analysis, 89–91
structure determination, of natural
products, 79–80
structure elucidation and analysis,
of isomers, 88–89
tissue-specific distribution, 80–81
xenobiotic metabolism
and degradation, 83
limitations, 62
liquid chromatography, 64–66
miniaturization
capNMR spectroscopy, 76
coupling capillary isotachophoresis, 77
thapsigargicin, 77–78
nuclear magnetic resonance spectroscopy,
66–68
operation modes
continuous-f low mode, 68–70
LC-SPE-NMR mode, 73–75
loop storage mode, 71–73
stopped-flow mode, 70–71
sample preparation, 63–64
Liquid core waveguide (LCW), 131–132
Liquid-phase chemiluminescence
detection, HPLC
chemiluminescene quantum yield, 222
indirect/sensitized chemiluminescene, 222
instrument components, 224–225
optimization, 225
post-column chemiluminescence detection,
225–226
reagents
lucigenin and acridinium esters, 230–232
luminol and related, 226–230
oxidant/enhancer combination, 242–243
peroxyoxalate, 232–238
potassium permanganate, 241–242
tris(2,2′-bipyridine)ruthenium(III),
238–241

Loop storage LC-NMR mode
features of, 72–73
hardware setup, 72
Luminol and related reagents, 226–230

M

Magnetic sector mass analyzer, 45
MALDI, see Matrix-assisted laser desorption
ionization (MALDI)
Mass analyzers
gas chromatography-mass spectrometry
ion trap, 46–47
magnetic sector, 45
quadrupole, 43–44
role of, 42
tandem mass spectrometry, 47–48
time-of-flight (TOF), 48–50
inductively coupled plasma mass
spectrometry
quadrupole, 165–166
sector field mass spectrometry, 166–167
time-of-flight mass spectrometry, 166
mass spectrometry
ion trap, 17–18
magnetic sector, 14–15
quadrupole, 15
tandem mass spectrometry, 18–20
time of flight, 15–17
Mass spectrometry
contamination, 23–24
data handling, 25–27
development of, 3
importance of, 2
ion enhancement/suppression solvent,
21–23
ionization energy and ionization efficiency,
8–9
mass analyzers
ion trap, 17–18
magnetic sector, 14–15
quadrupole, 15
tandem mass spectrometry, 18–20
time of flight, 15–17
mass spectral data
fragmentation, 11–14
isotopic information, 10–11
nitrogen rule, 9–10
rings and double bonds, 11–12
principle
APCI, 6–7
atmospheric pressure ionization, 5
chemical ionization (CI), 5
electron ionization (EI), 4–5
electrospray ionization, 5–6
fast atom bombardment (FAB), 7–8
inductively coupled plasma, 8

liquid chromatography column effluent,
3–4
MALDI, 6–7
quantitation, 25
resolution, 20–21
Matrix-assisted laser desorption ionization
(MALDI), 6–7
MDHPLC, *see* Multidimensional high-
performance liquid chromatography
(MDHPLC)
Metallothioneins, 179–180
Methamphetamine, electron ionization mass
spectra of, 38
Mie scattering process, 149–150
Mobile phase composition, ELSD parameters
mobile phase additives, 154
solvents, 153–154
Monodisperse aerosol generator (MAG), 128
Multidimensional high-performance liquid
chromatography (MDHPLC)
beneficial effects, coffee, 275–277
chemical fingerprinting, 252
2D chromatographic systems
comprehensive heart-cutting approach, 270
comprehensive separation, 269–270
data collection, 271–272
design, 270–271
heart cutting process, 269–270
peak matching, 274–275
peak retention time determination,
272–274
Decaffeinato coffee analysis, 279–281
description, 253–254
one-dimensional chromatograms, 278–279
orthogonality property
determination, 260–262
divergence measure, 260
low-molecular-weight polystyrenes,
266–267
mathematical tools, 262–265
peptides, 267–268
pharmaceuticals, 269
Ristretto coffee analysis, 279–281
sample dimensionality
characteristics, 254
chiral and achiral environment, 257
low-molecular-weight polymers, 254, 256
PAHs, 254–255
selective and nonselective displacements
alcohol ethoxylates, 259
combination types, 257–258
2D separation, low-molecular-weight
polystyrenes, 259
one liquid chromatographic
dimension, 258
selectivity factors, 257–258
unidimensional HPLC, 252–253

N

Nebulization process, ELSD
HPLC solvent results, 146–147
pneumatic nebulizers, 146, 148
Nitrogen rule, 9–10

O

Orthogonality property, MDHPLC
determination, 260–262
divergence measure, 260
low-molecular-weight polystyrenes, 266–267
mathematical tools
correlation matrix, 264–265
informational entropy, 263
normalization, retention data, 262–263
synentropy, 264
peptides, 267–268
pharmaceuticals, 269

P

PAHs, *see* Polynuclear aromatic hydrocarbons
(PAHs)
Peroxyoxalate reagents
amphetamines determination, 235–236
catecholamines determination, 235, 237
derivatizing reagents, 234–235
detection modes, 234
fluorophore, electronic excitation, 233
peroxyoxalate chemiluminescence, schematic
illustration, 232–233
Photon energy, 100
Piezo-actuated microdroplet deposition
device, 117
Pneumatic nebulizers, 146, 148
Polyacetylene, NMR spectra, 89–90
Polyethylene terephthalate (PET)
fragmentation pathways of, 13
mass spectrum of, 12
Polynuclear aromatic hydrocarbons (PAHs),
254–255
Potassium permanganate reagents,
241–242
Pulsed-amperometric detectors, 188

Q

Quadrupole mass analyzers
gas chromatography-mass spectrometry,
43–44
inductively coupled plasma mass
spectrometry, 165–166
mass spectrometry, 15
Quadrupole time of flight (QToF) mass
spectrometer, 19–20

R

Raman spectroscopy
    concentric flow interface, 137–138
    evaporative methods, 136
    flow cells, 129–130, 134–135
    instrumental factors, 108
    for nitroethane, 102–103
    solvent elimination method, 135–136
    UV-resonance Raman detection, 133–134
    waveguides, 131–132
Rayleigh scattering process, 149
Resonance Raman (RR) effect
    chemical enhancement, 107
    definition, 104
    field enhancement, 107
    laser excitation wavelength, 106
Reversed-phase HP LC–ICP MS, 172–173
Ruta graveolens, 80

S

Savitzky-Golay smoothing filter, 273
Segmented stream generator, 113–114
Separation science, *see* Mass spectrometry
Single-focusing magnetic sector mass
    analyzer, 45
Size-exclusion LC–ICP MS, 171–172
Soft ionization technique, *see* Chemical
    ionization (CI)
Solvent evaporation, ELSD, 148–149
Stable isotope labeling method, 81
Stopped-flow LC-NMR mode, 70–71
Surface enhanced Raman spectroscopy (SERS)
    anionicdyes, 138
    nitrophenols, 136
    silver sol, 137

T

Tandem mass spectrometry
    of domoic acid, 18–19
    triple quadrupole and quadrupole time
    of flight, 19–20
Thapsigargicin, LC-SPE-capNMR spectra of,
    77–78
Total ion current (TIC) chromatogram,
    33–34
    of kerosene sample, 51–52
    of lavender essential oil, 53–54
3-Trifluoromethylaniline(3-TFMA), 83
Triple quadrupole mass spectrometer, 19–20
Tris(2,2′-bipyridine)ruthenium(III) reagents
    chemiluminescent reduction, 238
    divinylsulfone, 240, 242
    pipecolic acid determination, 239
    preparation, 238

U

Ultrasonic nebulizer, 124–126

W

Windowless flow cell, 134–135

X

Xanthohumol, 86

Z

ZnSe, 111

Milton Keynes UK
Ingram Content Group UK Ltd.
UKHW021618071024
449327UK00020BA/1101

9 780367 452209